Streaming media server
design

~ ~~~ ~~~

The CD that accompanied this book has been replaced by
a web site that can be found at the following address:
http://www.phptr.com

Note that all references to the CD in the book now pertain
to the web site.

Streaming Media
Server Design

IMSC Press Multimedia Series

▶ Streaming Media Server Design

Ali E. Dashti, Seon Ho Kim, Cyrus Shahabi, and Roger Zimmermann

▶ Desktop Digital Video Production

Frederic Jones

▶ Touch in Virtual Environments:
Haptics and the Design of Interactive Systems

Edited by Margaret L. McLaughlin, João P. Hespanha, and Gaurav S. Sukhatme

▶ The MPEG-4 Book

Edited by Fernando Pereira and Touradj Ebrahimi

▶ Multimedia Fundamentals, Volume 1:
Media Coding and Content Processing

Ralf Steinmetz and Klara Nahrstedt

▶ Intelligent Systems for Video Analysis and Access Over the Internet

Wensheng Zhou and C. C. Jay Kuo

Integrated Media Systems Center

The Integrated Media Systems Center (IMSC), a National Science Foundation Engineering Research Center in the University of Southern California's School of Engineering, is a preeminent multimedia and Internet research center. IMSC seeks to develop integrated media systems that dramatically transform the way we work, communicate, learn, teach, and entertain. In an integrated media system, advanced media technologies combine, deliver, and transform information in the form of images, video, audio, animation, graphics, text, and haptics (touch-related technologies). IMSC Press, in partnership with Prentice Hall, publishes cutting-edge research on multimedia and Internet topics. IMSC Press is part of IMSC's educational outreach program.

Streaming Media Server Design

Ali E. Dashti
Seon Ho Kim
Cyrus Shahabi
Roger Zimmermann

PEARSON EDUCATION
PRENTICE HALL PROFESSIONAL TECHNICAL REFERENCE
UPPER SADDLE RIVER, NJ 07458
WWW.PHPTR.COM

Library of Congress Cataloging-in-Publication Data

Editorial/production supervision: *Mary Sudul*
Cover design director: *Jerry Votta*
Cover design: *Nina Scuderi*
Manufacturing manager: *Maura Zaldivar*
Acquisitions editor: *Bernard Goodwin*
Editorial assistant: *Michelle Vincenti*
Marketing manager: *Dan DePasquale*

Prentice Hall books are widely used by corporations and government agencies for training, marketing, and resale.
The publisher offers discounts on this book when ordered in bulk quantities. For more information, contact Corporate Sales Department, Phone: 800-382-3419; FAX: 201- 236-7141;
E-mail: corpsales@prenhall.com
Or write: Prentice Hall PTR, Corporate Sales Dept., One Lake Street, Upper Saddle River, NJ 07458.

Printed in the United States of America
10   9   8   7   6   5   4   3   2   1

ISBN 0-13-067038-3

Pearson Education LTD.
Pearson Education Australia PTY, Limited
Pearson Education Singapore, Pte. Ltd.
Pearson Education North Asia Ltd.
Pearson Education Canada, Ltd.
Pearson Educación de Mexico, S.A. de C.V.
Pearson Education—Japan
Pearson Education Malaysia, Pte. Ltd.

To my parents,
    my wife, Mona,
    and my daughter, Zaynab,
    for their love, support, and understanding throughout the times.
                    Ali Esmail Dashti

To my adorable wife, Eun Jin,
    and my cheerful sons, Nam Yun and Tae Hun.
                    Seon Ho Kim

To my beautiful wife, Parisa,
    and my lovely children, Layla and Kameron.
                    Cyrus Shahabi

To my wonderful wife Adele
    and my sweet daughter Kimberly.
                    Roger Zimmermann

CONTENTS

PREFACE

The Internet has grown from a military network (designed primarily for communication among major research and government institutions) to a phenomenon that has captured the world's attention and imagination over the past few years. This explosive interest in the Internet was due to the development of the browser for the World Wide Web by Marc Anderson while a graduate student at the University of Illinois. This new piece of software opened the door to new and novel ways of experiencing the Internet. At its core, the main technology that the browser offered was a user interface, which allowed the viewing of different media types, such as text, images, and animations. This glimpse at the possibility of delivering multimedia content, i.e., text, images, audio, video, animations, time series, etc., to users around the world made the browser an overnight success, and hence the Internet became a household name.

However, soon after the beginning of the Internet revolution (in the mid 1990s), it became apparent that the delivery of multimedia content requires more than just fancy user interfaces. Multimedia content can impose tremendous loads and constraints on the Internet's storage and retrieval infrastructure, and on networking infrastructure.

One of the main components of the storage and retrieval infrastructure for the delivery of multimedia data on the Internet is the *Streaming Media* (SM) server. These servers are designed to store and deliver SM content, such as video and audio streams, to hundreds (or thousands) of simultaneous users. For example, the Media nCUBE 3000 servers from nCUBE can support 20,000 MPEG-1 streams simultaneously. To support SM display, it is necessary to retrieve and display data at a prespecified rate; otherwise, the display will suffer from frequent disruptions and delays, termed hiccups. Moreover, SM much exceeds the resource demands of traditional data types and requires massive amounts of space and bandwidth for its storage and display. For example, an MPEG-2 stream requires between 3-15 megabits per second (Mb/s) bandwidth, which translates into 1.35-6.75 gigabytes (GB) of storage for a one-hour video clip. Storing thousands of such video clips can easily consume terabytes and even petabytes of storage. To support thousands of simultaneous users requires server bandwidths in excess of tens of thousands of megabits per second. To achieve these high bandwidth and massive storage requirements for multi-user SM servers, disk drives are commonly combined into disk clusters. Magnetic disks are usually the storage media of choice for such systems because of their high performance and moderate cost. With current technology and prices, other storage technologies are either slow (magneto-optical disks), provide limited random access (tapes), or have limited write capabilities (CD-ROM, DVD). However, other storage technologies can play an important role in special cases as discussed in this textbook.

This textbook is based on our research in the field of SM servers since the early 1990s. Its objective is to present an in-depth introduction to SM server design, with an emphasis on architectural and implementation aspects of the design. This textbook is designed for senior level undergraduate courses or first year graduate level courses, with a primary focus on the design and implementation of SM servers. Students need a strong background in operating systems and database systems.

The material in this textbook can be divided into two parts: 1) basics of SM server design, and 2) advanced topics in SM server design. Part I, Chapter 1 discusses the fundamentals of SM systems. Subsequently, in Chapter 2, we discuss the design of SM servers in the context of a single disk platform. Finally, in Chapter 3, we extend the design of the SM server to a multi-disk platform. In Part II, we start off by discussing low latency system design for a special class of applications, namely digital authoring tools, in Chapter 4. In Chapter 5, we extend the SM server design to adapt to heterogeneous disk platforms. In Chapter 6, we discuss fault tolerance issues in SM server design. In Chapter 7, we introduce the design of a hierarchical storage system for SM servers. In Chapter 8, we consider a distributed SM server design. In Chapter 9, we present a *Super Streaming* paradigm, where the delivery rate of the SM objects is higher than their consumption rate. In Chapter 10, we present the design and implementation of a second generation SM server, namely *Yima*. To stress the implementation aspects of SM server design, a scaled-down version of Yima SM server software, Yima-Personal Edition SM server, is included as a companion CD with this textbook. This server software is licensed through the University of Southern California's Integrated Media System Center (IMSC). We provide the complete source code of the personal edition server software with this textbook. This SM server can run on a single disk platform, using the Linux operating system, as described in Appendix A. Students can use the given source code to extend the SM server design, e.g., extend the server design to run on a multi-disk platform. We provide a set of exercises, however, students and instructors are free to experiment as they see necessary.

Supplementary Material

The home page for this book is at URL: streamingmedia.usc.edu, where instructors and students can find the following information online:

1. Lecture notes

2. Updates to the Yima-Personal Edition SM server software

3. Updates to all known errors in the book

4. Other relevant links

Instructors and students are advised to visit the above web page regularly to view important updates and/or register at the site to receive important updates automatically.

Acknowledgments

This book has grown out of our collective research at IMSC at the University of Southern California, and subsequently at the University of Denver (Seon Ho Kim) and Kuwait University (Ali Dashti). We thank IMSC and the Computer Science Department at the University of Southern California, the Computer Science Department at the University of Denver, and the Computer Engineering Department at Kuwait University for providing us with the necessary research environment and the necessary infrastructure to make our research in the field of SM servers possible. Moreover, Ali Dashti would like to thank Kuwait University Research Administration, where his work was supported by Kuwait University Research Grant No. [EE 05/00]. Cyrus Shahabi's research has been funded in part by NSF grants EEC-9529152 (IMSC ERC) and IIS-0082826, and unrestricted cash gifts from Okawa Foundation, Microsoft, and NCR. Roger Zimmermann's research has been supported in part by NSF grants EEC-9529152 (IMSC ERC) and IIS-0082826.

We owe special thanks to Professor Chrysostomos (Max) Nikias, the former IMSC director and current Dean of Engineering at the University of

Southern California, for his constant support, encouragement, and leadership. We are indebted to: Professor Michael Arbib, Professor Dennis McLeod, and Professor Shahram Ghandeharizadeh who have contributed generously to our research since the early 1990s. We are very grateful for the many wonderful graduate and undergraduate students who have contributed to our research and to this textbook. In particular, we would like to thank Farnoush Banaei-Kashani for his help in administrating the book account during its inception as well as helping tremendously with Chapters 8 and 9.

We also would like to thank the IMSC Press Editor Dr. Andy Tescher who initially proposed the idea of a textbook on SM servers, and to the publisher Mr. Bernard Goodwin, for his guidance and encouragements of the past two years.

The University of Southern California
February 2003

Chapter 1

FUNDAMENTALS OF STREAMING MEDIA SYSTEMS

Multimedia (MM) systems utilize audio and visual information, such as video, audio, text, graphics, still images, and animations to provide effective means for communication. These systems utilize multi-human senses in conveying information, and they play a major role in educational applications (such as e-learning and distance learning), library information systems (such as digital library systems), entertainment systems (such as Video-On-Demand and interactive TV), communication systems (such as mobile phone multimedia messaging), military systems (such as Advanced Leadership Training Simulation), etc. Due to the exponential improvements of the past few years in solid state technology (i.e., processor and memory) as well as increased bandwidth and storage capacities of modern magnetic disk drives, it has been technically feasible to implement these systems in ways we only could have dreamed about a decade ago.

A challenging task when implementing MM systems is to support the sustained bandwidth required to display *Streaming Media* (SM) objects, such as video and audio objects. Unlike traditional data types, such as records, text and still images, SM objects are usually large in size. For example, a two-hour MPEG-2 encoded movie requires approximately 3.6 gigabytes (GByte) of storage (at a display rate of 4 megabits per second (Mb/s)). Figure 1.1 compares the space requirements for ninety minute video clips encoded in different industry standard digital formats. Second, the isochronous nature of SM objects requires timely, real-time display of data blocks at a pre-specified rate. For example, the NTSC video standard requires that 30 video frames per second be displayed to a viewer. Any deviation from this real-time requirement may result in undesirable artifacts, disruptions, and jitters,

1

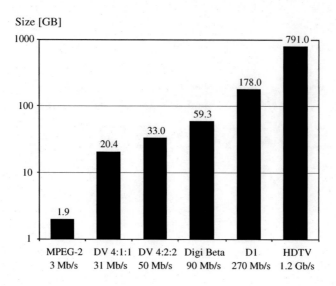

Figure 1.1. Storage requirements for a ninety minute video clip digitized in different industry standard encoding formats.

collectively termed *hiccups*.

In the remainder of this chapter, the following topics are covered. First, in Section 1.1, we introduce different display paradigms for SM objects. Next, in Section 1.2, we consider the overall SM system architecture. After that, in Section 1.3, we briefly overview data compression techniques. Subsequently, in Section 1.4, we describe a number of networking protocols that ensure real-time streaming over the Internet. Finally, in Section 1.5, we present an outline for the entire textbook.

1.1   Introduction to Streaming Media Display

Many applications that traditionally were the domain of analog audio and video are evolving to utilize digital audio and video. For example, satellite broadcast networks, such as *DirectTV*^{TM}, were designed from the ground up with a completely digital infrastructure. The proliferation of digital audio and video have been facilitated by the wide acceptance of standards for compression and file formats. Consumer electronics are also adopting these standards in products such as the digital versatile disk (DVD[1] and digital

---

[1]DVD is a standard for optical discs that feature the same form-factor as CD-ROMs but holds 4.7 GByte of data or more.

Figure 1.2. SM Display Paradigms.

VHS[2] (D-VHS)). Furthermore, increased network capacity for local area networks (LAN) and advanced streaming protocols (e.g., RTSP[3]) allow remote viewing of SM clips.

There are a number of possible display paradigms for SM objects, as shown in Figure 1.2. These display paradigms are defined according to:

1. The location of the SM object (local vs. remote), and

2. The capabilities of the network and the servers (none-streaming vs. streaming).

SM objects can either be displayed from a local machine (i.e., *local streaming*) or from a remote server. In the former approach, the SM objects are available locally in their entirety (e.g., stored on a hard disk, or available on DVDs and/or CD-ROMs). When the user requests the display of an object, the data blocks are retrieved from the local storage, passed to the *video-card/graphics-card* for decoding, and subsequently displayed to the user. Modern personal computers (PCs) and consumer electronic devices (such as Sony's *PlayStation*[TM] and Microsoft's *Xbox*[TM]) can display a single SM object without much difficulty. Local streaming does not require any guarantees from the remote servers and the network (for obvious reasons), and hence, it occupies the upper and lower left quadrants of Figure 1.2.

---

[2]Video Home System: a half-inch video cassette format.

[3]The Real Time Streaming Protocol is an Internet Engineering Task Force (IETF) proposed standard for control of SM on the Internet.

When the user requests the display of an object that resides on a remote server, the data blocks are retrieved from the remote server over a network (e.g., the Internet or a corporate intranet), and passed to the client for display. Depending on the capabilities of the server and the network, there are two alternative display paradigms: 1) *store-and-display* paradigm, and 2) *remote streaming* paradigm. With the store-and-display paradigm, the SM objects are first downloaded in their entirety from a remote server to a local storage before initiating the display process (which is similar to local streaming). This paradigm is favored when the server and/or the network cannot guarantee the isochronous nature of SM objects (upper right quadrant of Figure 1.2). With the remote streaming paradigm, the data blocks are retrieved from the remote server over a network and processed by the client as soon as they are received; data blocks are not written to the local storage (although they might be buffered). This paradigm is possible only when both the server and the network can guarantee the isochronous nature of the SM object (lower right quadrant of Figure 1.2). In the remainder of this book, we will be referring to remote streaming as *streaming*.

There are several advantages of streaming paradigm versus store-and-display paradigm:

1. No waiting for downloads (well, not much waiting).

2. No physical copies of the content are stored locally. reducing the possibility of copyright violations (if any).

3. No storage (or limited storage) requirements at the client side.

4. Support of live events.

On the other hand, there are a number of limitations to the streaming paradigm:

1. It requires real-time guarantees from the server and the network.

2. Lost or damaged packets (blocks) or missed deadlines may cause a hiccup in the display of the SM object.

It is important to note that streaming is not a progressive download, where the display starts as soon as enough of the data is buffered locally while the remainder of the file is being retrieved and stored on the local disk (i.e., at the end of the display the entire SM object is stored locally). Progressive download can be identified as a special case of the steaming paradigm (if the system can guarantee the hiccup-free display of the SM objects), or a special case of store-and-display paradigm.

In this book, we focus on the design of Steaming Media Servers (*SM Servers*) that guarantee the continuous display of SM objects, assuming that the network delivers the data blocks in a timely fashion. We ignore for the most part design issues related to SM encoding standards (such as MPEG[4] standards) and SM networking[5].

1.2   Streaming Media System Architecture

To support the streaming paradigm, servers and the network must guarantee the continuous display of the SM object without disruptions (i.e., hiccups). To illustrate, assume a SM object X that consists of n equi-sized blocks: X_0, X_1, ..., X_{n-1}, and resides on a SM server, as shown in Figure 1.3. There are some important observations when streaming this object:

1. The display time of each block is a function of the display requirements of each object and the size of the block. For example, if the display requirement of the object X is 4 Mb/s and the size of each block, X_i, is 1 MByte, then the display time of each block, X_i, is 2 seconds. However, it is important to note that SM objects can be encoded using either *constant bit-rate* (CBR) or *variable bit-rate* (VBR) schemes. As the name implies, the consumption rate of a CBR media is constant over time, while VBR streams use variable rates to achieve maximum compression. We will assume CBR media throughout the book to provide a focused discussion, unless indicated otherwise.

2. The retrieval time of each block is a function of the transfer rate of the SM server, and it is primarily related to the speed of the storage subsystem.

3. The delivery time of each block from the SM server to the display station is a function of the network speed, traffic, and protocol used.

Assume that a user requests the display of object X from the SM server, as depicted in Figure 1.3. Using the above information, the SM server schedules the retrieval of the blocks, while the network ensures the timely delivery

---

[4]Motion Picture Encoding Group (MPEG) (http://mpeg.telecomitalialab.com/) has standardized several audio and video compression formats, such as MPEG-1, MPEG-2, and MPEG-4.

[5]In this chapter, we present an overview of SM encoding standards and introduce a number of networking technologies that make SM possible over the Internet.

Figure 1.3. System Architecture.

Figure 1.4. Timing of retrieval, delivery, and display of object X.

of these blocks to the display station. The SM server stages a block of X (say X_i) from the disk into main memory and initiates its delivery to the display station (via the network). The SM server schedules the retrieval and the delivery of X_{i+1} prior to the completion of the display of X_i. This ensures a smooth transition between the two blocks in support of continuous display. This process is repeated until all blocks of X have been retrieved, delivered, and displayed. The periodic nature of the data retrieval and display process gives rise to the definition of a *time period* (T_p): it denotes the time required to display one data block. Note that the display time of a block is in general significantly longer than its retrieval and delivery time from the SM server to the display station. Thus, the SM server can be multiplexed among several displays.

The display station in turn might buffer enough data (i.e., blocks) in its memory, such that any delays in data delivery due to network or SM server delays are tolerated[6] (see Figure 1.3). To guarantee the continuous display of object X without any hiccups, it is critical that prior to the completion of the display of block X_i, block X_{i+1} is available in the buffer at the display station. For example, consider the retrieval times, delivery times, and the display times shown in Figure 1.4. Block X_1 is retrieved and delivered to the display station, and the display station delays the display of the block to reduce the probability of hiccups (e.g., due to network delays). For block X_1, the network delivers the block in a shorter time, and hence, the block is buffered for a longer period of time. If the network delay becomes longer than anticipated, as shown in Figure 1.4 for block X_2, then the display of the object might suffer a hiccup. Conversely, if the delivery of the blocks becomes faster than the display rate of the object, then the display station's buffer might overflow. In this case, the display station might have to either: 1) discard blocks, or 2) send a signal to the SM server to slow down or to pause the retrieval. The former will cause retransmissions of the blocks (increasing the network traffic) and extra load on the SM server, or it might cause hiccups in the display. The later will affect retrieval schedule at the SM server.

Due to the size and the isochronous characteristic of SM objects, the design of servers in support of SM has been different from that of conventional file servers and associated storage systems [175], [59]. The fundamental functionality of SM servers is a hiccup-free display of SM. However, just supporting a continuous display is not enough in the design of SM servers because many real applications, especially commercial ones such as movie-on-demand systems that concurrently service multiple users, require maximizing the performance of servers for cost-effective solutions. Thus, the following performance metrics are important: 1) the number of simultaneous displays that can be supported by a SM server, i.e., *throughput*, and 2) the amount of time elapsed from when a display request arrives at the system until the time the actual display is initiated by the system on behalf of this request, i.e., *startup latency*. Throughput, in general, is closely related to another important metric of SM servers, *cost per stream*. If a technique supports a higher throughput with fixed resources than others, it provides for a more

---

[6]Note that network and SM server delays can either be deterministic or otherwise.

cost-effective solution. Throughout this book, we compare alternative designs based on these metrics.

1.3   Data Compression

A ninety minute uncompressed video clip digitized in High Definition Television (HDTV) format at 1.2 gigabits per second (Gb/s) requires 791 GByte of storage. Even though modern storage devices, such as magnetic disk drives, provide large storage capacities, it might not be economical to store SM objects uncompressed. Moreover, the data transfer rate of magnetic disks is not high enough to retrieve multiple high bit rate objects in real time. A more serious problem arises when transferring a large number of SM objects over the network. Even though network speeds are rapidly increasing, it is not economically feasible to handle the simultaneous display of a large number of SM objects over existing networks.

To resolve these problems and to efficiently handle large number of SM objects, we need to compress these objects, where a smaller SM object requires less disk storage space and less network bandwidth. In the remainder of this section, we briefly overview data compression techniques.

1.3.1   Information vs. Data

Data is an individual fact or multiple facts, or a value, or a set of values. For example, a collection of digitally captured pixel values of an image is data. Information is data in a usable form, usually interpreted in a predefined way. For example, the content of an image is information. In general, information is meaningful to us and it is stored and handled as data in computer systems. Thus, the main goal of compression techniques is to reduce the size of data while maintaining as much information as possible.

A popular approach in multimedia data compression is *perceptual coding* that hides errors where humans will not see or hear them. Based on the studies of human hearing and vision to understand how we see/hear, perceptual coding exploits the limit of human perception. For example, human audio perception ranges between 20 Hz and 20 KHz but most voice sounds are in low frequencies (below 4 KHz). Thus, audio signals above 4 KHz are eliminated in telephone systems. Human visual perception is strongly influenced by edges and low frequency information such as big objects. Thus, many detailed patterns in an image can be ignored in the coding process. Perception coding takes advantage of this fact and encodes only those signals that will be perceived by humans, eliminating lots of imperceptible signals.

1.3.2   Coding Overview

Good compression algorithms should satisfy the following objectives:

- Achieve high compression ratios. It is obvious that a higher compression ratio is more beneficial.

- Ensure a good quality of the reconstructed information, i.e., it is important to maintain a good or acceptable quality while providing a high compression ratio.

- Minimize the complexity of the encoding and decoding process, i.e., if it takes too long to encode and/or decode the SM objects, then the usage of the algorithm might become limited. For real time encoding and decoding, coding and decoding processes must be fast.

Some other general requirements include independence of specific size and frame rate and support various data rates of an object.

Depending on the importance of the original information, there are two types of compression: *lossless compression* and *lossy compression*. Using lossless compression, the reconstructed information is mathematically equivalent to the original information (i.e., reconstruction is perfect). Thus, it does not lose or alter any information in the process of compression and decompression. For example, documents, databases, program executable files, and medical images, to name a few, do not tolerate a loss or alteration of information. Lossless compression reduces the size of data for these important types of information by detecting repeated patterns and compressing them. However, it achieves only a modest level of compression (about a factor of 5), which may not be sufficient for SM objects.

While lossless compression is desirable for information that cannot tolerate a loss of even a bit, some media types are not very sensitive to the loss of information. Human eyes usually do not detect minor color changes between the original image and the reconstructed one, and usually such small changes do not pose a problem in many practical applications. Thus, *lossy compression* achieves a very high degree of compression (compression ratios up to 200) by removing less important information such as imperceptible audio signals in music. Using lossy compression, reconstructed images may demonstrate some degradation in the quality of the image. The practical objective of lossy compression is to maximize the degree of compression while maintaining the quality of the information to be *virtually lossless*.

Entropy and Source Encoding

Entropy is a measure of amount of information. If N states are possible, each characterized by a probability p_i, with $\sum_{i=1}^{N} p_i = 1$, then $S = -\sum_{i=1}^{N} p_i log_2 p_i$ is the entropy which is the lowest bound on the number of bits needed to describe all parts of the system. It corresponds to the information content of the system. For example, when there are only two symbols, S1 (0.9) and S2 (0.1), the entropy is $0.9 log_2 0.9 - 0.1 log_2 0.1 = 0.47$. Another typical example is as follows. In an image with uniform distribution of gray-level intensity, i.e., $p_i = 1/256$, the number of bits needed to code each gray level is 8 bits. Thus, the entropy of this image is 8.

Entropy encoding ignores semantics of input data and compresses media streams by regarding them as sequences of bits. Run-length encoding and Huffman encoding are popular entropy encoding schemes. On the contrary, *source encoding* optimizes the compression ratio by considering media-specific characteristics. It is also called *sematic-based coding*. Many advanced schemes such as DPCM, DCT, FFT, and Wavelet fall into this category. In reality, most practical compression algorithms employ a hybrid of the source and entropy coding such that a source encoding is applied at the early stage of compression and then an entropy encoding is applied to results in an attempt to further reduce the data size.

Run-length Encoding

This is one of the simplest compression techniques. It replaces consecutive occurrences of a symbol with the symbol followed by the number of times it is repeated. For example, the string, "a a a a a" can be represented by the following two bytes, "5a". The first byte shows the number of occurrences of the character, and the second byte represents the character itself. This scheme is naturally lossless and its compression factor ranges from 1/2 to 1/5 depending on the contents of objects. This scheme is most useful where symbols appear in long runs: e.g., for images that have areas where the pixels all have the same value, cartoons for example. However, this may not be efficient for complex objects where adjacent symbols are all different.

Relative Encoding

This is a technique that attempts to improve efficiency by transmitting only the difference between each value and its predecessor. In an image, based on the fact that neighboring pixels may be changing slowly in many cases, one can digitize the value at a pixel by using the values of the neighboring pixels and encode the difference between two. A differential PCM coder (DPCM) quantizes and transmits the difference, $d(n) = x(n)x(n-1)$. The advantage of using difference $d(n)$ instead of the actual value $x(n)$ is the reduced number of bits to represent a sample. For example, the series of pixel values "60 105 161 129 78", can be represented by "60+45+56-32-51". By reducing the size of each value (in the example, 6 bits are enough to represent the difference while 8 bits are required to represent the original values), assuming the difference is far smaller than the original value, it can reduce the overall size of data. This scheme works well when the adjacent values are not greatly changing, such as voice signals. Furthermore, the transmitter can predict each value based on a mathematical model and transmit only the difference between the predicted and actual values, further reducing the size of required bits to represent a value (*predictive coding*).

Huffman Encoding

Huffman encoding is a popular compression technique that assigns variable length codes to symbols, so that the most frequently occurring symbols have the shortest codes. Thus, more frequently occurring values are assigned fewer bits exploiting the statistical distribution of the values within an object. To correctly decompress the encoded data, encoder and decoder must share the same codebook. Huffman coding is particularly effective where the data are dominated by a small number of symbols. Suppose that one wants to encode a source of N = 8 symbols: a,b,c,d,e,f,g,h. The probabilities of these symbols are: P(a) = 0.01, P(b) = 0.02, P(c) = 0.05, P(d) = 0.09, P(e) = 0.18, P(f) = 0.2, P(g) = 0.2, P(h) = 0.25. If we assign 3 bits per symbol ($N = 2^3 = 8$), the average length of the symbols is: $L = \sum_{i=1}^{8} 3P(i) = 3$ bits/symbol. The minimum average length we could ever achieve is equal to the entropy (according to Shannon's theorem): $S = -\sum_{i=1}^{8} P(i)log_2 P(i) = 2.5821$ bits/symbol $= 0.86 \times L$.

   The Huffman code assignment procedure is based on a binary tree structure. This tree is developed by a sequence of pairing operations in which the two least probable symbols are joined at a node to form two branches of a tree.

- Step 1. The list of probabilities of the source symbols are associated with the leaves of a binary tree.

- Step 2. Take the two smallest probabilities in the list and generate an intermediate node as their parent and label the branch from parent to one of the child nodes 1 and the branch from parent to the other child 0.

- Step 3. Replace the probabilities and associated nodes in the list by the single new intermediate node with the sum of the two probabilities. If the list contains only one element, quit. Otherwise, go to step 2.

It is very important to estimate the probability p_i, the relative frequency of the symbols. To decode the variable length codes, Huffman encoding uses prefix codes, which have the property that no codeword can be the prefix (i.e., an initial segment) of any other codeword. Note that there is no guarantee that the best possible codes always reduce the size of sources. In the worst case, there is no compression.

Transform Coding

In transform coding, using a mathematical transformation, the original data to be coded is converted into new data (transform coefficients) which is more suitable for compression. This process is reversible such that the original data can be recovered through inverse transformation. For example, Fourier transform and Cosine transform convert signals from space domain into frequency domain to obtain the frequency spectrum with which one can easily separate low frequency information.

Transform coefficients represent a proportion of energy contributed by different frequencies. In data compression using transform coding, it is important to choose a transformation so that only a subset of coefficients have significant values. In other words, energy is confined to a subset of important coefficients, known as energy compaction. Energy compaction is good for coding because we can consider only significant coefficients. If the number of significant coefficients is far smaller than the number of samples in original sequence, compression is possible. In practice, one can code significant coefficients accurately using a greater number of bits while allocating fewer or no bits to other less meaningful coefficients (which offer less perceptible information to humans). Moreover, many low energy coefficients can even be discarded through quantization. In practice, many algorithms apply transform coding at block level. For example, an N x N image is divided into several n x n blocks and each n x n block undergoes a reversible transformation.

1.3.3 JPEG

JPEG (Joint Photographic Expert Group) is an international compression standard for *continuous-tone* still images, both gray-scale and color. Its development was motivated by the fact that the compression ratio of lossless methods such as Huffman is not high enough for many applications such as

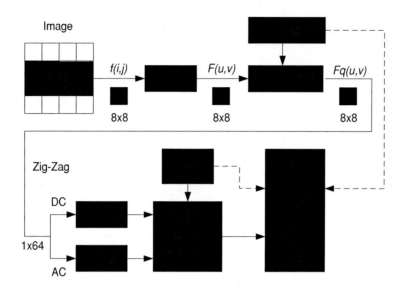

Figure 1.5. JPEG overview.

high quality gray-scale images, photographic images, and still video. JPEG achieves a high compression ratio by removing *spatial redundancy* (i.e., correlation between neighboring pixels) within an image, which is basically a lossy process.

Figure 1.5 shows the components and sequence of JPEG. JPEG utilizes discrete cosine transform (DCT), which is one of the best known transforms and is a close approximation to the optimal for a large class of images. DCT transforms data from a spatial domain to a frequency domain and separates high frequency information and low frequency information.

Discrete Cosine Transform (DCT)

As the first step in JPEG compression, an original image is partitioned into 8x8 pixel blocks and the algorithm independently applies DCT to each block. DCT transforms data from the spatial domain to the frequency domain in which they can be more efficiently encoded.

The definition of forward discrete cosine transform (FDCT) and inverse discrete cosine transform (IDCT) is as follows:

$$F(u,v) = \frac{1}{4}C(u)C(v)[\sum_{i=0}^{7}\sum_{j=0}^{7}f(i,j) * \cos\frac{(2i+1)u\pi}{16}\cos\frac{(2j+1)v\pi}{16}] \quad (1.1)$$

$$f(i,j) = \frac{1}{4}[\sum_{u=0}^{7} \sum_{v=0}^{7} C(u)C(v)F(u,v) * \cos \frac{(2i+1)u\pi}{16} \cos \frac{(2j+1)v\pi}{16}] \quad (1.2)$$

where: $C(u), C(v) = 1/\sqrt{2}$ for $u, v = 0$, $C(u), C(v) = 1$ otherwise.

Using FDCT, original pixel values, $f(i,j)$ where $0 \le i, j < 8$, are transformed into $F(u,v)$ named DCT coefficient. The output DCT coefficient matrix represents information of the 8x8 block in frequency domain. DCT enhances compression by concentrating most of the energy in the signal in the lower spatial frequencies. The left upper corner of the matrix represents low frequency information while the right lower corner represents high frequency information. IDCT restores original pixel values from DCT coefficients.

Quantization

The purpose of quantization is to achieve high compression by representing DCT coefficients with no greater precision than necessary by discarding information which is not visually significant to human perception. After FDCT, each of the 64 DCT coefficients is quantized: $F[u,v] = Round(F[u,v]/q[u,v])$. For example, assume a DCT coefficient, $101101 = 45$ (6 bits). If it is quantized by a quanta value, $q[u,v] = 4$, the original DCT value is truncated to 4 bits, $1011 = 11$, reducing the required number of bits to represent a DCT coefficient. Due to many-to-one mapping, quantization is a fundamentally lossy process, and is the principal source of lossiness in DCT-based encoders. There are two different types of quantization: *uniform* and *non-uniform*. Uniform quantization divides each $F[u,v]$ by the same constant N. Non-uniform quantization uses a quantization table from psychovisual experiments to exploit the limits of the human visual system.

Zig-Zag Sequence

After quantization, the 8x8 matrix of quantized coefficients is ordered into a one-dimensional array (1x64) using the zig-zag sequence (Figure 1.6). This is to facilitate entropy coding by placing low-frequency coefficients on the top of the vector (maps 8x8 to 1x64 vector).

Entropy Coding

In order to achieve additional compression losslessly by encoding the quantized DCT coefficients more compactly based on statistics, JPEG applies Differential Pulse Code Modulation (DPCM) on the DC component which is a measure of the average value of the 64 image samples. DPCM is useful because there is usually a strong correlation between the DC coefficients of

DC   AC$_{01}$ AC$_{07}$

AC$_{70}$ AC$_{77}$

Figure 1.6. Zig-zag sequence.

adjacent 8x8 blocks. Run Length Encoding (RLE) is applied on AC components which usually have lots of zeros. For further compression, Huffman coding is used for both DC and AC coefficients at the last stage.

JPEG Operating Modes

JPEG supports the following operation modes to meet the various needs of many applications:

- Sequential Mode:
  Each image component is encoded in a single left-to-right, top-to-bottom scan as explained in this section.

- Lossless Mode:
  A special case of the JPEG where indeed there is no loss. It does not use a DCT-based method. Instead, it uses a predictive (differential coding) method. Typical compression ratio is 2:1.

- Progressive Mode:
  Each image is coded in multiple scans with each successive scan refining the image until a desired quality of image is achieved. Display of the encoded image follows the same steps: a coarser image is displayed first, then more components are decompressed to provide a finer version of the image.

- Hierarchical Mode:

  An image is compressed to multiple resolution levels. This makes it possible to display a high resolution image on a low resolution monitor by accessing a lower resolution version without having to decompress the full version.

1.3.4   MPEG

JPEG achieves intra-frame compression for still images by exploiting the redundancy in images (spatial redundancy). This intra-frame compression is not enough for video because it doesn't consider inter-frame compression between successive frames. We need a different scheme for video to exploit both spatial redundancy and temporal redundancy. MPEG is a de facto compression standard for video, audio, and their synchronization. MPEG (Moving Picture Coding Experts Group) was established in 1988 to create standards for delivery of video and audio. MPEG achieves intra-frame encoding using DCT-based compression for the reduction of spatial redundancy (similar to JPEG). Its inter-frame encoding utilizes block-based motion compensation for the reduction of temporal redundancy. Specifically, MPEG uses bidirectional motion compensation.

Block-based Motion Compensation

To exploit the temporal redundancy between adjacent video frames, *difference coding* compares pixels in the current frame with ones in the previous frame so that only changed pixels are recorded. In this way, a fraction of the number of pixel values will be recorded for the current frame. In practice, pixels values are slightly different even with no movement of objects. This can be interpreted as lots of changes and result in less compression. Thus, difference coding needs to ignore small changes in pixel values and it is naturally lossy.

For more efficient compression, difference coding is done at the block level. Block-based difference coding receives a sequence of blocks rather than frames. For example, a 160 x 120 image (19200 pixels) can be divided into 300 8x8 blocks. If a block in the current frame is similar to the one in the same location of the previous frame, the algorithm skips it or stores only the difference. If they are different, a whole block of pixels is updated at once. Limitations of difference coding are obvious. It is useless where there is a lot of motion (few pixels unchanged). What if objects in the frame move fast? What if a camera itself is moving?

Motion compensation assumes that the current frame can be modelled as a translation of the previous frame. As in block-based difference coding, the current frame is divided into uniform non-overlapping blocks. Each block in the current frame to be encoded is compared to areas of similar size from the preceding frame in order to find an area that is similar, i.e., the best matching block. The relative difference in locations is known as the *motion vector*. The matching process can be based on *prediction* or *interpolation*. Because fewer bits are required to code a motion vector than to code an actual block, compression is achieved. Motion compensation is the basis of most video compression algorithms.

For further and better compression, *bidirectional motion compensation* can be used. Areas just uncovered in the current frame are not predictable from the past, but they can be predicted from the future. Thus, bidirectional motion compensation searches in both past and future frames to find the best matching block. Moreover, the effect of noise and errors can be reduced by averaging between previous and future frames. Bidirectional interpolation also provides a high degree of compression. However, it requires that frames be encoded and transmitted in a different order from which they will be displayed. In reality, because exact matching is not possible, it is a lossy compression.

Group of Pictures

MPEG defines a set of pictures to form a group of pictures (GOP) consisting of the following types (Figure 1.7):

- I frame: Intra-coded picture

- P frame: Unidirectionally predicted picture

- B frame: Bidirectionally predicted picture

I-frames are encoded using intra-frame compression that is similar to JPEG. Thus, they can be decoded without other frames. I-frames are used as access points for random access. P-frames are predicted frames with reference to a previous I or P frame. B-frames are bidirectional frames encoded using the previous and the next I/P frames.

MPEG Standards

MPEG-1 (ISO/IEC 11172) is a standard for storing and playing video on a single computer at low bit-rates up to 1.5 Mb/s. MPEG-2 (ISO/IEC 13818) is a standard for high quality video such as digital TV (HDTV and DVD). MPEG-2 builds upon MPEG-1 standard and supports both field prediction

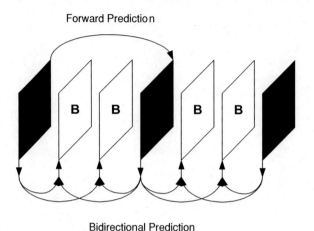

Figure 1.7. A group of pictures in MPEG.

and frame prediction (interlaced video format). While MPEG-2 aims at high quality video, MPEG-4 standard (ISO/IEC 14496) supports low bit-rate encoding of audio and video, user interactivity, and the special requirements for next generation broadcast services. MPEG-7 standard (ISO/IEC 15938) provides a set of standardized tools to describe multimedia content: metadata elements and their structure and relationships, that are defined by the standard in the form of Descriptors and Description Schemes. Officially, MPEG-7 is called *Multimedia Content Description Interface* [129]. MPEG-21 standard (ISO/IEC 21000) is being developed to define a multimedia framework in order to enable transparent and augmented use of multimedia resources across a wide range of networks and devices used by different communities.

Hierarchical Coding

Hierarchical coding in MPEG encodes images in a manner that facilitates access to images at different quality levels or resolutions. This makes the progressive transmission possible: Partial image information is transmitted in stages, and at each stage, the reconstructed image is progressively improved. Hierarchical coding is motivated by the need for transmitting images over low-bandwidth channels. Progressive transmission can be stopped either if an intermediate version is of satisfactory quality or the image is found to be of no interest. Hierarchical coding is also very useful in multi-use environments where applications need to support a number of display devices with different resolutions. It could optimize utilization of storage server and network resources.

MPEG Issues

The following issues are commonly considered in the design of streaming applications using MPEG:

- Avoiding propagation of errors. Due to the dependency among successive frames, errors such as missing frames or packets transfer over multiple frames. We can avoid this problem by sending an I-frame every once in a while (e.g., by setting a reasonable group of pictures).

- Bit-rate control. In general, complicated images result in a less compression and a higher bit-rate than simple images. To regulate the output bit-rate, a simple feedback loop based on buffer fullness is used. If the buffer is close to full, MPEG increases the quantization scale factor to reduce the size of data.

- Constant Bit Rate (CBR) vs. Variable Bit Rate (VBR). MPEG streams can be encoded either as constant bit rate or as variable bit rate. CBR approach is more appropriate for video broadcasting through fixed bandwidth channels while VBR supports fixed quality of images such as DVD better than CBR.

1.4   Delivery of Streaming Media Over Internet

Due to its ubiquitous existence, the Internet has become the platform of most networking activities including SM applications, where users require the integration of multimedia services. However, as a shared datagram network, the Internet is not naturally suitable for real-time traffic, such as SM. Dedicated links are not practical in many ways for transmitting SM data over the Internet. Moreover, because of the store-and-forwarding in routers and internetworking devices, Internet Protocol (IP) has some fundamental problems in transmitting real-time data. Thus, delivery of SM over the Internet requires special installation and new software development so that packets experience as little delay as possible. ATM is promising in transmitting real-time data because it supports high bandwidth connection-oriented transmission and various quality of service (QoS) for different applications. However, it is expensive and not widely available at user sites.

In general, the following issues must be resolved to stream multimedia data over the Internet.

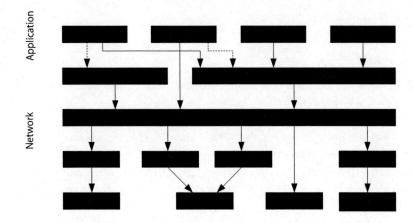

Figure 1.8. Network protocols for SM over the Internet.

- **High Bandwidth**
  The underlying hardware has to provide enough bandwidth to handle large size of multimedia data.

- **Multicast**
  Applications need to take into account multicast in order to reduce the traffic.

- **Guaranteed Bandwidth**
  There should be some mechanisms for real-time applications to reserve resources along the transmission path in order to guarantee QoS.

- **Out of Order Delivery**
  Applications need to take care of the timing issues so that audio and video data can be played back continuously with correct timing and synchronization. Applications also need to define standard operations for applications to manage the delivery and present the multimedia data.

This section explains several network protocols needed to support QoS over the Internet for many SM applications. Figure 1.8 shows their relations with other well-known protocols.

1.4.1   RTP

Media transport in many streaming applications is mainly implemented with RTP (Real-time Transport Protocol, RFC 1889) [148], which is an

IP-based transport protocol for audio and video conferences and other multi-participant real-time applications. It is a lightweight protocol without error correction or flow control functionality. Thus, it guarantees neither QoS nor resource reservation along the network path. RTP is also designed to work in conjunction with the auxiliary control protocol, RTCP, to get feedback on quality of data transmission and information about participants in the ongoing session. RTP is primarily designed for multicast of real-time data, but it can be also used in unicast. It can be used for one-way transport such as video-on-demand as well as interactive services such as Internet telephony.

Multimedia applications require appropriate timing in data transmission and playing back. To support real-time SM data transmission, RTP provides timestamping, sequence numbering, and other mechanisms. Through these mechanisms, RTP provides end-to-end transport for real-time data over the datagram network.

- *Timestamping* is the most important information for real-time applications. The sender sets the timestamp when the packet was sampled. The receiver uses the timestamp to reconstruct the original timing. It can be also used to synchronize different streams with timing properties, such as audio and video data in MPEG. However, RTP itself is not responsible for the synchronization. This has to be done at the application level.

- *Sequence numbers* are used to determine the correct order because UDP does not deliver packets in timely order. It is also used for packet loss detection. Notice that in some video formats, when a video frame is split into several RTP packets, all of them can have the same timestamp. So timestamp alone is not enough to put the packets in order.

- The *payload type identifier* specifies the payload format as well as the encoding/compression schemes. Using this identifier, receiving application knows how to interpret and play out the payload data. Example specifications include PCM, MPEG1/MPEG2 audio and video, et al. More payload types can be added by providing a profile and payload format specification. At any given time of transmission, an RTP sender can only send one type of payload, although the payload type may change during transmission, for example, to adjust to network congestion.

- *Source identification* allows the receiving application to know where the data is coming from. For example, in an audio conference, from the source identifier a user could tell who is talking.

RTP is typically run on top of UDP. RTP is primarily designed for multicast, the connection-oriented TCP does not scale well and therefore is not suitable. For real-time data, reliability is not as important as timely delivery. Even more, reliable transmission provided by retransmission as in TCP is not desirable. RTP and RTCP packets are usually transmitted using UDP/IP service.

1.4.2 RTCP

RTCP (RTP Control Protocol) is the control protocol designed to work in conjunction with RTP. In an RTP session, participants periodically send RTCP packets to convey feedback on quality of data delivery and information of membership:

- RR (receiver report): Receiver reports reception quality feedback about data delivery, including the highest packet number received, the number of packets lost, inter-arrival jitter, and timestamps to calculate the round-trip delay between the sender and the receiver.

- SR (sender report): Sender reports a sender information section, providing information on inter-media synchronization, cumulative packet counters, and number of bytes sent.

- SDES (source description items): Information to describe the sources.

- APP (application specific functions): It is now intended for experimental use as new applications and new features are developed.

Using the above messages, RTCP provides the following services to control data transmission with RTP:

- QoS monitoring and congestion control. RTCP provides feedback to an application about the quality of data distribution. Sender can adjust its transmission based on the receiver report feedback. Receivers can know whether a congestion is local, regional or global. Network managers can evaluate the network performance for multicast distribution.

- Source identification. RTCP SDES (source description) packets contain textual information called canonical names as globally unique identifiers of the session participants. They may include a user's name, telephone number, e-mail address and other information.

- Inter-media synchronization. RTCP sender reports contain an indication of real time and the corresponding RTP timestamp. This can be used in inter-media synchronization like lip synchronization in video.

- Control information scaling. RTCP packets are sent periodically among participants. In order to scale up to large multicast groups, RTCP has to prevent the control traffic from overwhelming network resources. RTP limits the control traffic to at most 5% of the overall session traffic. This is enforced by adjusting the RTCP generating rate according to the number of participants.

1.4.3   RTSP

RTSP (Real Time Streaming Protocol, RFC 2326) is a client-server multimedia presentation protocol to enable controlled delivery of streamed multimedia data over IP network. RTSP provides methods to realize commands (play, fast-forward, fast-rewind, pause, stop) similar to the functionality provided by CD players or VCRs. RTSP is an application-level protocol designed to work with lower-level protocols like RTP and RSVP to provide a complete streaming service over the Internet. It can act as a network remote control for multimedia servers and can run over TCP or UDP. RTSP can control either a single or several time-synchronized streams of continuous media.

RTSP provides the following operations:

- Retrieval of media from media server. The client can request a presentation description, and ask the server to setup a session to send the requested data.

- Invitation of a media server to a conference. The media server can be invited to the conference to play back media or to record a presentation.

- Adding media to an existing presentation. The server and the client can notify each other about any additional media becoming available.

In RTSP, each presentation and media stream is identified by an RTSP URL. The overall presentation and the properties of the media are defined in a presentation description file, which may include the encoding, language, RTSP URLs, destination address, port, and other parameters. The presentation description file can be obtained by the client using HTTP, e-mail or other means. RTSP aims to provide the same services for streamed audio and video just as HTTP does for text and graphics. But RTSP differs from HTTP in several aspects. First, while HTTP is a stateless protocol, an RTSP server has to maintain "session states" in order to correlate RTSP requests with a stream. Second, HTTP is basically an asymmetric protocol where the client issues requests and the server responds, but in RTSP both the media server and the client can issue requests. For example, the server can issue a request to set playback parameters of a stream.

1.4.4   RSVP

RSVP (Resource reSerVation Protocol) is the network control protocol that allows the data receiver to request a special end-to-end quality of service for its data flows. Thus, real-time applications can use RSVP to reserve necessary resources at routers along the transmission paths so that the requested bandwidth can be available when the transmission actually takes place. RSVP is a main component of the future Integrated Services Internet which can provide both best-effort and real-time service.

RSVP is used to set up reservations for network resources. When an application in a host (the data stream receiver) requests a specific quality of service (QoS) for its data stream, it uses RSVP to deliver its request to routers along the data stream paths. RSVP is responsible for the negotiation of connection parameters with these routers. If the reservation is set up, RSVP is also responsible for maintaining router and host states to provide the requested service. Each node capable of resource reservation has several local procedures for reservation setup and enforcement. Policy control determines whether the user has administrative permission to make the reservation (authentication, access control and accounting for reservation). Admission control keeps track of the system resources and determines whether the node has sufficient resources to supply the requested QoS.

RSVP is also designed to utilize the robustness of current Internet routing algorithms. RSVP does not perform its own routing; instead it uses underlying routing protocols to determine where it should carry reservation requests. As routing changes paths to adapt to topology changes, RSVP adapts its reservation to the new paths wherever reservations are in place.

1.5   Outline of the Book

The book is organized into the following chapters:

Chapter 2 concentrates on single disk platform SM server design, by presenting different techniques in support of a hiccup-free display. Even though a single disk server is not practical in many applications, however, it presents the fundamental concepts and techniques in designing SM servers.

Chapter 3 extends the discussion to multiple disk platform SM server design. It introduces possible design approaches: 1) cycle-based scheduling with round-robin data placement approach and 2) deadline-driven scheduling with unconstrained data placement approach. In addition, it presents a number of optimization techniques and an online reconfiguration process.

Chapter 4 deals exclusively with deadline-driven scheduling and unconstrained media placement approach. It quantifies the hiccup probability of this approach and presents techniques to reduce this probability.

Chapter 5 extends the discussion to a heterogenous disk platform. It presents a number of techniques that take advantage of the rapid development in disk storage devices.

Chapter 6 deals with fault-tolerance issues in SM servers. It presents techniques for homogenous disk platforms, then it extends the discussion to a heterogenous disk platform.

Chapter 7 is devoted to hierarchical storage system design for SM servers. It presents a pipelining technique to ensure the continuous display from tape jukeboxes and data placement techniques to improve the access time to tape resident SM objects.

In Chapter 8 and Chapter 9, the concept of distributed SM servers are introduced. Chapter 8 presents RedHi, a distributed SM server and its network components. Chapter 9 presents a super-streaming mechanism that takes advantage of a distributed system.

Chapter 10 presents a case study on the design of a second generation SM server, namely Yima, while Appendix A provides instructions on installing a personal version of Yima.

Chapter 2

SINGLE DISK PLATFORM SM SERVERS

The fundamental functionality of SM servers is a *hiccup-free display* of SM. However, just supporting a continuous display is not enough in the design of SM servers because many real applications, especially commercial ones such as movie-on-demand systems that concurrently service multiple users, require maximizing the performance of servers for cost-effective solutions. Thus, the following performance metrics are important in the design and implementation of SM servers:

1. *Throughput*, i.e., the number of simultaneous displays that can be supported by a SM server.

2. *Startup Latency*, i.e., the amount of time. elapsed from when a display request arrives at the system until the time the actual display is initiated by the system on behalf of this request.

Throughput, in general, is closely related to another important metric of SM servers, namely *cost per stream* (*CPS*). If a technique supports a higher throughput with fixed resources than others, it provides for a more cost-effective solution (i.e., lower CPS)[1].

Magnetic disk drives have been the choice of storage devices for the design of SM servers due to their high data transfer rate, large storage capacity, random access capability, and low price. Therefore, many studies have investigated hiccup-free display of SM using magnetic disk drives [9], [12], [13], [26], [59], [136], [138], [175], [183], [70].

This chapter describes different techniques in support of a hiccup-free display of SM with a single disk drive. Even though a single disk server is

---

[1]Throughout this textbook (in general) and this chapter (in particular) we compare alternative designs based on these metrics.

not practical in many applications, we still can describe many fundamental concepts and techniques in designing SM servers without introducing much higher complexity at the beginning. Most concepts and techniques can be straightforwardly applied to more complicated multi-disk platforms described in the later chapters.

In Section 2.1, we first describe the characteristics of modern disk drives and low-level SCSI programming interface to these drives. In Section 2.2, we present an overview of SM techniques discussed in this chapter. Next, in Section 2.3, we present three alternative disk scheduling techniques for SM display: Simple, SCAN, and GSS. Subsequently, in Section 2.4, we present a constrained data placement techniques, namely: **REgion BasEd bloCk Allocation** (REBECA). Finally, in Section 2.5, we describe four alternative techniques to support SM display using a single multi-zone disk: track pairing (TP), logical track (LT), FIXB, and VARB.

2.1 Modern Disk Drives

The magnetic disk drive technology has benefited from more than two decades of research and development. It has evolved to provide a low latency (in the order of milliseconds) and a low cost per megabyte (MByte) of storage (a few cents per MByte at the time of this writing in 2003). It has become commonplace with annual sales in excess of 30 billion dollars [124]. Magnetic disk drives are commonly used for a wide variety of storage purposes in almost every computer system. To facilitate their integration and compatibility with a wide range of host hardware and operating systems, the interface that they present to the rest of the system is well defined and hides a lot of the complexities of the actual internal operation. For example, the popular *SCSI* (Small Computer System Interface, see [8], [43]) standard presents a magnetic disk drive to the host system as a linear vector of storage blocks (usually of 512 bytes each). When an application requests the retrieval of one or several blocks the data will be returned after some (usually short) time, but there is no explicit mechanism to inform the application exactly how long such an operation will take. In many circumstances such a "best effort" approach is reasonable because it simplifies program development by allowing the programmer to focus on the task at hand instead of the physical attributes of the disk drive. However, for a number of data intensive applications, such as SM servers, exact timing information is crucial to satisfy the real-time constraints imposed by the requirement for a hiccup-free delivery of audio and video streams. Fortunately, with a model that imitates the internal operation of a magnetic disk drive it is possible to predict service

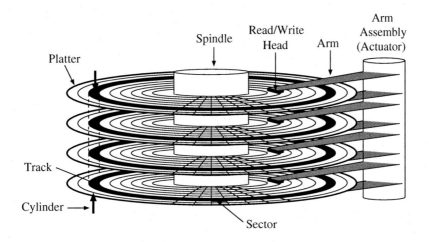

Figure 2.1. Disk drive internals.

times at the level of accuracy that is needed to design and configure SM server storage systems.

2.1.1 Internal Operation

We will first give an overview of the internal operation of modern magnetic disk drives. Next, we will introduce a model that allows an estimation of the service time of a disk drive. This will build the basis for our introduction of techniques that provide SM services on top of disk storage systems.

A magnetic disk drive is a mechanical device, operated by its controlling electronics. The mechanical parts of the device consist of a stack of platters that rotate in unison on a central spindle (see [139] for details). Presently, a single disk contains one, two, or as many as fourteen platters[2] (see Figure 2.1 for a schematic illustration). Each platter surface has an associated disk head responsible for reading and writing data. The minimum storage unit on a surface is a *sector* (which commonly holds 512 bytes of user data). The sectors are arranged in multiple, concentric circles termed *tracks*. A single stack of tracks across all the surfaces at a common distance from the spindle is termed a *cylinder*. To access the data stored in a series of sectors, the disk head must first be positioned over the correct track. The operation to reposition the head from the current track to the target track is termed a *seek*. Next, the disk must wait for the desired data to rotate under the head. This time is termed *rotational latency*.

---

[2]This is the case, for example, for the Elite series from Seagate Technology™, LLC.

Model	ST31200WD	ST32171WD	ST34501WD	ST336752LC
Series	Hawk 1LP	Barracuda 4LP	Cheetah 4LP	Cheetah X15
Manufacturer		Seagate Technology^{TM}, LLC		
Capacity C	1.006 GByte	2.061 GByte	4.339 GByte	34.183 GByte
Avg. transfer rate R_D	3.47 MB/s	7.96 MB/s	12.97 MB/s	49.28 MB/s
Spindle speed	5,400 rpm	7,200 rpm	10,033 rpm	15,000 rpm
Avg. rotational latency	5.56 msec	4.17 msec	2.99 msec	2.0 msec
Worst case seek time	21 msec	19 msec	16 msec	6.9 msec
Surfaces	9	5	8	8
Cylinders $\#cyl$	2,697	5,177	6,582	18,479
Number of Zones Z	23	11	7	9
Sector size	512 bytes	512 bytes	512 bytes	512 bytes
Sectors per Track ST	59 - 106	119 - 186	131 - 195	485 (avg.)
Sector ratio $\frac{ST_{Z_0}}{ST_{Z_{N-1}}}$	$\frac{106}{59} = 1.8$	$\frac{186}{119} = 1.56$	$\frac{195}{131} = 1.49$	
Introduction year	1990	1993	1996	2000

Table 2.1. Parameters for four commercial disk drives. Each drive was considered high performance when it was first introduced.

In their quest to improve performance, disk manufacturers have introduced the following state-of-the-art designs features into their most recent products:

- Zone-bit recording (ZBR). To meet the demands for a higher storage capacity, more data is recorded on the outer tracks than on the inner ones (tracks are physically longer towards the outside of a platter). Adjacent tracks with the same data capacity are grouped into *zones*. The tighter packing allows more storage space per platter compared with a uniform amount of data per track [173]. Moreover, it increases the data transfer rate in the outer zones (see the *sector ratio* in Table 2.1). Section 2.5 details how multi-zoning can be incorporated in SM server designs.

- Statistical analysis of the signals arriving from the read/write heads, referred to as *partial-response maximum-likelihood* (PRML) method [81], [177], [51]. This digital signal processing technique allows higher recording densities because it can filter and regenerate packed binary signals better than the traditional peak-detection method. For example, the switch to PRML in Quantum Corporation's Empire product series allowed the manufacturer to increase the per-disk capacity from 270 MBytes to 350 MBytes [51].

- High spindle speeds. Currently, the most common spindle speeds are 5,400 and 7,200 rpm. However, high performance disks use 10,000 and

15,000 rpm (see Table 2.1). All other parameters being equal, this
results in a data transfer rate increase of 39%, or 108% respectively,
over a standard 7,200 rpm disk.

Some manufacturers also redesigned disk internal algorithms to provide uninterrupted data transfer (e.g., to avoid lengthy thermal recalibrations) specifically for SM applications. Many of these technological improvements have
been introduced gradually with new disk generations entering the marketplace every year. To date (i.e., in 2003) disk drives of many different performance levels with capacities ranging from 20 up to 200 gigabytes are
commonly available[3].

The next section details disk modelling techniques to identify the physical
characteristics of a magnetic disk in order to estimate its service time. These
estimates have been applied to develop and implement Yima, a scalable
SM server [160] presented as a case study in Chapter 10. Our techniques
employ no proprietary information and are quite successful in developing a
methodology to estimate the service time of a disk.

2.1.2   Disk Drive Modelling

Disk drive simulation models can be extremely helpful to investigate performance enhancements or tradeoffs in storage subsystems. Hence, a number
of techniques to estimate disk service times have been proposed in the literature [178], [17], [139], [181], [75]. These studies differ in the level of detail
that they incorporate into their models. The level of detail should depend
largely on the desired accuracy of the results. More detailed models are
generally more accurate, but they require more implementation effort and
more computational power [55]. Simple models may assume a fixed time for
I/O, or they may select times from a uniform distribution. However, to yield
realistic results in simulations that imitate real-time behavior, more precise
models are necessary. The most detailed models include all the mechanical
positioning delays (seeks, rotational latency), as well as on-board disk block
caching, I/O bus arbitration, controller overhead, and defect management.
In the next few paragraphs we will outline our model and which aspects we
chose to include.

On-Board Cache

Originally, the small buffers of on-board memory implemented with the device electronics were used for speed-matching purposes between the media

---

[3]The highest capacity disk drives currently available with 200 GByte of storage space
are the WD Caviar 200 GB models from Western Digital Corporation.

data rate and the (usually faster) I/O bus data transfer rate. They have more recently progressed into dynamically managed, multi-megabyte caches that can significantly affect performance for traditional I/O workloads [139]. In the context of SM servers, however, data is retrieved mostly in sequential order and the chance of a block being re-requested within a short period of time is very low. Most current disk drives also implement aggressive read-ahead algorithms which continue to read data into the cache once the read/write head reaches the final position of an earlier transfer request. It is the assumption that an application has a high chance of requesting these data in the cache because of spatial locality. Such a strategy works very well in a single-user environment, but has a diminishing effect in a multi-programming context because disk requests from multiple applications may be interleaved and thus requests may be spatially disbursed[4]. In a SM server environment, a single disk may serve dozens of requests in a time period on behalf of different streams with each request being rather large in size. It has been our observation that disk caches today are still not large enough to provide any benefits under these circumstances. Hence, our model does not incorporate caching strategies.

Data Layout

To make a disk drive self-contained and present a clean interface to the rest of the system, its storage is usually presented to the host system as an addressable vector of data blocks (see Section 2.1). The logical block number (LBN) of each vector element must internally be mapped to a physical media location (i.e., sector, cylinder, and surface). Commonly, the assignment of logical numbers starts at the outermost cylinder, covering it completely before moving on to the next cylinder. This process is repeated until the whole storage area has been mapped.

This LBN-to-PBN mapping is non-contiguous in most disk drives because of techniques used to improve performance and failure resilience. For example, some media data blocks may be excluded from the mapping to store internal data needed by the disk's firmware. For defect management some tracks or cylinders in regular intervals across the storage area may be skipped initially by the mapping process to allow the disk firmware later to relocate sectors that became damaged during normal operation to previously unused areas.

As a result, additional seeks may be introduced when a series of logically

---

[4]This problem can be alleviated to some extent with the implementation of segmented caches.

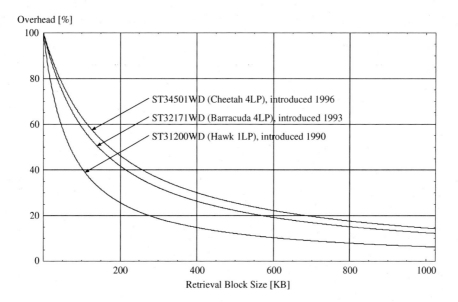

Figure 2.2. Mechanical positioning overhead as a function of the retrieval block size for three disk drive models. The overhead includes the seek time for the disk heads to traverse over half of the total number of cylinders plus the average rotational latency, i.e., $T_{Seek}(\frac{\#cyl}{2})$ (for details see Equation 2.2 and Table 2.2). The technological trends are such that the data transfer rate is improving faster than the delays due to mechanical positioning are reduced. Hence the wasteful overhead is increasing, relatively speaking, and seek-reducing scheduling algorithms are becoming more important.

contiguous blocks are retrieved by an application. It is possible to account for these anomalies when scheduling data retrievals, because most modern disk drives can be queried as to their exact LBN-to-PBN function. The effects of using accurate and complex mapping information versus a simple, linear approximation have been investigated and quantified in [180]. This study reports possible marginal improvements of less than 2% in combination with seek-reducing scheduling algorithms.

Scheduling Algorithms

Scheduling algorithms that attempt to reduce mechanical positioning delays can drastically improve the performance of magnetic disk drives. Figure 2.2 shows the seek and rotational latency overhead incurred by three different disk drive types when retrieving a data block that requires the read/write heads to move across half of the total number of cylinders. As illustrated, the overhead may waste a large portion of a disk's potential bandwidth. Furthermore, it is interesting to note that newer disk models spend a proportionally larger amount of time on wasteful operations (e.g., the Cheetah 4LP model is newer than the Barracuda 4LP, which in turn is newer than the Hawk 1LP). The reasons can be found in the technological trends which indicate that the media data rate is improving at a rate of approximately 40% per year while the mechanical positioning delays lag with an annual decrease of only 5%. Hence, seek-reducing algorithms are essential in conjunction with modern, high-performance disk drives.

More than 30 years of research have passed since Denning first analyzed the *Shortest Seek Time First* (SSTF) and the SCAN or *elevator*[5] policies [44], [138]. Several variations of these algorithms have been proposed in the literature, among them *Cyclical SCAN* (C-SCAN) [151], LOOK [113], and C-LOOK [180] (a combination of C-SCAN and LOOK). Specifically designed with SM applications in mind was the *Grouped Sweeping Scheme* (GSS) [183].

There are various tradeoffs associated with choosing a disk scheduling algorithm. For SM retrieval it is especially important that a scheduling algorithm is fair and starvation-free so that real-time deadlines can be guaranteed. A simplistic approach to guarantee a hiccup-free continuous display would be to assume the worst case seek time for each disk type, i.e., setting the seek distance d equal to the total number of cylinders, $d = \#cyl$. However, if multiple displays ($\mathcal{N}$) are supported simultaneously, then $\mathcal{N}$ fragments need to be retrieved from each disk during a time period. In the worst case scenario, the $\mathcal{N}$ requests might be evenly scattered across the disk surface. By ordering the fragments according to their location on the disk surface (i.e., employing the SCAN algorithm), an optimized seek distance of $d = \frac{\#cyl}{\mathcal{N}}$ can be obtained, reducing both the seek and the service time. This scheduling technique was employed in all our analytical models and simulation experiments.

Bus Interface

We base our observations on the SCSI I/O bus because it is most commonly available on today's high performance disk drives. The SCSI standard is both a bus specification and a command set to efficiently use that bus (see [8], [43]). The *host adapter* (also called *initiator*) provides a bridge between the system bus (based on any standard or proprietary bus; e.g., PCI[6]) and the

---

[5]This algorithm first serves requests in one direction across all the cylinders and then reverses itself.

[6]Peripheral Component Interconnect, a local bus specification developed for 32-bit or 64-bit computer system interfacing.

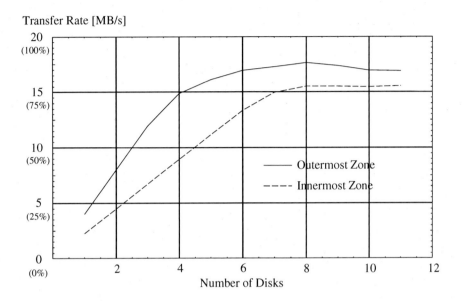

Figure 2.3. Example SCSI bus scalability with a Fast&Wide bus that can sustain a maximum transfer rate of 20 megabytes per second (MB/s). The graphs show read requests issued to both the innermost and outermost zones for varying numbers of disks. The bandwidth scales linearly up to about 15 MB/s (75%) and then starts to flatten out as the bus becomes a bottleneck.

SCSI I/O bus. The individual storage devices (also called *targets* in SCSI terminology) are directly attached to the SCSI bus and share its resources.

There is overhead associated with the operation of the SCSI bus. For example, the embedded controller on a disk drive must interpret each received SCSI command, decode it, and initiate its execution through the electronics and mechanical components of the drive. This controller overhead is usually less than 1 millisecond for high performance drives ([139] cite 0.3 to 1.0 msec, [181] measured an average of 0.7 msec, and [121] reports 0.5 msec for a cache miss and 0.1 msec with a cache hit). As we will see in the forthcoming sections, SM retrieval requires large block retrievals to reduce the wasted disk bandwidth. Hence, we consider the controller overhead negligible (less than 1% of the total service time) and do not include it in our model.

The issue of the limited bus bandwidth should be considered as well. The SCSI I/O bus supports a peak transfer rate of 10, 20, 40, 80, 160, or 320 MB/s depending on the SCSI version implemented on both the controller and all the disk devices. Current variations of SCSI are termed Fast, Fast&Wide, Ultra, Ultra&Wide, Ultra2&Wide, Ultra160, or Ultra320 respectively. If the aggregate bandwidth of all the disk drives attached to a SCSI bus exceeds

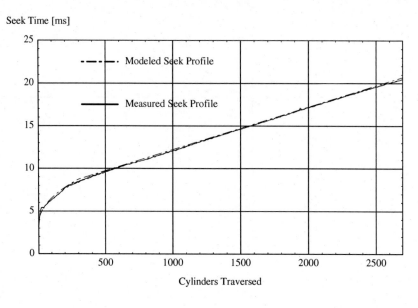

Figure 2.4. Example measured and modelled seek profile for a disk type ST31200WD.

its peak transfer rate, then bus contention occurs and the service time for the disks with lower priorities[7] will increase. This situation may arise with today's high performance disk drives. For example, SCSI buses that implement the wide option (i.e., 16-bit parallel data transfers) allow up to 15 individual devices per bus, which together can transfer data in excess of 600 MB/s. Furthermore, if multiple disks want to send data over this shared bus, arbitration takes place by the host bus adapter to decide which request has higher priority. In an earlier study we investigated these scalability issues and concluded that bus contention has a negligible effect on the service time if the aggregate bandwidth of all attached devices is less than approximately 75% of the bus bandwidth (see Figure 2.3 [75]). We assume that this rule is observed to accomplish a balanced design.

Mechanical Positioning Delays and Data Transfer Time

The disk service time (denoted $T_{Service}$) is composed of the data transfer time (desirable) and the mechanical positioning delays (undesirable), as in-

---

[7]Every SCSI device has a fixed priority that directly corresponds to its SCSI ID. This number ranges from 0 to 7 for narrow (8-bit) SCSI. The host bus adapter is usually assigned ID number 7 because this corresponds to the highest priority. The wide SCSI specification was introduced after the narrow version. Consequently, the priorities in terms of SCSI IDs are 7 ... 0, 15 ... 8, with 7 being the highest and 8 the lowest.

Metric	Disk Model			
	Hawk 1LP ST31200WD	Barracuda 4LP ST32171WD	Cheetah 4LP ST34501WD	Cheetah X15 ST336752LC
Seek constant c_1 [msec]	$3.5 + 5.56^a$	$3.0 + 4.17^a$	$1.5 + 2.99^a$	$1.5 + 2.0^a$
Seek constant c_2 [msec]	0.303068	0.232702	0.155134	0.029411
Seek constant c_3 [msec]	$7.2535 + 5.56^a$	$7.2814 + 4.17^a$	$4.2458 + 2.99^a$	$2.4 + 2.0^a$
Seek constant c_4 [msec]	0.004986	0.002364	0.001740	0.000244
Switch-over point z	300 cyl.	600 cyl.	600 cyl.	924 cyl.
Total size #cyl	2,697 cyl.	5,177 cyl.	6,578 cyl.	18,479 cyl.

[a]Average rotational latency based on the spindle speed: 5,400 rpm, 7,200 rpm, 10,033 rpm, and 15,000 rpm respectively.

Table 2.2. Seek profile model parameters for four commercially available disk models.

troduced earlier in this chapter. For notational simplicity in our model, we will subsume all undesirable overhead into the expression for the seek time (denoted T_{Seek}). Hence, the composition of the service time is as follows:

$$T_{Service} = T_{Transfer} + T_{Seek} \qquad (2.1)$$

The transfer time ($T_{Transfer}$) is a function of the amount of data retrieved and the data transfer rate of the disk: $T_{Transfer} = \frac{B}{R_D}$. The seek time ($T_{Seek}$) is a non-linear function of the number of cylinders traversed by the disk heads to locate the proper data sectors. A common approach (see [139], [181], [75], [121]) to model such a seek profile is a first-order-of-approximation with a combination of a square-root and linear function as follows:

$$T_{Seek}(d) = \begin{cases} c_1 + (c_2 \times \sqrt{d}) & \text{if } d < z \text{ cyl.} \\ c_3 + (c_4 \times d) & \text{if } d \geq z \text{ cyl.} \end{cases} \qquad (2.2)$$

where d is the seek distance in cylinders. In addition, the disk heads need to wait on average half a platter rotation—once they have reached the correct cylinder—for the data to move underneath the heads. Recall that this time is termed rotational latency and we include it in our model of T_{Seek} (see c_1 and c_3 in Table 2.2). Every disk type has its own distinct seek profile. Figure 2.4 illustrates the measured and modelled seek profile for a Seagate ST31200WD disk drive (shown without adding the rotational latency) and Table 2.2 lists the corresponding constants used in Equation 2.2. These empirical results have been obtained with the embedded differential Fast&Wide SCSI I/O interface of an HP 9000 735/125 workstation.

```
int scsi_inquiry (int handle)
    /* This procedure will send a SCSI INQUIRY command to the specified  */
    /* device and print the vendor and product names as well as the      */
    /* product revision. The file handle must be obtained by opening a    */
    /* generic SCSI device, e.g., handle = fileno(fopen("/dev/sga", "w+")); */
{

    int           status, i;
    unsigned char  buffer[1124];
    unsigned char *cmd;
    unsigned char *pagestart;
    unsigned char  tmp;

    memset(buffer, '\0'; 1124);

    *((int *)buffer) = 0;                /* length of input data */
    *(((int *)buffer) + 1) = 1024;     /* length of output buffer */
    cmd = (char *)(((int *)buffer) + 2);

    /* For the exact input and output formats of the SCSI INQUIRY command */
    /* please see the SCSI specifications and clause 8.2.5 online at:      */
    /* http://www.danbbs.dk/~dino/SCSI/SCSI2-08.html#8.2.5                 */

    cmd[0] = 0x12;      /* SCSI INQUIRY command code */
    cmd[1] = 0x00;      /* lun=0, evpd=0 */
    cmd[2] = 0x00;      /* page code = 0 */
    cmd[3] = 0x00;      /* (reserved) */
    cmd[4] = 0xff;      /* allocation length */
    cmd[5] = 0x00;      /* control */

    status = ioctl(handle, 1 /* SCSI_IOCTL_SEND_COMMAND */, buffer);
    if (status) {
        printf("ioctl(SCSI_IOCTL_SEND_COMMAND) status\t= %d\n", status);
        return status;
    }

    pagestart = buffer + 8;
    printf("SCSI: Inquiry Information\n");

    tmp = pagestart[16];
    pagestart[16] = 0;
    printf("%s%s\n", "Vendor:              ", pagestart + 8);
    pagestart[16] = tmp;
    tmp = pagestart[32];
    pagestart[32] = 0;
    printf("%s%s\n", "Product:             ", pagestart + 16);
    pagestart[32] = tmp;
    printf("%s%s\n", "Revision level:      ", pagestart + 32);
    printf("\n");

    return status;
}
```

Figure 2.5. Sample code fragment to send the SCSI command INQUIRY to the controller electronics of a SCSI device under Linux. The code utilizes the Linux SCSI Generic (sg) driver. More information is available online at http://gear.torque.net/sg/.

2.1.3   Low-Level SCSI Programming Interface

Information about disk characteristics such as the seek profile or the detailed zone layout is not generally available. Hence, it is necessary to device methods to measure these parameters. The SCSI standard specifies many useful commands that allow device interrogation. However, SCSI devices are normally controlled by a device driver, which shields the application programmer from all the low-level device details and protects the system from faulty programs. In many operating systems, for example Linux and Windows NT™, it is possible to explicitly send SCSI commands to a device through specialized system calls. Great care must be taken when issuing such commands because the usual error detection mechanisms are bypassed.

The format to initiate a SCSI command is operating system dependent. For example, Linux utilizes variations of the `ioctl()`, `read()` and `write()` system calls while Windows NT provides the advanced SCSI programming interface (ASPI)[8]. The syntax and semantics of the actual SCSI commands are defined in the SCSI standard document [8]. Three types of SCSI commands can be distinguished according to their command length: 6, 10, and 12 bytes. The first byte of each command is called *operation code* and uniquely identifies the SCSI function. The rest of the bytes are used to pass parameters. The standard also allows vendor-specific extensions for some of the commands, which are usually described in the vendor's technical manuals of the device in question.

Figure 2.5 shows a sample code fragment that illustrates how a SCSI command is sent to a device under Linux. In the example, the SCSI command `INQUIRY` is transmitted to the device to obtain vendor and product information. The shown SCSI command is 6 bytes in length and its command code is 0x12. Each SCSI command has its own input and output data format that is described in the SCSI standard document [8]. SCSI commands to read and write data exist together with a host of other functions. For example, to translate a linear logical block address into its corresponding physical location of the form ⟨head/cylinder/sector⟩ one may use the `SEND DIAGNOSTICS` command to pass the translate request to the device (the translate request is identified through the `XLATE_PAGE` code in the parameter structure). By translating a range of addresses of a magnetic disk drive, track and cylinder sizes as well as zone boundaries can be identified.

---

[8]ASPI was defined by Seagate Technology, LLC.

2.2   Overview of SM Techniques

A basic approach to support continuous display in a single disk SM server is to divide an SM object into equi-sized blocks. Each block is a unit of retrieval and is stored contiguously. For example, a SM object X is divided into n equi-sized blocks: $X_0, X_1, X_2, \ldots, X_{n-1}$. The size of blocks, the display time of a block, and the time to read a block from a disk drive can be calculated as a function of display bandwidth requirement of an object, the number of maximum simultaneous displays that a disk drive can support, and the physical disk characteristics such as the data transfer rate. Upon the request for the SM object X, the system stages the first block of X, i.e., X_0, from the disk into main memory and initiates its display. Prior to the completion of the display of X_0, the system stages the next block X_1 from the disk into main memory to provide for a smooth transition and a hiccup-free display. This process is repeated until all blocks of X are displayed. This process introduces the concept of a *time period* (T_p), which denotes the time to display a block. For example, the display time of one 0.5 MByte block of a SM object encoded with 4 Mb/s is one second $(T_p = 1 \text{ sec})$.

In general, the display time of a block is longer than its retrieval time from a disk drive. Thus, the bandwidth of a disk drive can be multiplexed among multiple simultaneous displays accessing the same disk drive. For example, with the 4 Mb/s of display bandwidth requirement of MPEG-2 encoded objects, a disk drive with the 80 Mb/s of data transfer rate can support up to 20 simultaneous displays. This is the ideal case when there is no overhead in disk operation. However, in reality, a magnetic disk drive is a mechanical device and incurs a delay when required to retrieve data. This delay consists of: 1) *seek time* to reposition the disk head from the current track to the target track, and 2) *rotational latency* to wait until the data block arrives under the disk head. These are wasteful operations that prevent a disk drive from transferring data. Both their number of occurrence and duration of each occurrence must be reduced in order to maximize the number of simultaneous displays supported by a disk drive. Thus, the performance of SM servers significantly depends on the physical characteristics of magnetic disk drives such as data transfer rate, seek times, and rotational latency. For example, it is obvious that we can increase the throughput of a server using disk drives having a higher data transfer rate.

Another important physical characteristic of a magnetic disk drive is its zones: a disk consists of several zones with each providing a different storage capacity and transfer rate. Zone-bit recording (ZBR) is an approach utilized by disk manufactures to increase the storage capacity of magnetic disks, as described in Section 2.1.1. This technique groups adjacent disk cylinders into zones [139], [121]. Tracks are longer towards the outer portions of a disk platter as compared to the inner portions, hence, more data may be recorded in the outer tracks when the maximum linear density, i.e., bits per inch, is applied to all tracks. A zone is a contiguous collection of disk cylinders whose tracks have the same storage capacity, i.e., the number of sectors per track is constant in the same zone. Hence, outer tracks have more sectors per track than inner zones. Different disk models have a different number of zones. For example, a disk may consist of 7 zones while another one may consist of 9 zones.

Different zones provide a different transfer rate because: 1) the storage capacity of the tracks for each zone is different, and 2) the disk platters rotate at a fixed number of revolutions per second. Assuming a constant recording density in tracks, typical disks provide the maximum data transfer rates from the outermost zones and the minimum ones from the innermost zones. One can observe a significant difference in data transfer rates between the minimum and maximum (more than 50%).

This chapter introduces two important components of a SM server design: *scheduling* of the disk bandwidth and *placement* of data blocks. Traditionally, SM servers employ the *scheduling* of the available disk bandwidth to guarantee a continuous display and to maximize the throughput by minimizing the wasteful work of disk drives. Among various possible disk scheduling algorithms such as first-come-first-serve, shortest-seek-first, SCAN, elevator [44], and earliest-deadline-first (EDF) [168], [104], two scheduling approaches are widely utilized: *deadline-driven* and *cycle-based*. A technique for real-time scheduling of I/O tasks is a deadline-driven approach and it can be applied for the scheduling of SM data retrievals [138]. With this approach, a request for a block is tagged with a deadline that ensures a continuous display. Disks are scheduled to service requests with EDF. A limitation of this approach is that seek times may not be optimized because the sequence of block retrievals are determined by deadlines.

A cycle-based scheduling technique [70], [126], [170] is an approach exploiting the periodic nature of SM display. A time period is partitioned into a number of time slots ($\mathcal{N}$) such that the duration of a slot is long enough to retrieve a block from a disk. The number of slots denotes the number of simultaneous displays supported by the system. During a given time period (or a cycle), the system retrieves up to $\mathcal{N}$ blocks, only one block for each display. Block requests for the next cycle are not issued until the current cycle ends. During the next cycle, the system retrieves the next blocks for displays in a cyclic manner. For example, assuming that the system can support up to three block retrievals during a time period, suppose that the system retrieves X_i, Y_j, and Z_k for a given time period. During the next time period, it retrieves X_{i+1}, Y_{j+1}, and Z_{k+1}.

Another way to reduce the worst seek time is to control the location of data blocks. Based on the observation of the typical sequential access pattern of continuous data blocks, one can place adjacent data blocks in the way that minimize the seek time between two retrievals.

A number of studies have investigated techniques to support a hiccup-free display of SM using magnetic disk drives with a single zone [9], [12], [13], [26], [59], [135], [136], [138], [175], [183], [70]. These studies assume a fixed data transfer rate for a disk drive. If a system designer elects to use one of these techniques with multi-zone disks, the system is forced to use the minimum data transfer rate of the zones for the entire disk in order to guarantee a continuous display of video objects. This approach results in a significant reduction of data transfer rate by wasting the transfer rates of outer zones. A couple of studies strive to cure this limitation by modelling a multi-zone disk drive as a logical single-zone disk. Track Pairing (TP) [15] provides logical tracks with an identical storage capacity by pairing two physical tracks (see Section 2.5.1). Logical Track (LT) [85] constructs a logical track by combining one physical track from each zone (see Section 2.5.2). Then, one can construct a SM server based on logical single-zone disks using one of traditional continuous display techniques. We introduce two alternative techniques, FIXB and VARB, that harness the average transfer rate of zones while ensuring a continuous display.

2.3   Disk Scheduling for SM Display

2.3.1   Simple Technique

We make the following simplifying assumptions:

1. The system is configured with a fixed amount of memory and a single disk drive. The disk drive has a fixed data transfer rate (R_D).

2. The objects that constitute the SM server belong to a single media type and require a fixed bandwidth for their display (R_C).

3. $R_D > R_C$.

4. A multi-user environment requiring simultaneous display of objects to different users.

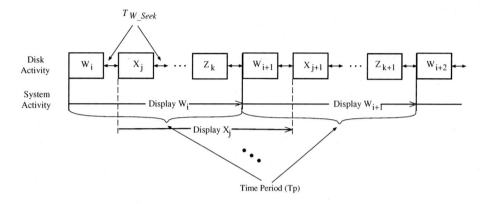

Figure 2.6. Continuous display with Simple.

To support continuous display of an object X, it is partitioned into n equi-sized blocks: X_0, X_1, ..., X_{n-1}, where n is a function of the block size ($\mathcal{B}$) and the size of X. We assume a block is laid out contiguously on the disk and is the unit of transfer from disk to main memory. The time required to display a block is defined as a *time period* (T_p):

$$T_p = \frac{\mathcal{B}}{R_C} \qquad (2.3)$$

With this technique, when an object X is referenced, the system stages X_0 in memory and initiates its display. Prior to completion of a time period, it initiates the retrieval of X_1 into memory in order to ensure a continuous display. This process is repeated until all blocks of an object have been displayed.

To support simultaneous displays of several objects, a time period is partitioned into fixed-size slots, with each slot corresponding to the retrieval time of a block from the disk drive. The number of slots in a time period defines the number of simultaneous displays that can be supported by a disk drive. Figure 2.6 demonstrates the concept of time period and time slots. Each box represents a block reading time. Assuming that each block is stored contiguously on the surface of the disk, the disk incurs a seek every time it switches from one block of an object to another. We denote this as T_{W_Seek} and assume that it includes the average rotational latency time of the disk drive. We will not discuss rotational latency further because it is a constant added to every seek time.

Since the blocks of different objects are scattered across the disk surface, a simple technique (*Simple* [137]) should assume the maximum seek time (i.e.,

$Seek(\#cyl)^9$) when multiplexing the bandwidth of the disk among multiple displays. Otherwise, a continuous display of each object cannot be guaranteed. As in Figure 2.6, the following equation should be satisfied to support $\mathcal{N}_{simple}$ simultaneous display.

$$\sum^{\mathcal{N}_{simple}} (\text{a block reading time} + \text{a maximum seek time}) \leq T_p \qquad (2.4)$$

$$\mathcal{N}_{simple} \left(\frac{\mathcal{B}}{R_D} + Seek(\#cyl) \right) \leq T_p \qquad (2.5)$$

For a fixed block size, the maximum number of simultaneous displays that a disk can support is:

$$\mathcal{N}_{simple} = \left\lfloor \frac{R_D}{R_C} \frac{\mathcal{B}}{(\mathcal{B} + R_D \times Seek(\#cyl))} \right\rfloor \qquad (2.6)$$

The optimal block size to support $\mathcal{N}_{simple}$ can be obtained when the left side of Eq. 2.5 equals to T_p:

$$\mathcal{B}_{simple} = \frac{R_C \times R_D}{R_D - \mathcal{N}_{simple} \times R_C} \times \mathcal{N}_{simple} \times Seek(\#cyl) \qquad (2.7)$$

When $\$MB$ and $\$DISK$ denote the price of main memory, $ per MByte, and the price of a single disk drive, respectively, the cost per stream (CPS) of this technique is:

$$CPS_{simple} = \frac{\mathcal{N}_{simple} \times \mathcal{B}_{simple} \times \$MB + \$DISK}{\mathcal{N}_{simple}} \qquad (2.8)$$

The maximum startup latency observed by a request with this technique is:

$$\ell_{simple} = T_p = \frac{\mathcal{B}_{simple}}{R_C} \qquad (2.9)$$

This is because a request might arrive a little too late to employ the empty slot in the current time period. Note that ℓ_{simple} is the maximum startup latency (the average latency is $\frac{\ell_{simple}}{2}$) when the number of active users is $\mathcal{N}_{simple} - 1$. If the number of active displays exceeds $\mathcal{N}_{simple}$ then Eq. 2.9 should be extended with appropriate queuing models.

---

[9] We define $Seek(\#cyl)$ as a *full stroke* that is the time to reposition the disk head from the outermost cylinder to the innermost one.

2.3.2   Disk Bandwidth Utilization and Performance

Seek is a wasteful operation that minimizes the number of simultaneous displays supported by a disk. To retrieve $\mathcal{N}$ blocks, the disk performs $\mathcal{N}$ seeks during a time period, where $T_{W\_Seek} = Seek(d)$. Hence, the percentage of time that disk performs wasteful work can be quantified as: $\frac{\mathcal{N} \times Seek(d)}{T_p} \times 100$, where d is the maximum distance between two blocks retrieved consecutively ($d = \#cyl$ with Simple). By substituting T_p from Eq. 2.3, we obtain the percentage of wasted disk bandwidth:

$$wasteful = \frac{\mathcal{N} \times Seek(d) \times R_C}{\mathcal{B}} \times 100 \qquad (2.10)$$

By reducing this percentage, the system can support a higher number of simultaneous displays. We can manipulate two factors to reduce this percentage: 1) decrease the distance traversed by a seek, and/or 2) increase the block size. A limitation of increasing the block size is that it results in a higher memory requirement. However, if one can decrease the duration of the seek time, then the same number of simultaneous displays can be supported with smaller block sizes because the size of a block is proportional to $Seek(d)$ for a given $\mathcal{N}$ (see Eq. 2.7). This will save some memory. Briefly, for a fixed number of simultaneous displays, as the duration of the worst seek time decreases (increases) the size of the block shrinks (grows) proportionally with no impact on throughput. This impacts the amount of memory required to support $\mathcal{N}$ displays. For example assume: $Seek(\#cyl) = 17$ msec, $R_D = 68$ Mb/s, $R_C = 4$ Mb/s, and $\mathcal{N} = 15$. From Eq. 2.7, we compute a block size of 1.08 MByte that wastes 12% of the disk bandwidth. If a display technique reduces the worst seek time by a factor of two, then the same throughput can be maintained with a block size of 0.54 MByte, reducing the amount of required memory by a factor of two and maintaining the percentage of wasted disk bandwidth at 12%. Moreover, the startup latency reduces from 2.16 seconds to 1.08 seconds.

2.3.3   SCAN

One approach to reduce the worst seek time is scheduling of the disk bandwidth for multiple block retrievals in a time period. One can apply a SCAN algorithm [168], [131] for the block retrievals during a time period. The system sorts the order of block retrievals during a time period based on the location of blocks in a disk. The movement of the disk head to retrieve the blocks during a time period abides by the SCAN algorithm, in order to reduce the incurred seek times among retrievals. However, a hiccup may happen if

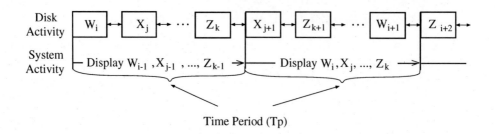

Figure 2.7. Continuous display with SCAN.

the system initiates the display of a block immediately after its retrieval as in Simple. This is because the time elapsed between two consecutive block retrievals can be greater than a time period. In order to prevent hiccups, the displays of all the blocks retrieved during the current time period must start at the beginning of the next time period. Figure 2.7 demonstrates a continuous display with SCAN. The blocks $W_i, X_j, ..., Z_k$ are retrieved during the first time period. The displays of these blocks are initiated at the beginning of the next time period.

Eq. 2.5 and 2.7 still hold with SCAN but with a reduce seek time, $T_{W\_Seek} = Seek(\frac{\#cyl}{N_{SCAN}})$. Thus, the block size to support N_{SCAN} simultaneous displays with SCAN is:

$$\mathcal{B}_{SCAN} = \frac{R_C \times R_D}{R_D - N_{SCAN} \times R_C} \times N_{SCAN} \times Seek\left(\frac{\#cyl}{N_{SCAN}}\right) \qquad (2.11)$$

SCAN requires two buffers for a display because a block is displayed from one buffer while the next block is being retrieved from the disk into the other buffer. Thus, the cost per stream is:

$$CPS_{SCAN} = \frac{N_{SCAN} \times 2 \times \mathcal{B}_{SCAN} \times \$MB + \$DISK}{N_{SCAN}} \qquad (2.12)$$

The maximum startup latency happens when a request arrives just after a SCAN begins in the current time period and the retrieval of the first block is scheduled at the end of the next time period. Thus, it is:

$$\ell_{SCAN} = 2 \times T_p = 2 \times \frac{\mathcal{B}_{SCAN}}{R_C} \qquad (2.13)$$

2.3.4   Grouped Sweeping Scheme (GSS)

A more general scheduling technique is *Grouped Sweeping Scheme* [182], GSS. GSS groups $\mathcal{N}$ active requests of a time period into g groups. This divides

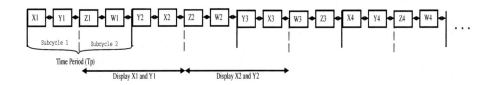

Figure 2.8. Continuous display with GSS.

a time period into g *subcycles*, each corresponding to the retrieval of $\lceil \frac{N}{g} \rceil$ blocks. Across the groups there is no constraint on the disk head movement. To support the SCAN policy within a group, GSS shuffles the order that the blocks are retrieved. For example, assuming X, Y, and Z belong to a single group, the sequence of the block retrieval might be X_1 followed by Y_4 and Z_6 (denoted as $X_1 \rightarrow Y_4 \rightarrow Z_6$) during one time period, while during the next time period it might change to $Z_7 \rightarrow X_2 \rightarrow Y_5$. In this case, the display of (say) X might suffer from hiccups because the time elapsed between the retrievals of X_1 and X_2 is greater than one time period. To eliminate this possibility, [182] suggests the following display mechanism: the displays of all the blocks retrieved during subcycle i start at the beginning of subcycle $i + 1$. To illustrate, consider Figure 2.8 where $g = 2$ and $N = 4$. The blocks X_1 and Y_1 are retrieved during the first subcycle. The displays are initiated at the beginning of subcycle 2 and last for two subcycles. Therefore, while it is important to preserve the order of groups across the time periods, it is no longer necessary to maintain the order of block retrievals in a group.

The maximum startup latency observed with this technique is the summation of one time period (if the request arrives when the empty slot is missed) and the duration of a subcycle ($\frac{T_p}{g}$):

$$\ell_{gss} = T_p + \frac{T_p}{g} \tag{2.14}$$

By comparing Eq. 2.14 with Eq. 2.9, it may appear that GSS results in a higher latency than Simple. However, this is not necessarily true because the duration of the time period is different with these two techniques due to a choice of different block size. The duration of a time period is a function of the block size.

To compute the block size with GSS, we first compute the total duration of time contributed to seek times during a time period. Assuming $\lceil \frac{N_{gss}}{g} \rceil$ blocks retrieved during a subcycle are distributed uniformly across the disk surface, the disk incurs a seek time of $Seek(\frac{\#cyl}{N_{gss}})$ (i.e., $T_{W\_Seek} = Seek(\frac{\#cyl}{\frac{N_{gss}}{g}})$) between every two consecutive block retrievals. Since N_{gss} blocks are retrieved during a time period, the system incurs $\mathcal{N}_{gss}$ seek times in addition to $\mathcal{N}_{gss}$ block retrievals during a time period, i.e., $T_p = \frac{\mathcal{N}_{gss}\mathcal{B}_{gss}}{R_D} + \mathcal{N}_{gss} \times$ $Seek(\frac{\#cyl \times g}{\mathcal{N}_{gss}})$. By substituting T_p from Eq. 2.3 and solving for $\mathcal{B}_{gss}$, we obtain:

$$\mathcal{B}_{gss} = \frac{R_C \times R_D}{R_D - \mathcal{N}_{gss} \times R_C} \times \mathcal{N}_{gss} \times Seek\left(\frac{\#cyl \times g}{\mathcal{N}_{gss}}\right) \qquad (2.15)$$

By comparing Eq. 2.15 with Eq. 2.7, observe that the bound on the distance between two blocks retrieved consecutively is reduced by a factor of $\frac{g}{\mathcal{N}_{gss}}$, noting that $g \leq \mathcal{N}_{gss}$.

   GSS is a generalization of different techniques. Observe that $g = \mathcal{N}$ simulates Simple (by substituting g with $\mathcal{N}$ in Eq. 2.15, it reduces to Eq. 2.7). And $g = 1$ simulates SCAN.

2.4   Constrained Data Placement

This study introduces a **RE**gion **B**as**E**d blo**C**k **A**llocation (REBECA) mechanism that reduces T_{W_Seek} by: 1) partitioning the disk space into $\mathcal{R}$ regions, and 2) forcing the system to retrieve the block of $\mathcal{N}$ active requests from a single region. By reducing the worst seek time, some of the disk bandwidth is freed to retrieve additional blocks per time period, providing for a higher number of simultaneous displays (throughput). However, with a fixed amount of memory, the latency increases as the number of regions ($\mathcal{R}$) is increased. Hence, it is crucial to use a planner to determine a value for the configuration parameters of a single disk multimedia server to realize a pre-specified throughput and latency time requirement.

   The seek time is a function of the distance that the disk head travels from its current track to the track that contains the referenced block. Hence, the worst possible seek time depends on the longest distance between the two blocks that could potentially be retrieved after each other. For example, assume the jth block of object X (X_j) should be retrieved after the ith block of object W (W_i) as in Figure 2.7. If the blocks of an object are assigned to the available disk space in a random manner then the worst seek time (T_{W_Seek}) depends on the distance between the first and the last cylinder of the disk (*longest_d*). However, if the placement of X_j and W_i are controlled such that their distance is at most d, where $d < longest\_d$, then T_{W_Seek} is reduced. By reducing the seek time (wasteful work), the disk can spend more of its time transferring data (useful work), resulting in a higher throughput.

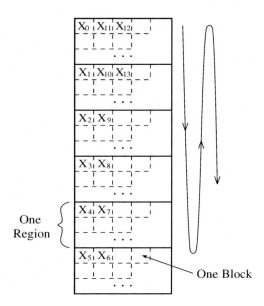

Figure 2.9. REBECA.

REBECA increases the throughput using the above observation. Assuming that $\mathcal{N}$ blocks of $\mathcal{N}$ different objects can be retrieved and displayed in one time period (T_p), its design is as follows. First, REBECA partitions the disk space into $\mathcal{R}$ regions. Next, successive blocks of an object X are assigned to the regions in a zigzag manner as shown in Figure 2.9. The zigzag assignments follows the efficient movement of disk head as in the elevator algorithm [169]. To display an object, the disk head moves *inward* (see Figure 2.10) until it reaches the innermost region and then it moves *outward*. This procedure repeats itself once the head reaches the out-most region on the disk. This minimizes the movement of the disk head required to simultaneously retrieve $\mathcal{N}$ objects because the display of each object abides by the following rules:

1. The disk head moves in one direction (either *inward* or *outward*) at a time.

2. For a given time period, the disk services those displays that correspond to a single region (termed *active region*, R_{active}).

3. In the next time period, the disk services requests corresponding to either $R_{active} + 1$ (*inward* direction) or $R_{active} - 1$ (*outward* direction). The only exception is when R_{active} is either the first or the last region.

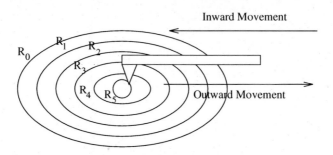

Figure 2.10. Disk head movement.

In these two cases, R_{active} is either incremented or decremented after two time periods because the consecutive blocks of an object reside in the same region. For example, in Figure 2.9, X_5 and X_6 are both allocated to the last region and R_{active} changes its value after two time periods. This scheduling paradigm does not waste disk space (an alternative assignment/schedule that enables R_{active} to change its value after every time period would waste 50% of the space managed by the first and the last region). Note that for two regions (i.e., $\mathcal{R} = 2$) the above scheduling paradigm is not necessary and the blocks should be assigned in a round-robin manner to the regions.

4. Upon the arrival of a request referencing object X, it is assigned to the region containing X_0 (say R_X).

5. The display of X does not start until the active region reaches X_0 ($R_{active} = R_X$) **and** its direction corresponds to that required by X. For example, X requires an *inward* direction if X_1 is assigned to $R_X + 1$ and *outward* if $R_X - 1$ contains X_1 (assuming that the organization of regions on the disk is per Figure 2.10).

To compute the worst seek time with REBECA, let b denote the number of blocks per region. The worst seek time during one time period is a function of b, because the blocks within a region are at most b blocks apart. However, across the time periods the worst seek time is a function of $2 \times b$. To observe consider Figure 2.7. The last scheduled object during time period i (Z_k in Figure 2.7) can be $2 \times b$ blocks apart from the first scheduled object during time period $i + 1$ (W_{i+1} in Figure 2.7). This is because Z_k and W_{i+1} reside in two different regions. Hence, the worst seek time across the time periods is the time required for the disk head to skip $2 \times b$ blocks. However, without REBECA, in the worst case the two blocks can be $\mathcal{R} \times b$ blocks apart, where

$\mathcal{R} \times b$ is the total number of blocks in the disk drive. Note that even for $\mathcal{R} = 2$ the total seek time is reduced. This is because the seek time within a time period is still a function of b for $\mathcal{N} - 1$ requests and $2 \times b$ for the last one.

Introducing regions to reduce the seek time increases the average latency time observed by a request. This is because during each time period the system can initiate the display of only those objects that correspond to the active region and whose assignment direction corresponds to that of the current direction of the arm. To illustrate this, consider Figure 2.11. In Figure 2.11.a, Y is stored starting with R_2, while the assignment of both X and Z starts with R_0. Assume that the system can support three simultaneous displays ($\mathcal{N} = 3$). Moreover, assume a request arrives at time T_1, referencing object X. This causes region R_0 to become active. Now, if a request arrives during T_1 referencing object Y, it cannot be serviced until the third time period even though sufficient disk bandwidth is available (see Figure 2.11.b). Its display is delayed by two time periods until the disk head moves to the region that contains Y_0 (R_2).

In the worst case, assume: 1) a request arrives referencing object Z when $R_{active} = R_0$, 2) both the first and the second block of object Z (Z_0 and Z_1) are in region 0 ($R_Z = R_0$) and the head is moving *inward*, and 3) the request arrives when the system has already missed the empty slot in the time period corresponding to R_0 to retrieve[10] Z_0. Hence, $2 \times \mathcal{R} + 1$ time periods are required before the disk head reaches R_0, in order to start servicing the request. This is computed as the summation of: 1) $\mathcal{R} + 1$ time periods until the disk head moves from R_0 to the last region, and 2) $\mathcal{R}$ time periods until the disk head moves from the last region back to R_0 in the reverse direction. Hence, the maximum latency time (ℓ) is computed as

$$\ell = \begin{cases} (2 \times \mathcal{R} + 1) \times T_p & \text{if } \mathcal{R} > 2 \\ (2 \times T_p) & \text{if } \mathcal{R} = 2 \\ T_p & \text{if } \mathcal{R} = 1 \end{cases} \qquad (2.16)$$

Note that ℓ is the maximum latency time (the average latency is $\frac{\ell}{2}$) when the number of active users is less than $\mathcal{N}$; otherwise, Eq. 2.16 should be extended with appropriate queuing models.

An interesting observation is that the computed latency time (ℓ in Eq. 2.16) is not observed for **recording** of *live*[11] objects. That is, if $\mathcal{N}$ sessions

---

[10] An intelligent scheduling policy might prevent this scenario.

[11] Recording a live session is similar to taping a live football game. In this case, a video camera or a compression algorithm is the producer and the disk drive is the consumer.

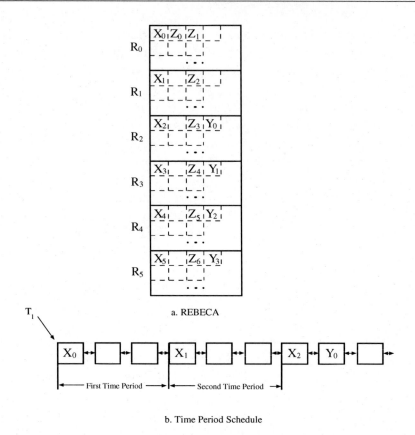

a. REBECA

b. Time Period Schedule

Figure 2.11. Latency Time.

of multimedia objects are recorded live, the transfer of each stream from memory to the disk can start immediately. This is because the first block of an object X can be stored starting with any region. Hence, it is possible to start its storage from the active region (i.e. $R_{active} \leftarrow R_X$).

In summary, partitioning the disk space into regions using REBECA is a tradeoff between throughput and latency time.

2.5   SM Display with Multi-zone Disks

Over the past few years, magnetic recording areal density has increased at the rate of 60% per year from increases in both bits per inch (bpi) and tracks per inch (tpi) [121], [82], [33]. Bits per inch (called *linear density*) shows how many bits can be stored on a track, i.e., the number of sectors on a track. To meet the demands for a higher storage capacity, disk drive manufacturers

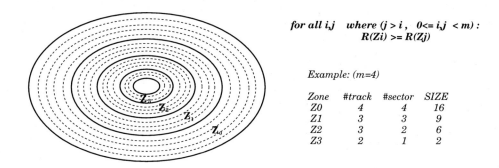

for all i,j   where (j > i,   0<= i,j < m):
$$R(Zi) >= R(Zj)$$

Example: (m=4)

Zone	#track	#sector	SIZE
Z0	4	4	16
Z1	3	3	9
Z2	3	2	6
Z3	2	1	2

Figure 2.12. An example of zone configuration.

have invariably introduced disks with *zones*. A zone is a contiguous collection of disk cylinders whose tracks have the same storage capacity, i.e., the number of sectors per track is constant in the same zone (ZBR technique). Tracks are longer towards the outer portions of a disk platter as compared to the inner portions, hence, more data can be recorded in the outer tracks. This is because that the linear density should be maintained near the maximum that the disk drive can support. Thus the amount of data stored on each track should scale with its length.

While ZBR increases the storage capacity of the disk, it produces a disk that does not have a single data transfer rate. The multiple transfer rates are due to: 1) the variable storage capacity of the tracks, and 2) a fixed number of revolutions per second for the platters. Assuming m zones in a disk drive, we denote i^{th} zone from the outermost zone as Z_i, i.e., Z_0 is the outermost zone and Z_{m-1} is the innermost one. We also denote the data transfer rate of zone Z_i as $R(Z_i)$. Then, $R(Z_i) \geq R(Z_j)$ for all i, j where $j > i$ and $0 \leq i, j < m$. For example, Figure 2.12 shows a disk drive with four zones. The outermost zone, Z_0, has four tracks of the size of four sectors. The innermost zone, Z_3, has two tracks with the track size of one sector.

In this section, we discuss alternative designs to support a continuous display of SM objects using a single disk with m zones: Z_0, Z_1, ..., Z_{m-1}. The transfer rate of zone i is denoted as $\mathcal{R}(Z_i)$. Z_0 denotes the outermost zone with the highest transfer rate: $\mathcal{R}(Z_0) \geq \mathcal{R}(Z_1) \geq ... \geq \mathcal{R}(Z_{m-1})$. We will extend these techniques to a multi-disk architecture in Section 5.3. To describe these techniques, we first make a number of simplifying assumptions. Subsequently, we will relax these assumptions one by one. These assumptions are as follows:

1. A single media type with a fixed display bandwidth, $\mathcal{R}_C$.

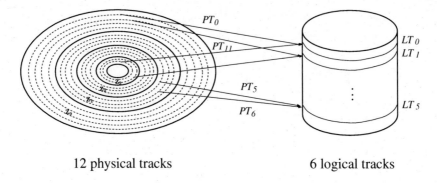

12 physical tracks　　　　　　　　6 logical tracks

Figure 2.13. Track pairing (TP).

2. A constant bit rate. That is, the display bandwidth $\mathcal{R}_C$ of a single object is fixed for the entire display duration of the object. This will be relaxed in Section 2.5.3.

3. The display bandwidth is less than or equal to the average transfer rate of the disk, $\mathcal{R}_C \leq \frac{\sum_{i=0}^{m-1} \mathcal{R}(Z_i)}{m}$.

2.5.1　Track Pairing (TP)

A straightforward way to model a multi-zone disk drive is to construct logical tracks that have the same size and identical data transfer rate regardless of the location of data in a multi-zone disk drive. Assuming a single disk drive with $\#TR$ physical tracks, track pairing (TP) [15] constructs a logical track by pairing two physical tracks such as pairing the outermost track ($\mathcal{PT}_0$) with the innermost track ($\mathcal{PT}_{\#TR-1}$), working itself toward the center of the disk platter. The result is a logical disk drive that consists of $\frac{\#TR}{2}$ logical tracks that have the same storage capacity. Thus, a logical track $\mathcal{LT}_i$ consists of two physical tracks, $\mathcal{PT}_i$ and $\mathcal{PT}_{\#TR-i-1}$, where $0 \leq i < \frac{\#TR}{2}$. Figure 2.13 shows the mapping between physical tracks and logical tracks in TP.

We can guarantee the same storage capacity of logical tracks with the assumption that the storage capacity of physical tracks increases linearly from the innermost track to the outermost track. However, in real multi-zone disk drives, there is no such linear increase in the storage capacity of tracks. Assuming m zones, there are only m different sizes of physical tracks because the storage capacity of tracks in the same zone is identical. To enforce the identical storage capacity of logical tracks in TP, the system limits the storage capacity of logical tracks by the smallest one among all logical tracks. Therefore, the fragmented data in a logical track is wasted.

For example, Seagate Cheetah 4LP disk drive (see Table 2.1) with 4 GByte storage capacity consists of 56,782 physical tracks. We can construct 28,391 logical tracks by TP. The storage capacity of the smallest logical track is 0.146 MByte and the largest one is 0.158 MByte. Thus, the storage capacity of a logical track is determined as 0.146 MByte ($sizeof(\mathcal{LT}) = 0.146$ MByte) and 181 MByte (4.2%) of disk space is wasted in this disk.

Using this logical single-zone disk drive, one can apply Simple in support of hiccup-free display. A block consists of one or multiple logical tracks:

$$\mathcal{B}_{TP} = n \times sizeof(\mathcal{LT}) \tag{2.17}$$

Eq. 2.5 can be modified to support $\mathcal{N}_{TP}$ simultaneous displays with TP.

$$\mathcal{N}_{TP}(2 \times n \times rot + 2 \times Seek(\#cyl)) \leq T_p \tag{2.18}$$

where rot is one revolution time required to read a physical track. For example, one revolution time is 8.33 msec for a disk drive with the rotation speed of 7,200 revolutions per minute. Note that only one intra-block seek is considered even though a block consists of multiple logical tracks. This is because we can eliminate unnecessary seeks by constructing a block with adjacent logical tracks[12]. Thus, for a given block size, the maximum throughput is:

$$\mathcal{N}_{TP} = \left\lfloor \frac{1}{R_C} \times \frac{\mathcal{B}_{TP}}{2 \times n \times rot + 2 \times Seek(\#cyl)} \right\rfloor \tag{2.19}$$

Then, the cost per stream and the maximum startup latency are:

$$CPS_{TP} = \frac{\mathcal{N}_{TP} \times \mathcal{B}_{TP} \times \$MB + \$DISK}{\mathcal{N}_{TP}} \tag{2.20}$$

$$\ell_{TP} = T_p = \frac{\mathcal{B}_{TP}}{R_C} \tag{2.21}$$

TP may suffer from a long intra-block seek time because a block consists of at least two separate physical tracks. For example, in order to retrieve Y_1 in Figure 2.14, the system performs the maximum intra-block seek because it retrieves the outermost track and the innermost track. Combined with the maximum inter-block seek time assumed in Simple, the portion of wasteful work might be significant. Applying SCAN to TP (TP_SCAN) prevents this limitation because one block retrieval can be separated in two

---

[12]A disk drive minimizes inter-track seek time with track skewing and read-ahead buffer techniques when it reads adjacent tracks [139].

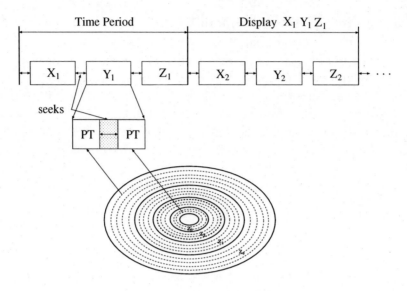

Figure 2.14. Continuous display with TP.

track retrievals and their retrieval order can be shuffled to minimize seek time among track retrievals. To determine the maximum throughput with TP_SCAN, the following equation should be satisfied:

$$\mathcal{N}_{TP\_SCAN} = \left\lfloor \frac{1}{R_C} \times \frac{n \times sizeof(\mathcal{LT})}{2 \times n \times rot + 2 \times Seek\left(\frac{\#cyl}{2\mathcal{N}_{TP\_SCAN}}\right)} \right\rfloor \quad (2.22)$$

The cost per stream and the maximum startup latency with TP_SCAN are:

$$CPS_{TP\_SCAN} = \frac{\mathcal{N}_{TP\_SCAN} \times 2 \times n \times sizeof(\mathcal{LT}) \times \$MB + \$DISK}{\mathcal{N}_{TP\_SCAN}}$$

$$(2.23)$$

$$\ell_{TP\_SCAN} = 2 \times T_p \quad (2.24)$$

2.5.2 Logical Track (LT)

LT [85] is an alternative approach to construct equi-sized logical tracks from various-sized physical tracks of a multi-zone disk drive. LT constructs a logical track with a set of physical tracks, each from a distinct zone provided

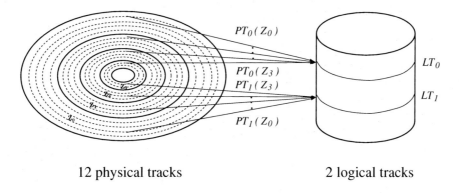

12 physical tracks   2 logical tracks

Figure 2.15. LT for a single disk drive.

by a disk drive. Thus, assuming m zones in a disk, a logical track consists of a set[13] of physical tracks:

$$\mathcal{LT}_i = \{ \; \mathcal{PT}_i(Z_j) \; : \; 0 \le j < m \; \} \tag{2.25}$$

where $\mathcal{PT}_i(Z_j)$ denotes i^{th} physical track in zone Z_j of the disk. The value of i is bounded by the zone with the fewest physical tracks, i.e., $0 \le i < Min[NT(Z_j)]$, where $NT(Z_j)$ is the number of physical tracks in the zone j of the disk dive. Figure 2.15 shows the mapping between physical tracks and logical tracks in LT.

Ideally, if there are the same number of tracks in all zones, there would be no wasted disk space with LT. However, different zones consist of different number of physical tracks (see Table 2.1). In order to enforce the same storage capacity of a logical track, LT wastes disk space because the zone with the fewest physical tracks determines the total number of logical tracks in a disk. In particular, this technique eliminates the physical tracks of those zones that have more than $NT_{min} = Min[NT(Z_j)]$, i.e., $\mathcal{PT}_k(Z_j)$ with $NT_{min} \le k < NT(Z_j)$, $0 \le j < m$, are eliminated. For example, Seagate Cheetah 4LP disk drive (see Table 2.1) with 4 GByte storage capacity consists of 7 zones. The innermost zone has the smallest number of physical tracks, 5421, while the outermost zone has 11617. Thus, 6196 tracks in zone 0 must be eliminated. LT wastes 1.5 GByte (35%) of disk space with this disk model. Note that the amount of waste disk space with LT depends on the physical zone characteristics of a disk drive. However, in general, LT wastes far more disk space than TP.

---

[13]We use the set notation, { : }, to refer to a collection of tracks from different zones of several disk drives. This notation specifies a variable before the colon and the properties that each instance of the variable must satisfy after the colon.

Similar to TP, LT provides equi-sized logical tracks with a single data transfer rate such that one can apply the continuous display techniques such as Simple and SCAN. A block consists of one or multiple logical tracks:

$$\mathcal{B}_{LT} = n \times sizeof(\mathcal{LT}) \tag{2.26}$$

Eq. 2.5 can be modified to support $\mathcal{N}_{LT}$ simultaneous displays with LT.

$$\mathcal{N}_{LT}\left(m \times n \times rot + m \times Seek\left(\frac{\#cyl}{m}\right)\right) \leq T_p \tag{2.27}$$

where rot is one revolution time required to read a single physical track. Thus, for a given block size, the maximum throughput is;

$$\mathcal{N}_{LT} = \left\lfloor \frac{1}{R_C} \times \frac{n \times sizeof(\mathcal{LT})}{m \times n \times rot + m \times Seek\left(\frac{\#cyl}{m}\right)} \right\rfloor \tag{2.28}$$

Then, the maximum startup latency is:

$$\ell_{LT} = T_p = \frac{\mathcal{B}_{LT}}{R_C} \tag{2.29}$$

The increased number (m) of intra-block seeks may result in a poor performance. Applying SCAN to LT (LT_SCAN) reduces seek times as in TP_SCAN. The maximum throughput with LT_SCAN is;

$$\mathcal{N}_{LT\_SCAN} = \left\lfloor \frac{1}{R_C} \times \frac{n \times sizeof(\mathcal{LT})}{m \times n \times rot + m \times Seek\left(\frac{\#cyl}{m\mathcal{N}_{LT\_SCAN}}\right)} \right\rfloor \tag{2.30}$$

The maximum startup latency with LT_SCAN is;

$$\ell_{LT\_SCAN} = 2 \times T_p \tag{2.31}$$

2.5.3   FIXB and VARB

We start by describing the two techniques based only on constrained data placement. The two proposed techniques partition each object X into f blocks: $X_0, X_1, X_2, ..., X_{f-1}$. The first, termed FIXB, renders the blocks equi-sized. With the second technique, termed VARB, the size of a block depends on the transfer rate of its assigned zone. The advantage of FIXB is its ease of implementation. There are two reasons for this claim. First, a fixed block size simplifies the implementation of a file system that serves as

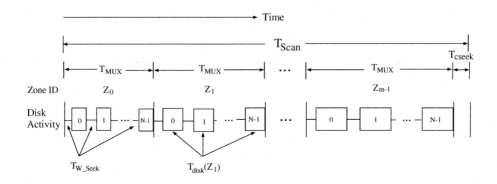

Figure 2.16. T_{Scan} and its relationship to T_{MUX}

the interface between memory and magnetic disk drive. Second, the design and implementation of a memory manager with fixed frames (the size of a frame corresponds to the size of a block) is simpler than one with variable sized frames. This is particularly true in the presence of multiple displays that result in race conditions when competing for the available memory[14]. While VARB might be more complicated to implement, it requires a lower amount of memory and incurs a lower latency as compared to FIXB when the bandwidth of the disk drive is a scarce resource. FIXB and VARB share many common characteristics. In Section 2.5.3, we describe FIXB. Subsequently, Section 2.5.3 describes VARB and its differences as compared to FIXB. In Sec. 2.5.3, we extend both techniques to employ disk scheduling for further minimization of seek time.

Fixed Block Size, FIXB

With this technique, the blocks of an object X are rendered equi-sized. Let $\mathcal{B}$ denote the size of a block. The system assigns the blocks of X to the zones in a round-robin manner starting with an arbitrary zone. FIXB configures the system to support a fixed number of simultaneous displays, $\mathcal{N}$. This is achieved by requiring the system to scan the disk in one direction, say starting with the outermost zone moving inward, visiting one zone at a time and multiplexing the bandwidth of that zone among $\mathcal{N}$ block reads. Once the disk arm reads $\mathcal{N}$ blocks from the innermost zone, it is repositioned to the outermost zone to start another sweep of the zones. The time to perform one

---

[14]We have participated in the design, coding, and debugging of a multi-user buffer pool manager with a fixed block size for a relational storage manager. We explored the possibility of extending this buffer pool manager to support variable block size, and concluded that it required significant modifications.

such a sweep is denoted as T_{Scan}. The system is configured to produce and display an identical amount of data per T_{Scan} period. The time required to read $\mathcal{N}$ blocks from zone i, denoted $T_{MUX}(Z_i)$, is dependent on the transfer rate of zone i. This is because the time to read a block ($T_{disk}(Z_i)$) during one $T_{MUX}(Z_i)$ is a function of the transfer rate of a zone.

Figure 2.16 shows T_{Scan} and its relationship with $T_{MUX}(Z_i)$ for m zones. During each T_{MUX} period, $\mathcal{N}$ active displays might reference different objects. This would force the disk to incur a seek when switching from the reading of one block to another, termed[15] $T_{W\_Seek}(Z_i)$. The computation of $T_{W\_Seek}(Z_i)$ depends on the disk scheduling technique employed and hence discussed in Sec. 2.5.3. At the end of a T_{Scan} period, the system observes a long seek time (T_{cseek}) attributed to the disk repositioning its arm to the outermost zone. The disk produces m blocks of X during one T_{Scan} period ($m \times \mathcal{B}$ bytes). The number of bytes required to guarantee a hiccup-free display of X during T_{Scan} should either be lower than or equal to the number of bytes produced by the disk. This constraint is formally stated as:

$$\mathcal{R}_C \times (T_{cseek} + \sum_{i=0}^{m-1} T_{MUX}(Z_i)) \leq m \times \mathcal{B} \qquad (2.32)$$

The amount of memory required to support a display is minimized when the left hand side of Eq. 2.32 equals its right hand side.

During a T_{MUX}, $\mathcal{N}$ blocks are retrieved from a single zone, Z_{Active}. In the next T_{MUX} period, the system references the next zone $Z_{(Active+1) \bmod m}$. When a display references object X, the system computes the zone containing X_0, say Z_i. The transfer of data on behalf of X does not start until the active zone reaches Z_i. One block of X is transferred into memory per T_{MUX}. Thus, the retrieval of X requires f such periods. (The display of X may exceed $\sum_{j=0}^{f-1} T_{MUX}(Z_{(i+j) \bmod m})$ seconds as described below.) The memory requirement for displaying object X varies due to the variable transfer rate. This is best illustrated using an example. Assume that the blocks of X are assigned to the zones starting with the outermost zone, Z_0. If Z_{Active} is Z_0 then this request employs one of the idle $T_{disk}(Z_0)$ slots to read X_0. Moreover, its display can start immediately because the outermost zone has the highest transfer rate. The block size and $\mathcal{N}$ are chosen such that the data accumulates in memory when accessing outermost zones and decreases when reading data from innermost zones on behalf of a display (see Figure 2.17). In essence, the system uses buffers to compensate for the low transfer rates

---

[15]$T_{W\_Seek}(Z_i)$ also includes the maximum rotational latency time.

of innermost zones using the high transfer rates of outermost zones, harnessing the average transfer rate of the disk. Note that the amount of required memory reduces to zero at the end of one T_{scan} in preparation for another sweep of the zones.

The display of an object may not start upon the retrieval of its block from the disk drive. This is because the assignment of the first block of an object may start with an arbitrary zone while the transfer and display of data is synchronized relative to the outermost zone, Z_0. In particular, if the assignment of X_0 starts with a zone other than the outermost zone (say Z_i, $i \neq 0$) then its display might be delayed to avoid hiccups. The duration of this delay depends on: 1) the time elapsed from retrieval of X_0 to the time that block X_{m-i} is retrieved from zone Z_0, termed $T_{accessZ_0}$, and 2) the amount of data retrieved during $T_{accessZ_0}$. If the display time of data corresponding to item 2 ($T_{display(m-i)}$) is lower than $T_{accessZ_0}$, then the display must be delayed by $T_{accessZ_0} - T_{display(m-i)}$. To illustrate, assume that X_0 is assigned to the innermost zone Z_{m-1} (i.e., $i = m - 1$) and the display time of each of its block is 4.5 seconds, i.e., $T_{display(1)} = 4.5$ seconds. If 10 seconds elapse from the time X_0 is read until X_1 is read from Z_0 then the display of X must be delayed by 5.5 seconds relative to its retrieval from Z_{m-1}. If its display is initiated upon retrieval, it may suffer from a 5.5 second hiccup. This delay to avoid a hiccup is shorter than the duration of a T_{scan}. Indeed, the maximum latency observed by a request is T_{scan} when the number of active displays is less than[16] $\mathcal{N}$:

$$\ell = T_{Scan} = T_{cseek} + \sum_{i=0}^{m-1} T_{MUX}(Z_i) \tag{2.33}$$

This is because at most $\mathcal{N} - 1$ displays might be active when a new request arrives referencing object X. In the worst case scenario, these requests might be retrieving data from the zone that contains X_0 (say Z_i) and the new request arrives too late to employ the available idle slot. (Note that the display may not employ the idle slot in the next T_{MUX} because Z_{i+1} is now active and it contains X_1 instead of X_0.) Thus, the display of X must wait one T_{scan} period until Z_i becomes active again.

One can solve for the block size by observing from Figure 2.16 that $T_{MUX}(Z_i)$ can be defined as:

$$T_{MUX}(Z_i) = \mathcal{N} \times \left(\frac{\mathcal{B}}{\mathcal{R}(Z_i)} + T_{W\_Seek}(Z_i) \right) \tag{2.34}$$

---

[16]When the number of active displays exceeds $\mathcal{N}$ then this discussion must be extended with appropriate queuing models.

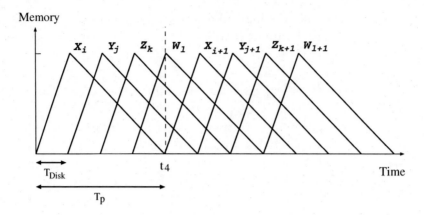

Figure 2.17. Memory required on behalf of a display.

Substituting this into Eq. 2.32, the block size is defined as:

$$\mathcal{B} = \frac{\mathcal{R}_C \times (T_{cseek} + (\mathcal{N} \times \sum_{i=0}^{m-1} T_{W\_Seek}(Z_i)))}{m - \mathcal{R}_C \times \sum_{i=0}^{m-1} \frac{\mathcal{N}}{\mathcal{R}(Z_i)}} \qquad (2.35)$$

Observe that FIXB wastes disk space when the storage capacity of the zones is different. This is because once the storage capacity of the smallest zone is exhausted then no additional objects can be stored as they would violate a round-robin assignment[17].

Variable Block Size, VARB

VARB is similar to FIXB except that it renders the duration of $T_{MUX}(Z_i)$ independent of the zone's transfer rate. This is achieved by introducing variable block size where the size of a block, $\mathcal{B}(Z_i)$, is a function of the transfer rate of a zone. This causes the transfer time of each block, T_{disk} to be identical for all zones (i.e., $T_{disk} = \frac{\mathcal{B}(Z_i)}{\mathcal{R}(Z_i)} = \frac{\mathcal{B}(Z_j)}{\mathcal{R}(Z_j)}$ for $0 \leq (i, j) < m$). Similar to FIXB, the blocks of an object are assigned to the zones in a round-robin manner and the concept of T_{Scan} is preserved. This means that the blocks of an object X are no longer equi-sized. The size of a block X depends on the zone it is assigned to. However, the change in block size requires a slight modification to the constraint that ensures a continuous display:

$$\mathcal{R}_C \times (T_{cseek} + \sum_{i=0}^{m-1} T_{MUX}(Z_i)) \leq \sum_{i=0}^{m-1} \mathcal{B}(Z_i) \qquad (2.36)$$

---

[17]Unless the number of blocks for an object is less than m. We ignored this case from consideration because video objects are typically very large.

The duration of $T_{MUX}(Z_i)$ is now independent of the transfer rate of a zone but still is a function of Z_i due to the incurred seek time (see Sec. 2.5.3). $T_{MUX}(Z_i)$ is now defined as:

$$T_{MUX}(Z_i) = \mathcal{N} \times (T_{disk} + T_{W\_Seek}(Z_i)) \qquad (2.37)$$

Substituting this into Eq. 2.36, the size of a block for a zone Z_i is defined as:

$$\mathcal{B}(Z_i) = \mathcal{R}(Z_i) \times \frac{\mathcal{R}_C \times \mathcal{N} \times \sum_{i=0}^{m-1} T_{W\_Seek}(Z_i) + \mathcal{R}_C \times T_{cseek}}{\sum_{i=0}^{m-1} \mathcal{R}(Z_i) - \mathcal{R}_C \times \mathcal{N} \times m} \qquad (2.38)$$

Similar to FIXB, VARB employs memory to compensate for the low bandwidth of innermost zones using the high bandwidth of the outermost zones. This is achieved by reading more data from the outermost zones. Moreover, the display of an object X is synchronized relative to the retrieval of its block from the outermost zone and may not start immediately upon retrieval of X_0. VARB wastes disk space when $\frac{size(Z_i)}{\mathcal{R}(Z_i)} \neq \frac{size(Z_j)}{\mathcal{R}(Z_j)}$ for $i \neq j$, and $0 \leq (i,j) < m$. The amount of wasted space depends on the zone that accommodates the fewest blocks. This is because the blocks of an object are assigned to the zones in a round-robin manner and once the capacity of this zone is exhausted the storage capacity of other zones cannot be used by other video objects.

Disk Scheduling

In Section 2.5.3, we termed $T_{W\_Seek}(Z_i)$ as the maximum seek time incurred between the retrieval of two blocks residing in a single zone. $T_{W\_Seek}(Z_i)$ is an important factor in computing the block size, duration of T_{MUX}, and latency time with each technique. However, the computation of $T_{W\_Seek}(Z_i)$ depends on the disk scheduling technique employed. In this section we describe the computation of $T_{W\_Seek}(Z_i)$ assuming *Grouped Sweeping Scheme* [183], or GSS, as our disk scheduling technique employed within the zones. The reason that we chose GSS is because GSS is the most general disk scheduling technique and as we show by the end of this section, other disk scheduling techniques become a special case of GSS.

The seek time is a function of distance travelled by the disk arm. Let $Seek(k)$ (see [139] for details) denote the time required for the disk arm to travel k cylinders to reposition itself from cylinder i to cylinder $i+k$ (or $i-k$). Hence, assuming a disk with $\#cyl$ cylinders, $Seek(1)$ and $Seek(\#cyl)$ denote the time required to reposition the disk arm between two adjacent cylinders, and the first and the last cylinder of the disk, respectively. Consequently, the maximum time required to seek among the cylinders of a zone i with $\#zcyl_i$ cylinders is $Seek(\#zcyl_i)$. If no disk scheduling is employed within the zones, $T_{W\_Seek}(Z_i)$ is simply $Seek(\#zcyl_i)$.

GSS groups $\mathcal{N}$ active requests of a T_{MUX} into g groups. This divides a T_{MUX} into g subcycles, each corresponding to the retrieval of $\lceil \frac{\mathcal{N}}{g} \rceil$ blocks. The movement of the disk head to retrieve the blocks within a group abides by the SCAN algorithm, in order to reduce the incurred seek time in a group. Across the groups there is no constraint on the disk head movement. To support the SCAN policy within a group, GSS shuffles the order that the blocks are retrieved. For example, assuming X, Y, and W belong to a single group, the sequence of the block retrieval might be X_1 followed by Y_4 and W_6 (denoted as $X_1 \rightarrow Y_4 \rightarrow W_6$) during the first subcycle of $T_{MUX}(Z_i)$, while during the first subcycle of $T_{MUX}(Z_{i+1})$ it might change to $W_7 \rightarrow X_2 \rightarrow Y_5$. In this case, the display of (say) X might suffer from hiccups because the time elapsed between the retrievals of X_1 and X_2 is greater than one $T_{MUX}(Z_i)$. To eliminate this possibility, [183] suggests the following display mechanism: the displays of all the blocks retrieved during subcycle i start at the beginning of subcycle $i + 1$. To illustrate, consider Figure 2.8 where $g = 2$ and $\mathcal{N} = 4$. The blocks X_1 and Y_1 are retrieved during the first subcycle. The displays are initiated at the beginning of subcycle 2 and last for two subcycles. Therefore, while it is important to preserve the order of groups across the T_{MUX}s, it is no longer necessary to maintain the order of block retrievals in a group.

To estimate $T_{W\_Seek}(Z_i)$ with GSS, we assume $\lceil \frac{\mathcal{N}}{g} \rceil$ blocks retrieved during a subcycle are distributed uniformly across the zone surface. Hence, the disk incurs a seek time of $Seek(\frac{\#zcyl_i}{\frac{\mathcal{N}}{g}})$ between every two consecutive block retrievals. That is,

$$T_{W\_Seek}(Z_i) = Seek \left(\frac{\#zcyl_i}{\frac{\mathcal{N}}{g}} \right) \tag{2.39}$$

Note that since $T_{W\_Seek}(Z_i)$ with GSS is now reduced as compared to no disk scheduling, all the parameters in FIXB and VARB that were a function of $T_{W\_Seek}(Z_i)$ such as block size, duration of T_{MUX}, and latency time are now reduced. In addition, observe that GSS with $g = \mathcal{N}$ simulates the no disk scheduling scenario while GSS with $g = 1$ is the SCAN disk scheduling. GSS with $g = 1$ is the SCAN disk scheduling.

Chapter 3

MULTIPLE DISK PLATFORM SM SERVERS

A single disk server is not appropriate for most applications for several reasons. First, its storage capacity might not be large enough to store all required objects. Second, the bandwidth of a single disk is limited to a small number of simultaneous displays. Third, the bandwidth requirements of an object might exceed that of a single disk drive. Hence, multi-disk architectures are introduced to resolve these limitations.

3.1 Overview

Assuming SM servers with multiple disk drives, the data blocks are assigned to the disks in order to distribute the load of a display evenly across the disks. Thus, data placement can affect the continuous display and performance of SM servers in conjunction of the scheduling techniques. There are two well-known approaches to assign blocks of an object across multiple disk drives: *constrained* and *unconstrained*. A typical example of constrained data placement is *round-robin* [12], [67], [70]. As suggested by its name, this technique assigns the blocks of an object to the disks in a round-robin manner, starting with an arbitrarily chosen disk. Assuming d disks in the system, if the first block of an object X is assigned to disk d_i, j^{th} block of X is assigned to disk $d_{(i+j-1) \ mod \ d}$ where $0 \leq i, j < d$. An example of unconstrained data placement is *random* [116], [172], [68] which assigns data blocks to disk drives using a random number generator.

Based on scheduling and data placement techniques, we can classify continuous display techniques on multi-disk SM servers. There are four possible approaches: 1) cycle-based and round-robin, 2) deadline-driven and random, 3) cycle-based and random, and 4) deadline-driven and round-robin.

Cycle-based, Round-robin

Many studies [12], [70], [126], [175], [105] investigated the combination of cycle-based scheduling and round-robin data placement. With this approach, one block is retrieved from each disk drive for each display in every time period. Thus, assuming d disk drives in the system, data retrieval for a display cycles through all d disks in d successive time periods, following the round-robin data placement in a lock-step manner. The system load should be distributed across disk drives to prevent formation of bottlenecks. This load can be balanced by intentionally delaying the retrieval of the first block of requested object whenever a bottleneck is formed on a disk drive (see Section 3.2 for details). Due to the harmony of round-robin data placement and periodic cycle-based data retrieval, this approach provides a deterministic service guarantee for a hiccup-free display of a SM object once its retrieval is initiated. This approach maximizes the utilization of disk bandwidth by distributing the load of a display across the disks evenly. Thus, the system throughput scales linearly as a function of the number of disk drives in the system. The drawback of this approach is that the startup latency also scales[1] because the system might delay the initiation of data retrievals of objects. Thus, this approach is suitable to the applications that require a high throughput and can tolerate a long startup latency such as movie-on-demand systems.

Deadline-driven, Random

A few studies [138], [97], [172], [68] have investigated the approach with deadline-driven scheduling and random data placement. By controlling the deadlines for block retrievals, this approach can provide a shorter startup latency than the cycle-based and round-robin approach. Hence, this approach is more appropriate to the applications requiring a short start latency such as a digital editing system. However, this approach may suffer from the statistical variation of the number of block retrievals in a disk drive. Due to the nature of random data placement, a disk might receive more than its fair share of requests. A formation of bottleneck on a disk drive may result in the violation of deadlines set forth on requested blocks, causing some displays to incur hiccups. This hiccup probability might be significant depending on the system load.

---

[1]The worst startup latency linearly scales as a function of the number of disks while the average startup latency increases sub-linearly.

Cycle-based, Random

The approach based on a cycle-based scheduling and random data placement is discarded from further considerations because of the following drawbacks. First, this approach cannot provide the deterministic service guarantee as with the cycle-based and round-robin approach due to a random placement of data. Second, this approach cannot provide a short startup latency as with the deadline-driven and random methods because of the cycle-based scheduling. Third, this approach results in a low utilization of disk bandwidth because requests might be distributed unevenly across the disks and those disks that finish a cycle early remain idle until other disks finish (based on the assumption that requests for the next blocks are issued at the end of the current cycle).

Deadline-driven, Round-robin

Similarly, we will not consider the approach with a deadline-driven scheduling and round-robin data placement. With this approach, once a bottleneck is formed on a disk drive, it will reoccur repeatedly due to the round-robin placement of data and the sequential data retrievals of SM displays. For example, when a bottleneck is formed on disk d_0, it will reoccur most probably on the adjacent disk d_1 that has the next blocks of displays participating in the formation of the bottleneck on d_0. Thus, the formation of bottleneck reoccurs across disks in a round-robin manner. The bottleneck is resolved only when one or more participating displays terminate. Assuming a movie-on-demand system, a bottleneck could last for the entire display time of a movie, say 2 hours, in the worst case. With random data placement, the bottleneck is resolved quickly because the system load will be redistributed based on the random placement of data.

3.2   Cycle-based, Round-robin

To support continuous display of an object X, it is partitioned into n equisized blocks: X_0, X_1, ..., X_{n-1}, where n is a function of the block size ($\mathcal{B}$) and the size of X. A *time period* (T_p) is defined as the time required to display a block:

$$T_p = \frac{\mathcal{B}}{R_C} \qquad (3.1)$$

With a multi-disk platform consisting of D disks, the workload of a display should be evenly distributed across the D disks in order to avoid the formation of bottlenecks. Striping [12], [67], [175] is a technique to accomplish

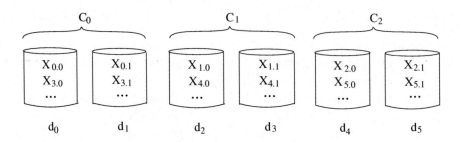

Figure 3.1. Round-robin Data Placement.

this objective. This technique partitions the D disks into C clusters of disks with each cluster consisting of d disks: $C = \lfloor \frac{D}{d} \rfloor$. Next, it assigns the blocks of object X to the clusters in a round-robin manner. The first block of X is assigned to an arbitrarily chosen disk cluster. Let $A(j, O_i)$ be the location of the jth block of object O_i.

$$A(j, O_i) = \begin{cases} \lfloor rand() \times d \rfloor, & j = 0 \\ (A(0, O_i) + j - 1) \ mod \ d, & 0 \leq j < n \end{cases} \tag{3.2}$$

where $rand()$ generates a random number that is uniformly distributed between 0 and 1. For example, in Figure 3.1, a system consisting of six disks is partitioned into three clusters, each consisting of two disk drives. The assignment of the blocks of X starts with cluster 0. This block is declustered [72] into two fragments: $X_{0.0}$ and $X_{0.1}$.

When a request references object X, the system stages X_0 from cluster C_0 in memory and initiates its display. Prior to completion of a time period, it initiates the retrieval of X_1 from cluster C_1 into memory in order to ensure a continuous display. This process is repeated until all blocks of an object have been displayed. To support simultaneous display of several objects ($R_C < R_D$), a time period is partitioned into fix-sized slots, with each slot corresponding to the retrieval time of a block from a cluster. The number of slots ($\mathcal{N}$) in a time period defines the maximum number of simultaneous displays supported by a cluster. With C clusters, because a cluster supports $\mathcal{N}$ simultaneous displays in a time period and the system accesses C clusters concurrently in the same time period, the system maintains $\mathcal{N} \times C$ time slots in a time period. (It is trivial to compute $\mathcal{N}$, $\mathcal{B}$, and T_p, see [64] for details.) We conceptualize a set of slots supported by a cluster in a time period as a group. Each group has a unique identifier. To support a continuous display in a multi-cluster system, a request maps onto one group and the individual groups visit the clusters in a round-robin manner (Figure 3.2). If group G_5

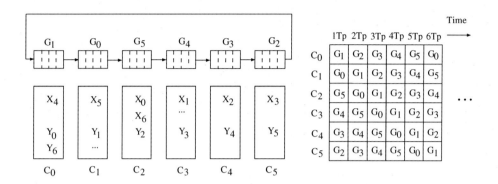

Figure 3.2. Rotating groups.

accesses cluster C_2 during a time period, G_5 would access C_3 during the next
time period. During a given time period, the requests occupying the slots
of a group retrieve blocks that reside in the cluster that is being visited by
that group.

Assuming a first-come-first-served policy, upon the arrival of a request,
the system assigns it to a group that services this request until the end of its
display.[2] With a request referencing object X, the system first determines
the group (say G_x) which is currently accessing the cluster where the first
block of X resides. If G_x has an available time slot, the system assigns this
request to G_x to initiate the retrieval on behalf of this request. Otherwise, a
failure has occurred and the system checks whether the next coming group
has an empty time slot. This is repeated until the system finds an available
group. Due to the rotation of groups, there is a fair chance for a request to
be assigned to a specific group (regardless of the location of the first block
of each object and their access frequency).

Therefore, if there are C clusters (or groups) in the system and each
cluster (or group) can support $\mathcal{N}$ simultaneous displays then the maximum
throughput of the system is $m = \mathcal{N} \times C$ simultaneous displays. The maximum startup latency is $T_p \times C$ because: 1) groups are rotating (i.e., playing
musical chairs) with the C clusters using each for a T_p interval of time, and
2) at most $C - 1$ failures might occur before a request can be activated
(when the number of active displays is fewer than $\mathcal{N} \times C$). Thus, both the
system throughput and the maximum startup latency scale linearly. Note
that system parameters such as blocks size, time period, throughput, etc., for
a cluster can be computed using display techniques such as REBECA [64]

---

[2]With the request migration scheme of Section 3.4.1, a request can be migrated to
another group during its display.

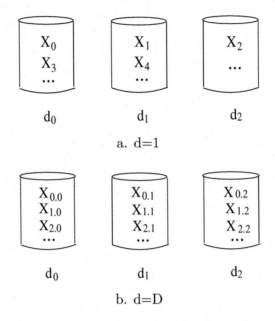

Figure 3.3. Two extreme stripe sizes (D=3).

and GSS [182]. These display techniques are local optimizations that are orthogonal to the optimization techniques proposed by this study.

Given a system with a fixed amount of resources (disk and memory), the number of disks that constitute a cluster have a significant impact on the overall performance of a system. To illustrate, assume a system that consists of 20 disks with 320 megabyte of memory. The impact of the number of disks per cluster on the system performance is best explained by considering the extreme choice of values for d, i.e., $d = 1$ and $d = D$. For each, we describe the scalability characteristics of the system as a function of additional resources (disk drive and memory) in a system. We assume a system consisting of one disk and 16 megabytes of memory as our base and focus on a media type with a bandwidth requirement of 4 Mb/s because the trends are identical for the other media types. The amount of memory (i.e., 16 megabytes) was chosen such that 20% of the disk bandwidth is wasted when the system is configured to maximize the number of displays (with $R_C = 4$ Mb/s). Next, we quantify the maximum throughput and latency time of the system with each technique when both the number of disk drives and the amount of memory increases by a factor of i, $1 \leq i \leq 20$. For example, when $i = 4$, the system consists of four disks and 64 megabyte of memory.

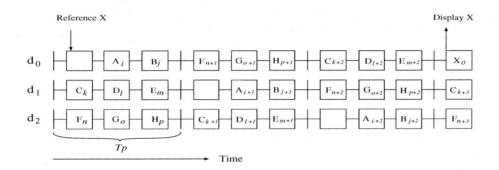

Figure 3.4. Maximum latency time with striping.

3.2.1 $d = 1$

When $d = 1$, the blocks of an object X are assigned to the D disk drives in a round-robin manner. The assignment of X_0 starts with an arbitrary disk. Assuming a system with three disk drives, Figure 3.3a demonstrates the assignment of blocks of X with this choice of value. The definition of a time period and its derivation are identical to Equation 3.1. The total number of simultaneous displays supported by the system is: $N \times D$. Hence, throughput of the system (i.e., maximum number of displays) scales linearly as a function of additional resources in the system. However, its maximum latency also scales linearly (see Figure 3.5). To demonstrate this, assume that each disk in Figure 3.3a can support three simultaneous displays. Assume that eight displays are active and that the assignment of object X starts with d_0 (X_0 resides on d_0). If the request referencing object X arrives too late to utilize the idle slot of d_0, it must wait three (i.e., D) time periods before the idle slot on d_0 becomes available again (see Figure 3.4). Hence, the maximum latency time is $T_p \times D$.

The effective disk bandwidth of the system is a function of $\mathcal{B}$, T_{W_Seek}, and D. It is defined as:

$$B_{Disk} = D \times R_D \times \frac{\mathcal{B}}{\mathcal{B} + (T_{W\_Seek} \times R_D)} \qquad (3.3)$$

The percentage of wasted disk bandwidth is quantified as:

$$\frac{(D \times R_D) - B_{Disk}}{D \times R_D} \times 100 \qquad (3.4)$$

Thus, the wasted disk bandwidth is a constant as a function of additional resources, see Figure 3.5. This explains why the throughput of the system scales linearly.

3.2.2 $d = D$

When $d = D$, each display utilizes the total bandwidth of D disks. Logically, D disk drives are conceptualized as a single disk. Each block of an object is dispersed across the D disk drives. For example, in Figure 3.3b, block X_0 is declustered across the 3 disks. Each fragment of this block is denoted as $X_{0,i}, 0 \leq i < D$. This approach enables the system to display an object using the design of Section 2.3. The only difference is that D disks are activated during each time slot (instead of one disk). The derivation of a time period, the maximum latency and required amount of memory are identical to those of Section 2.3. The effective bandwidth of the disks in the system is defined as:

$$B_{Disk} = D \times R_D \times \frac{B}{B + (T_{W\_Seek} \times D \times R_D)} \qquad (3.5)$$

The amount of wasted disk bandwidth is computed using Equation 3.4. (This equation is not equivalent to Equation 3.3 because $D \times R_D$ appears in the denominator of Equation 3.5.)

As compared to $d = 1$, the system wastes the bandwidth of the disk drives as a function of additional resources, see Figure 3.5. This is because the amount of data transferred per disk drive decreases as a function of additional disks. The amount of memory needed to render the percentage of wasted disk bandwidth a constant far exceeds that of a linear growth. This causes a sub-linear increase in throughput, see Figure 3.5. By supporting fewer users, the duration of a time period remains relatively unchanged, enabling the system to provide a lower latency time. This demonstrates the tradeoff between throughput and latency time of a system. It is important to note that if enough memory was provided to enable the number of users to scale linearly as a function of additional disk drives, its maximum latency would have been identical to that of $d = 1$.

3.2.3   Expected Startup Latency

It might be pessimistic to consider the maximum latency as the latency that a request experiences before being serviced. This section quantifies the characteristics of latency and develops a probabilistic approach to determine the expected startup latency of a request.

In traditional file servers, a multi-disk platform cannot be modelled as a multi-server queuing system because not all servers are identical: upon the arrival of a request, it should be assigned to the disk that contains the required data (and not an arbitrarily chosen disk). However, due to a

a. Throughput.

b. Maximum latency time.

c. Wasted disk bandwidth.

Figure 3.5. d=1 vs. d=D ($R_C = 4$ Mb/s).

random distribution of the first blocks of objects across disks and a round-robin access pattern, a request can be assigned to a time slot of any group in our approach. Assuming d=1, we can conceptualize our striping system as a queuing system with m ($m = d\mathcal{N}$) identical servers where a server corresponds to a time slot (and not a disk). Hence, we can compute the probability, $p(k)$, that there are k busy servers in the system at a given point in time by applying a queuing model. For example, with a Poisson arrival pattern and an exponential service time, the probability of k busy servers in an m server loss system is [99]:

$$p(k) = Prob\{k \text{ busy servers in the system}\} = \frac{(\lambda/\mu)^k/k!}{\sum_{k=0}^{m}(\lambda/\mu)^k/k!} \qquad (3.6)$$

where λ and μ are the arrival rate of requests and the service rate of the server (1/average service time) respectively. Note that Eq. 3.6 could be different for a different queueing model and our approach is independent to it.

When a request for an object X arrives at time t, the system determines the disk that contains the first block of X (say $A(0, X)$) and the group (say G_t) currently accessing this disk ($d_{A(0,X)}$). If G_t has at least one available slot and there remains enough time to retrieve a block till the end of the time period, the request is assigned to a slot in G_t and its retrieval is initiated. If the time slots of G_t are exhausted (occupied by other requests), the request cannot be served by this group (*failure*). Next, the system checks the availability of slots in $G_{(t+1) \bmod d}$. (Note that in contrast to how the disks are numbered, the numbering of the groups is descending, see Figure 3.2.) If the time slots of this group are also fully exhausted, the system checks the next group, $G_{(t+2) \bmod d}$. This procedure is repeated until a group with an idle slot is found (*success*). Hence, a request might have several failures before being assigned to a specific group in the system, assuming the first-come-first-serve policy for activating requests. This may result in a longer startup latency for some requests because several time periods might pass before the assigned group reaches the disk where X_0 resides. Traditionally, double buffering has been widely used to absorb the variance of block retrieval time [182], [67], [175]. The idea is as follows: while a buffer is being consumed from memory, the system fills up another buffer with data. The system can initiate display after the first buffer is filled and when the next time period begins, i.e., the next block is requested. Assuming a request is issued every time period (as in most round-based techniques [64], [183], [70],

[126]), the startup latency is defined as:

$$L(i) = (i + 0.5) \cdot T_p \qquad (3.7)$$

where i is the number of failures that a request experiences before a success.

Let $p_f(i, k)$ be the probability that a request has i failures before a success when there are k busy servers in the system. For a given k, the probability that a request experiences i failures before a success is:

$$p_f(i,k) = \begin{cases} \dfrac{\dbinom{m - i \cdot \mathcal{N}}{k - i \cdot \mathcal{N}}}{\dbinom{m}{k}} & if \quad i = \lfloor \frac{k}{\mathcal{N}} \rfloor \\[3em] \dfrac{\dbinom{m - i \cdot \mathcal{N}}{k - i \cdot \mathcal{N}} - \dbinom{m - (i+1) \cdot \mathcal{N}}{k - (i+1) \cdot \mathcal{N}}}{\dbinom{m}{k}} & for \quad other \quad i \quad values \end{cases}$$

$$(3.8)$$

where $0 \leq k < m$ and $0 \leq i \leq \lfloor \frac{k}{\mathcal{N}} \rfloor$.

Let a random variable L define the latency for a request with i failures. The probability that a request has a latency of L is the summation of the probability of i failures conditioned by all k values. Hence, the expected latency is:

$$E[L] = \sum_{k=0}^{m-1} \sum_{i=0}^{\lfloor \frac{k}{\mathcal{N}} \rfloor} p(k) \cdot p_f(i, k) \cdot L(i) \qquad (3.9)$$

Note that $L(i)$ could be different from Eq. 3.7 when one adopts different buffering and scheduling schemes. That is independent to this equation.

3.3   Deadline-driven, Random

Another way to distribute the system load across the clusters is to employ a random placement of data across the clusters [116], [172] (Figure 3.6). With deadline-driven (DD) approach, each block request is tagged with a deadline and the disk services requests using an earliest-deadline-first (EDF) policy. Note that round-robin (RR) uses an object-based retrieval while random uses a block-based retrieval, thus, a request with RR is for an entire object while with random, it is for a block of an object. Thus, a request with RR is for an entire object while with DD, it is for a block of an object.

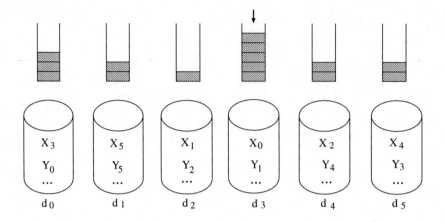

Figure 3.6. Random placement.

With random, the system may suffer from a statistical variation of the number of block requests in a cluster. Even though all the objects are displayed with a constant rate, the probability that a cluster receives a higher number of requests in a given time than other clusters might be significant. Thus, the time to retrieve a block might be greater than a time period, resulting in a hiccup. For example, assume that Figure 3.6 demonstrates a load distribution in a system consisting of six disks at a certain time. Each disk has its own queue and requests remain in queues until being serviced. For a simple discussion, assume that each block request is tagged with the same deadline, a T_p, and each disk drive can support up to three block retrievals during a T_p. Then, all requests can be serviced in a T_p but the last two requests in the queue of disk d_3. The deadlines of these two block requests are violated and hiccups happen.

We can employ a queueing model to quantify the expected hiccup probability with DD. With a Poisson arrival process and a deterministic service process, this is the probability that a request remains in the system more than the duration of a T_p (including waiting time in the queue and the service time) [172]. In particular, when a new request finds $\mathcal{N}_{sys}$ or more waiting requests in the queue of a disk, this request cannot be serviced in T_p and will experience a hiccup.

Suppose there are d disks in a system and each disk supports a maximum of $\mathcal{N}_{disk}$ requests in a T_p. When there are k active requests in the system, each disk receives k/d requests on the average per T_p. This is because blocks are randomly distributed across disks and a request accesses a specific disk with a probability of $1/d$. Using a M/D/1 queueing model [6], the probability that a request spends time less than or equal to t in the system can be calculated using the queue length distribution, p_n:

$$P[\omega \leq t] = \sum_{n=0}^{j-1} p_n \qquad (3.10)$$

where $js \leq t < (j+1)s$, $j = 1, 2, ...$ and ω is the random variable describing the total time a request spends in the queueing system, and

$$p_0 = 1 - \rho \qquad (3.11)$$

$$p_1 = (1 - \rho)(e^\rho - 1) \qquad (3.12)$$

$$p_n = (1 - \rho) \sum_{j=1}^{n} \frac{(-1)^{n-j}(j\rho)^{n-j-1}(j\rho + n - j)e^{j\rho}}{(n-j)!} \quad \text{for} \quad n = 2, 3, ... \qquad (3.13)$$

where s is the average service time, ρ is the system utilization (load), and λ is the arrival rate. Then, the probability of hiccups is:

$$P[hiccup] = 1 - P[\omega \leq t] \quad \text{when} \quad t = T_p = s \times \mathcal{N}_{disk} \qquad (3.14)$$

The average startup latency with random placement can be defined by the average time in the queueing system (average queueing time plus service time):

$$E[L] = \bar{\omega} = \frac{\rho W_s}{2(1 - \rho)} \qquad (3.15)$$

3.3.1   Functionality versus Performance

Based on the analytical models in Section 3.2 and 3.3, Figure 3.7.a shows an example of the average startup latency of two approaches, RR and random. The x axis of Figure 3.7.a shows the system load (utilization) and the y axis denotes the average startup latency in seconds. In all cases, the average startup latency with RR is significantly higher than that with random.[3] With a high system utilization this difference becomes larger. The knee of the curves prior to a rapid increase is 85% with RR and 95% with random.

---

[3]Results from various configurations showed similar trend in startup latency between RR and random.

Startup Latency (sec)

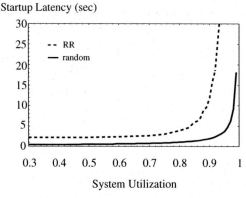

(a) Average startup latency.

Prob. of Hiccup

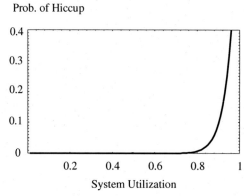

(b) Prob. of hiccup with random.

Figure 3.7. RR vs. random.

While the results of Figure 3.7.a argue in favor of random, there is a fundamental difference in servicing continuous displays between RR and random.
With RR, once a request is accepted, a hiccup-free display for the request
is guaranteed until the end of its display. However, with random, there is
a possibility of hiccups. Because hiccup-free display of continuous media is
an important functionality, the hiccup probability should be minimized with
random.

Figure 3.7.b shows the probability of hiccups with random as a function
of the system utilization. With a utilization higher than 80%, the quality of
display would suffer due to a high probability of hiccup. Thus, the maximum
utilization of a server that employs random should be less than 80%.

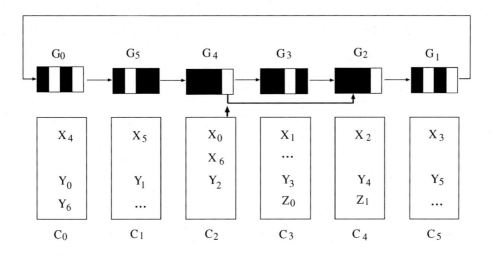

Figure 3.8. Load balancing.

In general, random is more appropriate to latency-sensitive applications than RR. RR is more appropriate if an application can tolerate a higher startup latency. For example, if 10 seconds of average startup latency is tolerable, RR can reach up to 90% system utilization while random realizes an 80% system utilization. However, if the average startup latency is required to be less than 2 seconds, random is the right choice.

In Chapter 4, we will discuss several techniques to reduce the hiccup probability with the deadline driven and random approach.

3.4   Optimization Techniques

3.4.1   Request Migration

By migrating one or more requests from a group with zero idle slots to a group with many idle slots, the system can minimize the possible latency incurred by a future request. For example, in Figure 3.8, if the system migrates a request for X from G_4 to G_2 then a request for Z is guaranteed to incur a maximum latency of one time period. Migrating a request from one group to another increases the memory requirements of a display because the retrieval of data falls ahead of its display. Migrating a request from G_4 to G_2 increases the memory requirement of this display by three buffers. This is because when a request migrates from G_4 to G_2 (see Figure 3.8), G_4 reads X_0 and sends it to the display. During the same time period, G_3 reads X_1 into a buffer (say, B_0) and G_2 reads X_2 into a buffer (B_1). During the next time period, G_2 reads X_3 into a buffer (B_2) and X_1 is displayed from memory buffer B_0. (G_2 reads X_3 because the groups move one cluster to the right at the end of each time period to read the next block of active displays occupying its servers.) During the next time period, G_2 reads X_4 into a memory buffer (B_3) while X_2 is displayed from memory buffer B_1. This round-robin retrieval of data from clusters by G_2 continues until all blocks of X have been retrieved and displayed.

With this technique, if the distance from the original group to the destination group is B then the system requires $B + 1$ buffers. However, because a request can migrate back to its original group once a request in the original group terminates and relinquishes its slot (i.e., a time slot becomes idle), the increase in total memory requirement could be reduced and become negligible.

3.4.2   Object Replication

Full Replication (FR)

To reduce the startup latency of the system, one may replicate objects. The simplest way is to replicate entire objects in the database so that all blocks have the same number of replicas. Let the original copy of an object X be its primary copy, X^P. All other copies of X are termed its secondary copies. The system may construct r secondary copies for object X. Each of its copies is denoted as X^i where $1 \leq i \leq r$. The number of instances of X is the number of copies of X, $r + 1$ (r secondary plus one primary). Assuming two instances of an object, by starting the assignment of X^1 with a disk different than the one containing the first block of its primary copy (X_0^P), the maximum startup latency incurred by a display referencing X can be reduced by one half. This also reduces the expected startup latency. The assignment of the first block of each copy of X should be separated by a fixed number of disks in order to maximize the benefits of replication. Assuming that the primary copy of X is assigned starting with an arbitrary disk (say d_i contains X_0^P), the assignment of secondary copies of X is as follows. The assignment of the first block of copy X^j should start with disk $(d_i + \frac{jd}{r+1})$ mod d. For example, if there are two secondary copies of object Y (Y^1, Y^2) assuming its primary copy is assigned starting with disk d_0. Y_0^1 is assigned starting with disk d_2 while Y_0^2 is assigned starting with disk d_4 when $d = 6$.

With two instances of an object, the expected startup latency for a request referencing this object can be computed as follows. To find an available slot, the system simultaneously checks two groups corresponding to the two different disks that contain the first blocks of these two instances. A failure happens only if both groups are full, reducing the number of failures for a request. The maximum number of failures before a success is reduced to $\frac{d}{2}$ due to two simultaneous searching of groups in parallel. Therefore, the probability of i failures in a system with each object having two instances is identical to that of a system consisting of $\frac{d}{2}$ disks with $2\mathcal{N}$ servers per disk. A request would experience a lower number of failures with more instances of objects.

This also reduces the expected startup latency. The assignment of the first block of each copy of O_i should be separated by a fixed number of disks in order to maximize the benefits of replication. Let A_i^j denote the disk that stores the first block of jth replica of object O_i. Then, assuming R_i copies of O_i, from Eq. 3.2,

$$A_i^j = \left\{ \begin{array}{ll} A(0, O_i^0), & j = 0 \\ (A_i^0 + \frac{j \times d}{R_i}) \bmod d, & 1 \leq j < R_i \end{array} \right. \tag{3.16}$$

The location of an object, O_i, with R_i replicas can be represented by the set T_i:

$$T_i = \{A_i^j\}, \ 0 \leq j < R_i \tag{3.17}$$

For example, in a six-disk system, if there are two secondary copies of object O_i (O_i^1 and O_i^2) assuming its primary copy, O_i^0 is assigned starting with disk d_0. O_i^1 is assigned starting with disk d_2 while O_i^2 is assigned starting with disk d_4. Thus, $T_i = \{0, 2, 4\}$.

With two instances of an object, the expected startup latency for a request referencing this object can be computed as follows. To find an available server, the system simultaneously checks two groups corresponding to the two different disks that contain the first blocks of these two instances. A failure happens only if both groups are full, reducing the number of failures for a request. The maximum number of failures before a success is reduced to $\lfloor \frac{k}{2\mathcal{N}} \rfloor$ due to two simultaneous searching of groups in parallel. Therefore, the probability of i failures in a system with each object having two instances is identical to that of a system consisting of $\frac{d}{2}$ disks with $2\mathcal{N}$ servers per disk. A request would experience a lower number of failures with more instances of objects. With j instances of an object in the system, the probability of a request referencing this object to observe i failures is:

$$
p_{f_j}(i,k) =
\begin{cases}
\dfrac{\dbinom{m-j\cdot i\cdot \mathcal{N}}{k-j\cdot i\cdot \mathcal{N}}}{\dbinom{m}{k}} & if\ \ i=\lfloor \frac{k}{j\cdot \mathcal{N}}\rfloor \\[4ex]
\dfrac{\dbinom{m-j\cdot i\cdot \mathcal{N}}{k-j\cdot i\cdot \mathcal{N}}-\dbinom{m-j\cdot(i+1)\cdot \mathcal{N}}{k-j\cdot(i+1)\cdot \mathcal{N}}}{\dbinom{m}{k}} & for\ \ other\ \ i\ \ values
\end{cases}
\tag{3.18}
$$

where $0 \le i \le \lfloor \frac{k}{j\cdot \mathcal{N}}\rfloor$. Hence, the expected startup latency of requests that reference an object with j instances is:

$$
E[L_j] = \sum_{k=0}^{m-1} \sum_{i=0}^{\lfloor \frac{k}{j\cdot \mathcal{N}}\rfloor} p(k)\cdot p_{f_j}(i,k)\cdot L(i)
\tag{3.19}
$$

Selective Replication (SR)

Object FR greatly increases the storage requirement of an application. One important observation in real applications is that objects may have different access frequencies. For example, in a video-on-demand system, more than half of the active requests might reference only a handful of recently released movies [2]. [34] models the empirical distribution of video rental frequency using a Zipf distribution. By replicating frequently referenced objects more times than less popular ones, i.e., selectively determine the number of replicas of an object based on its access frequency, we could significantly reduce the startup latency without a dramatic increase in storage space requirement of an application.

The optimal number of secondary copies per object is based on its access frequency and the available storage capacity. The formal statement of the problem is as follows. Assuming n objects in the system, let S be the total amount of disk space for these objects and their replicas. Let R_j be the optimal number of instances for object j, S_j to denote the size of object j and F_j to represent the access frequency (%) of object j. The problem is to determine R_j for each object j $(1 \le j \le n)$ while satisfying $\sum_{j=1}^{n} R_j\cdot S_j \le S$.

There exist several algorithms to solve this problem [93]. A simple one known as the Hamilton method computes the number of instances per object j based on its frequency by calculating the quota for an object $(F_j \times S)$. It

Step 0: Let $R_j = l_j$ for $j = 1, 2, ..., n$ and
$minsize = Min[S_j]$, $maxsize = Max[S_j]$,
(l_j is a lower bound of object j)

Step 1: Compute $\frac{F_j}{d(R_j)}$ for all j.
Find index j' having $Max[\frac{F_j}{d(R_j)}]$

Step 2: Let $rem = S - \sum_{j=1}^{n} S_j \times R_j$
If $rem < minsize$, Then output R and stop.
If $rem \geq S_{j'}$, Then $R_{j'} = R_{j'} + 1$
Else find index j'' which has $Max[\frac{F_j}{d(R_j)}]$
among those satisfying $S_j \leq rem$ and
let $R_{j''} = R_{j''} + 1$
Return to step 1

Figure 3.9. Divisor method to compute the number of replicas per object.

rounds the remainder of the quota to compute R_j. However, this method suffers from two paradoxes, namely, the Alabama and Population paradoxes. Generally speaking, with these paradoxes, the Hamilton method may reduce the value of R_j when either S or F_j increases in value. The divisor methods provide a solution free of these paradoxes (see Figure 3.9). For further details and proofs of this method, see [93]. Using a divisor method named Webster ($d(R_j) = R_j + 0.5$), we classify objects based on their instances. Therefore, objects in a class have the same instances. The expected startup latency in this system with n objects is:

$$E[L] = \sum_{i=1}^{n} F_i \cdot E[L_{R_i}] \tag{3.20}$$

where $E[L_{R_i}]$ is the expected startup latency for object having R_i instances.

Partial Replication (PR)

FR and SR replicate all blocks in an object. Considering the size of large SM objects and a bounded amount of available space for replication, replicating only the first small portion of each object several times could greatly reduce the extra space requirement while providing a much shorter startup latency. For example, we can replicate only the first 10 blocks of an object X when the number of blocks of X is 100. The placement of blocks follows the same way

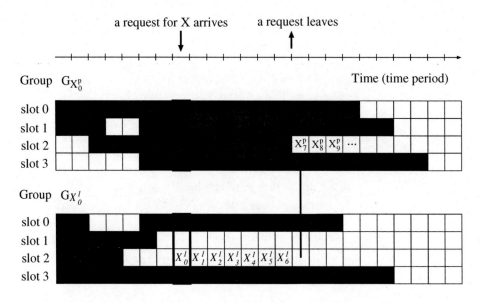

Figure 3.10. Partial replication technique.

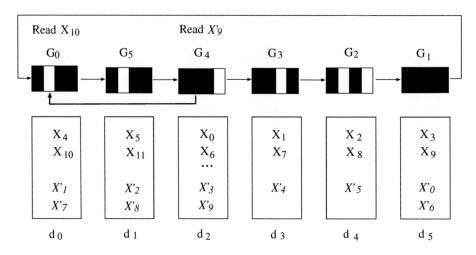

Figure 3.11. Request migration in the partial replication technique.

in the previous replication techniques. The assignment of requests is similar as in FR and SR but the system allocates a new request into the group which is currently accessing the disk where the first block of the primary copy of the requested object resides (G_{X^P}) whenever it is possible. In other words, when both $G_{X_0^P}$ and $G_{X_0^1}$ have at least one empty slot, a new request should be assigned to $G_{X_0^P}$. However, if $G_{X_0^P}$ is full and $G_{X_0^1}$ has empty slots, a request goes to $G_{X_0^1}$. Only when both have no empty slot, a request experiences a failure. This provides a same impact on the startup latency as in FR with two copies per object. The only difference is that a request assigned to $G_{X_0^1}$ should be relocated to $G_{X_0^P}$ before it reaches the last block of the partially replicated copy (tenth block in the previous example). The newly relocated request will retrieve the primary copy until the end of its display. For example, in Figure 3.10, a request for X arrives and is assigned to $G_{X_0^1}$ because $G_{X_0^P}$ is full. For the next seven time periods, the request is serviced in $G_{X_0^1}$ until $G_{X_0^P}$ releases a slot when a display ends. Then, the request is relocated from $G_{X_0^1}$ to $G_{X_0^P}$ and is serviced until the end of display. With PR, if we replicate the first 10% of an object 10 times, it takes two times of the original storage requirement (the size of the primary copy). As we discussed in the previous section, this can greatly reduce the startup latency as if the system has ten full copies of the object requiring ten times of the original storage requirement. However, there is a chance that a request assigned to the partially replicated copy could not be relocated to primary copy until the display reaches the end of partially replicated copy. Then, hiccups happen.

If a disk drive can support $\mathcal{N}$ simultaneous displays and s is average display time (service time) of objects, the service rate of a disk drive (a group) is $\mathcal{N}/s$. Hence, ideally, if we replicate the first $s/\mathcal{N}$ portion of an object, no hiccup happens. However, due to the statistical variation of time when requests arrive at and leave the system, there exists a probability that a request experiences hiccup in this technique. To reduce this probability, we should replicate more portion than $s/\mathcal{N}$, resulting in a higher space requirement.

Request migration with temporary buffering can eliminate this hiccup problem. In Figure 3.11, assume that a request in G_4 is accessing a partially replicated copy of X while G_1 is accessing the primary copy of X. A hiccup happens when G_1 does not have any idle slots until G_4 reaches to the last block of a secondary copy. Request migration utilizes buffers to avoid a potential hiccup situation. For example, while G_4 accesses the last block of the secondary copy X_9, G_0 reads the block X_{10} from the disk drive d_0 and stores it into a temporary buffer in the same time period. From the next time period on, G_0 retrieves the remaining blocks sequentially until the group G_1 releases a time slot. Then the scheduler migrates the request to the group G_1 and the temporary buffers are freed. Hence, hiccups don't happen even though G_1 is full, as long as there exists at least one available slot in the entire groups. Note that there is a tradeoff between the amount of required buffers and the number of hiccups. Extra buffers increase system cost per

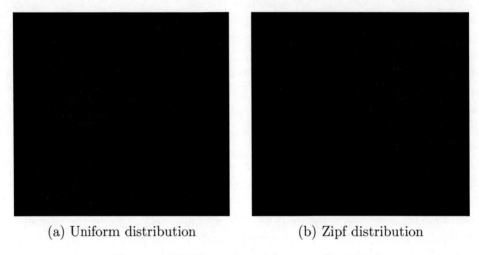

(a) Uniform distribution   (b) Zipf distribution

Figure 3.12. Two access frequency distributions.

display.

Partial Selective Replication (PSR)

This is a hybrid approach of PR and SR discussed in the previous subsections. By taking advantage of utilizing skewed access frequencies of an application and reduced storage requirement of PR, this approach determines the number of partially replicated secondary copies of an object based on its access frequency. Modifying the divisor method for partial replication is trivial and straightforward.

Comparison

We compare four replication techniques using simulation studies. Assuming two different access frequency models: 1) uniform and 2) skewed (Zipf), the average startup latency of each technique is quantified as a function of available extra disk storage space. In these experiments, we assumed that the entire database was disk resident. A server was configured with twelve Seagate Cheetah ST39103LW/LC disks. Each disk has 9 GBytes of disk space and 80 Mb/s of data transfer rate. We assumed that each video clip was encoded using MPEG2 with a 4 Mb/s of display rate. The database consisted of 30 1-hour video clips with a total of 54 GByte disk space. A video clip consisted of 3600 blocks with one second of time period. For FR and SR, the entire 3600 blocks were replicated while first 720 blocks (20%)

were replicated for PR and PSR.

We assumed a uniform distribution of access frequency for 30 video clips, and also analyzed a skewed distribution (Zipf) of access as a simplified model of real access frequencies (Figure 3.12). The bandwidth of each disk can support sixteen simultaneous displays ($\mathcal{N}$=16). Hence, the maximum throughput of this configuration (12 disks) is 192 simultaneous displays. We assumed that request arrivals to the system followed a Poisson pattern with an arrival rate of $\lambda = 0.05$req/sec for a 95% system utilization. Upon the arrival of a request, if the scheduler fails to find an idle slot in the system then this request is rejected.

Figure 3.13 shows the quantified average startup latency from our simulations using both uniform and Zipf access distributions. The x axis represents the storage space of a system as a multiple of database size. For example, 1.1 on x axis means an extra space of 10% of DB size. The y axis shows resulting average startup latency. Note that, when $x = 1$, it shows an average startup latency without any replication. Due to statistical variance in request arrivals, the system could reach to its maximum capacity, 192 simultaneous displays, then newly arrived requests were rejected during such a peak time. The rejection rate was 2.5% on the average for all experiments.

As we increase extra disk space for replication, the average startup latency decreases as a function of it. PR and PSR provides the shortest startup latency with both uniform and Zipf distribution. In the best case, when $x = 2$, 79% and 77% of reduction in the average startup latency were observed with uniform and Zipf distribution, respectively. FR and PR show steps downward because they cannot create more number of replicas when the increase of extra space is not enough to replicate all objects in database. However, SR and PSR continuously decrease as extra space grows.

One observation is that the average startup latency can be significantly reduced even with a small amount of extra space. In Figure 3.13.b, with only 20% increase of storage requirement, SR, PR, and PSR provide 45%, 59%, and 56% of reduction, respectively, comparing to one without replication (2.39 seconds). This implies that the system does not require a huge amount of extra space to achieve a smaller startup latency to meet the latency criteria of many applications. While PR and PSR provide the shortest startup latency, they require more memory space because of increased number of buffers for request migration. Thus, for a cost-effective solution, SR works fine without any increase in system cost, especially for applications with highly skewed access frequency.

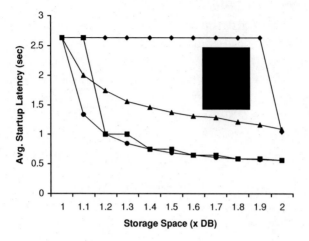

(a) Uniform access distribution

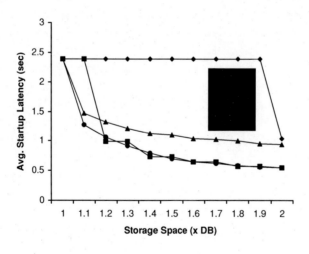

(b) Zipf access distribution

Figure 3.13. Average startup latency ($\rho = 0.95$).

3.5   Online Reconfiguration

In real applications, access frequency varies over time in many reasons. Because the effectiveness of the selective replication is sensitive to the change of access frequency, one may need to periodically reconfigure the number of instances of objects and data placement of them to maximize the system performance. In general, it requires time and extra disk bandwidth to relocate even some parts of huge multimedia objects while servicing user requests. This may result in a significant degradation of the system performance during reconfiguration process. However, while this performance degradation is temporary during reconfiguration, it could be permanent without reconfiguration. Thus, we assume that a system performs a periodic reconfiguration such as every midnight or every week.

3.5.1   Reconfiguration Process

We can represent the current configuration of data placement as a set with n elements:

$$C = \{(O_i, R_i, \{A_i^j\})\} \ where \ 0 \leq i < n, \ 0 \leq j < R_i, 1 \leq R_i < n \quad (3.21)$$

Then, formally, reconfiguration process is a function from the current configuration C to a new configuration C':

$$C = \{(O_i, R_i, \{A_i^j\})\} \Rightarrow C' = \{(O_i', R_i', \{A'_i^j\})\} \quad (3.22)$$

Our proposed reconfiguration process works as indicated in Figure 3.14. In step 1, the system gathers necessary information for reconfiguration including the object list and their updated access frequency. In step 2, the number of instances is calculated using selective replication as described in Figure 3.9. In step 3, the location of individual blocks is determined using Eq. 3.2 and 3.16. At this stage, the system can begin relocating data blocks based on the data placement.

For dynamic reconfiguration, the system needs to gather logging access information and update access frequency to each object in the database. Then, according to a preset schedule, it performs a dynamic reconfiguration by executing step 4 and 5. There are two different cases in reconfiguration. First, there could be changes only in access frequency of existing objects. No new object is introduced and no change in the database. In this case, reconfiguration may happen based on the tradeoff analysis between the overhead of reconfiguration process and the performance enhancement after it (step 5). For example, assuming a slight change in access frequency and a minor enhancement in performance, expensive reconfiguration won't be justified. The overhead of reconfiguration process can be quantified from the number of data blocks to be moved (say, S_m) during the process, which can be quantified using an algorithm in Figure 3.15.b. Then, we can calculate the percentage

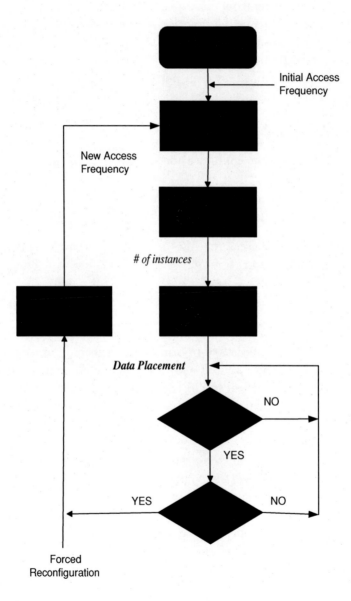

Figure 3.14. Reconfiguration Process.

Step 0: Sort O_i in descending order based on F_i
for $0 \leq i < n$

Step 1: Let $F_{del} = \sum_{j=0}^{k-1} F_{n-j-1}$
for k least popular objects
Delete k objects, O_i, $(n - k \leq i < n)$
Add k new objects, O'_i, with F'_i $(n - k \leq i < n)$
Let $F_{add} = \sum_{j=0}^{k-1} F'_{n-j-1}$.

Step 2: For all i, where $0 \leq i < n - k$
Compute $F'_i = F_i \times \frac{1-F_{add}}{1-F_{del}}$

Step 3: Output F'_i and stop.

a. Adjust access probability with new objects

Step 0: Compute $T'_i = \{A'^j_i\}$ for each O_i
based on F'_i and R'_i $(0 \leq i < n)$

Step 1: Let $S_m = 0$
For all O_i $(0 \leq i < n)$
For all j $(0 \leq j < R'_i)$
If A'^j_i is NOT an element of T_i
$S_m = S_m + S_i$

Step 2: Output S_m and stop.

b. Compute the overhead

Figure 3.15. Algorithms for reconfiguration process.

of wasted disk bandwidth ($W_{overhead}$) during a predefined reconfiguration period (T_{reconf} in seconds) as the overhead, $W_{overhead} = (100 \cdot S_m)/(m \cdot T_{reconf})$. When $E[L]$ is the expected startup latency with old configuration C and $E[L']$ is the one with new configuration C', the performance gain is defined as the percentage reduction, $G = (100 \cdot (E[L] - E[l']))/E[L]$. Then, the system commits reconfiguration based on a cost function only when the gain is greater than the cost, i.e., $G > c \times W_{overhead}$, where c is a cost coefficient set by the system administrator.

The other case happens when there is a change in the database such as adding, deleting, and replacing a number of objects. This case is a forced reconfiguration committed by the administrator, regardless of the cost. When one introduces a new object, its access probability should be determined (estimated) rather than obtained from the history. Figure 3.15.a shows an algorithm to adjust the access probability of existing objects when k $(0 < k \leq n)$ least popular objects are replaced in the database. Step 6 in Figure 3.14

Object ID (O_i)	Access Freq. (%)(F_i)	No. of Instances (R_i)
O_0	10.29	9
O_1	9.33	8
O_2	8.77	7
O_3	7.74	7
O_4	7.14	6
O_5	6.91	6
O_6	4.16	4
O_7	3.41	3
O_8	3.21	3
O_9	3.12	3
O_{10}	3.10	3
O_{11}	2.56	2
O_{12}	1.84	2
O_{13}	1.74	1
O_{14}	1.73	1
O_{15}	1.65	1
O_{16}	1.55	1
O_{17}	1.53	1
O_{18}	1.41	1
O_{19}	1.32	1
O_{20}	1.28	1
O_{21}	1.19	1
O_{22}	1.01	1
O_{23}	0.97	1
O_{24}	0.93	1
O_{25}	0.93	1
O_{26}	0.92	1
O_{27}	0.88	1
O_{28}	0.79	1
O_{29}	0.77	1
O_{30}	0.71	1
O_{31}	0.68	1
O_{32}	0.57	1
O_{33}	0.56	1
O_{34}	0.42	1
O_{35}	0.40	1
O_{36}	0.39	1
O_{37}	0.38	1
O_{38}	0.38	1
O_{39}	0.38	1
O_{40}	0.37	1
O_{41}	0.33	1
O_{42}	0.33	1
O_{43}	0.30	1
O_{44}	0.29	1
O_{45}	0.28	1
O_{46}	0.27	1
O_{47}	0.26	1
O_{48}	0.23	1
O_{49}	0.22	1

a. Week 1

Object ID (O'_i)	Access Freq. (%)(F'_i)	No. of Instances (R'_i)
O'_0 (New)	14.43	12
O'_1 (New)	9.73	8
O'_2 (O_0)	7.98	6
O'_3 (O_2)	6.88	6
O'_4 (O_2)	6.74	5
O'_5 (O_2)	5.62	5
O'_6 (O_4)	5.31	4
O'_7 (New)	4.90	4
O'_8 (O_5)	4.45	4
O'_9 (O_6)	3.09	2
O'_{10} (O_9)	2.52	2
O'_{11} (O_8)	2.46	2
O'_{12} (O_7)	2.12	2
O'_{13} (O_{11})	1.94	2
O'_{14} (O_{10})	1.81	1
O'_{15} (O_{14})	1.30	1
O'_{16} (O_{13})	1.27	1
O'_{17} (O_{12})	1.15	1
O'_{18} (O_{17})	1.08	1
O'_{19} (O_{15})	1.01	1
O'_{20} (O_{16})	0.88	1
O'_{21} (O_{25})	0.78	1
O'_{22} (O_{22})	0.77	1
O'_{23} (O_{19})	0.77	1
O'_{24} (O_{18})	0.77	1
O'_{25} (O_{20})	0.74	1
O'_{26} (O_{21})	0.73	1
O'_{27} (O_{24})	0.72	1
O'_{28} (O_{23})	0.63	1
O'_{29} (O_{26})	0.59	1
O'_{30} (O_{28})	0.58	1
O'_{31} (O_{31})	0.55	1
O'_{32} (O_{29})	0.53	1
O'_{33} (O_{27})	0.51	1
O'_{34} (O_{30})	0.48	1
O'_{35} (O_{33})	0.43	1
O'_{36} (O_{32})	0.35	1
O'_{37} (New)	0.31	1
O'_{38} (O_{36})	0.31	1
O'_{39} (O_{38})	0.28	1
O'_{40} (O_{42})	0.27	1
O'_{41} (O_{40})	0.27	1
O'_{42} (O_{37})	0.26	1
O'_{43} (O_{34})	0.26	1
O'_{44} (O_{35})	0.26	1
O'_{45} (New)	0.25	1
O'_{46} (O_{41})	0.25	1
O'_{47} (O_{39})	0.25	1
O'_{48} (O_{44})	0.22	1
O'_{49} (O_{43})	0.21	1

b. Week 2

Figure 3.16. Statistics of top 50 video rentals in the United States.

executes this adjustment and produces a new access frequency. The system resumes the process to accommodate the change.

3.5.2   Experiments

In all experiments, we assumed that the entire database consisting of 50 video objects was disk resident. The selection of objects and their access frequencies were based on real world statistics from a WWW page [42], maintained by *The Internet Movie Database*, which provides ranks and revenues of weekly top 50 video rentals in the United States. Without loss of generality, we converted video rental revenue of an object into its access frequency. We used statistics for ten weeks (say, week 1 to 10) between the third week of February, 2002 and the third week of April, 2002. For example, the first two columns of tables in Figure 3.16 show the statistics from the rental revenue. We assumed constant-bit-rate compressed video objects with a fixed one-hour length in all our experiments.

First, we performed a series of simulations to obtain the expected startup latency and compared them with analytical results. We assumed a system consisting of 12 disks ($d = 12$). The bandwidth of each disk can support ten simultaneous displays ($\mathcal{N}$=10). Hence, the maximum throughput of this configuration was 120 simultaneous displays ($m = 120$). We assumed two Poisson arrival rates ($\lambda = 0.0317$/sec for a system load, $\rho = 0.95$, and $\lambda = 0.0267$/sec for $\rho = 0.8$) for user requests. Upon the arrival of a request, if the scheduler fails to find an idle slot in the system (d failures), then this request is rejected (m server loss system). Figure 3.17 presents the distribution of number of failures from both the analytical and simulation results. The results demonstrated that the proposed equations model well the server's behavior producing very close values. We performed the same experiments for different configurations and obtained similar results. Note that we assumed that each object has only a primary copy (no replication) for these experiments.

Second, we quantified the startup latency in a system with 12 disks while varying the number of maximum simultaneous displays of a disk ($\mathcal{N}$) from 1 to 12. By varying the value of $\mathcal{N}$ with a constant data transfer rate of a disk, we model different display rate requirements of objects. A lower $\mathcal{N}$ value means that the system is configured with objects requiring a higher display rate (more disk bandwidth). In this experiment, we assumed two configurations: one with no replication and the other with replicas using the selective replication based on the access frequency from week 2. For the selective replication, we assumed a total storage capacity of 100 copies for equi-size 50 objects. Based on the quantified access frequency, we determined the number of replicas using the divisor method in Figure 3.9. The third column of table in Figure 3.16.b shows the exact number of instances of each object. For each configuration, we quantified startup latency using both analytical model and simulation. Figure 3.18 shows results from this experiment. Startup latency with the selective replication was always far shorter than that with no replication. Figure 3.18 also demonstrated that the results from the analytical model were always slightly lower than those

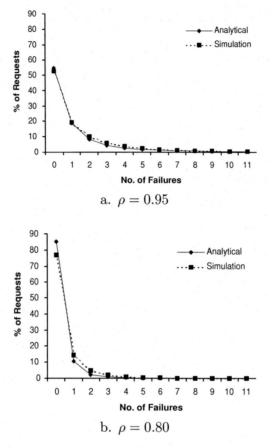

a. $\rho = 0.95$

b. $\rho = 0.80$

Figure 3.17. Distribution of number of failures $(d = 12, N = 10)$.

from the simulation but the trends were exactly matched. This implies that we would not have any problem when we use the analytical model to make a decision in the algorithm (Figure 3.15.b) in order to commit a reconfiguration. Another observation is that startup latency depends on the $\mathcal{N}$ value. The larger $\mathcal{N}$ is, the shorter the expected startup latency. This is because the probability that all slots are exhausted in a group decreases as a function $\mathcal{N}$ resulting in a smaller number of failures.

Finally, we performed nine forced reconfigurations during a ten week period based on the statistics from The Internet Movie Database. Every week, we recalculated the number of instances of objects using selective replication technique with the updated access frequency. Then, we quantified the overhead of moving data blocks to new locations according to the new data

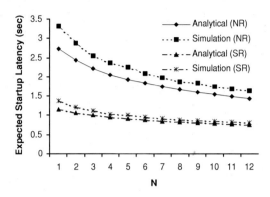

Figure 3.18. Expected startup latency ($\rho = 0.95$, $d = 12$).

placement. Consequently, a new expected startup latency was quantified. Due to the lack of space, we reported only one detailed reconfiguration from the first week to the second week (see Figure 3.16).

Figure 3.19.a shows the percentage reduction in expected startup latency, which was calculated using the difference between latencies before the reconfiguration and after it. Note that a latency before the reconfiguration means the one quantified using the previous week's configuration with the current week's access frequency representing the system performance just before the configuration happens. On the average, we could achieve a 17.0% reduction in expected startup latency when we performed a reconfiguration. We observed that the impact of reconfiguration was more significant when a bigger change in access frequency was introduced. For example, from week 1 to week 2 (see Figure 3.16.b), two very popular videos were introduced and ranked as the top two, resulting in the largest reduction (28%) in our experiments. However, from week 7 to week 8, newly introduced videos could not attract people's attention and ranked low, resulting in the smallest reduction (5.6%). This observation fits to the rationale of selected replication based on popularity.

Figure 3.19.b shows the percentage of disk bandwidth wasted during the 4-hour reconfiguration process. Our results show that, on the average, 18.6% of disk bandwidth was wasted. In other words, when we reserve 20% of total disk bandwidth for reconfiguration, reconfiguration takes 3.7 hours on the average. Considering the fact that system load significantly decreases during the night in most applications, this result demonstrates that reconfiguration is feasible without affecting the overall performance of the system.

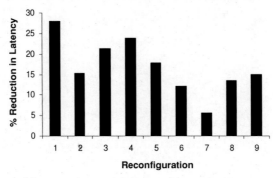

a. % reduction in expected startup latency

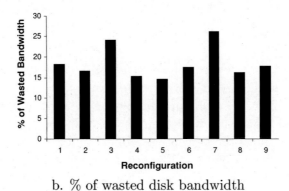

b. % of wasted disk bandwidth

Figure 3.19. Reconfiguration process, ($\rho = 0.95$, $m = 120$, $d = 12$).

3.6 Supporting Multi-zone Disks

In this chapter, we discussed techniques to support multiple disk platform SM servers with single zone disks. In reality, modern disk drives consist of multiple zones as illustrated in Section 2.5. Assuming logical single zone disks created using techniques in Section 2.5 and Section 5.3, we can straightforwardly apply techniques in this chapter to multiple multi-zone disk platform. A homogeneous collection of multi-zone disks is a special case of a collection of heterogeneous multi-zone disks in Section 5.3 when the number of disk types is one. Thus, the details can be found in section 5.3.

Chapter 4

DEADLINE-DRIVEN SCHEDULING & UNCONSTRAINED DATA PLACEMENT

This chapter analyzes the performance of a multi-disk CM server utilizing the deadline-driven scheduling and random data placement (DD). Section 4.1 investigates the distribution of block retrieval times and its impact on the hiccup probability. We propose a taxonomy of deadline-driven approaches in Section 4.2. Section 4.2.1 introduces a technique to reduce both the hiccup probability and the startup latency. We evaluate and compare our proposed techniques in Section 4.3.

4.1 Hiccup Probability

DD assigns blocks across disks in a random manner. Each block request is tagged with a deadline. When a disk receives more than its fair share of block requests at an instance in time, it become a bottleneck for the entire system, increasing block retrieval time (ω) which consists of service time (block reading time) and waiting time in a disk queue. In this section, we assume that a block is contiguously stored in a disk. Thus, assuming a fixed block size, the block retrieval time varies depending on the location of a block in a multi-zone disk. We measure block retrieval time and describe the relation between block retrieval time and hiccup probability.

Traditionally, double buffering (Figure 4.3.a) has been widely used to absorb the variance of block retrieval time[182], [67], [175]. The idea is as follows: while a buffer is being consumed from memory, the system fills up another memory frame with data. The system can initiate display after the

Probability [ω < t]

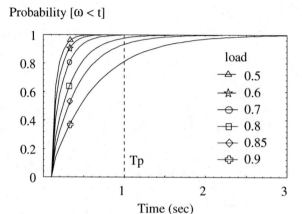

Time (sec)

Figure 4.1. Block retrieval time.

Load	$\bar{\omega}$	Max. ω	Probability		
(ρ)	(msec)	(msec)	$\omega > 1T_p$	$\omega > 2T_p$	$\omega > 3T_p$
0.50	151.2	1138	0.000030	0.0	0.0
0.55	162.5	1229	0.000070	0.0	0.0
0.60	176.7	1304	0.000160	0.0	0.0
0.65	195.5	1367	0.000460	0.0	0.0
0.70	220.6	1441	0.001570	0.0	0.0
0.75	256.1	2036	0.006430	0.000020	0.0
0.80	308.3	2621	0.018990	0.000450	0.0
0.85	397.9	3415	0.055720	0.002630	0.000200
0.90	597.6	4286	0.177530	0.023490	0.003140

Table 4.1. Examples of retrieval time distributions.

first buffer is filled and a request for the next one is issued.

In order to provide a rough idea about how the system load and hiccup probability are related, Figure 4.1 presents typical results of block retrieval time with DD. In these simulations, we assumed that: 1) a block consisted of 30 frames and the duration of a time period was one second long ($T_p = 1$), 2) block requests followed the Poisson arrival pattern, and 3) a constant service time of 100 msec based on a simple single zone disk model with a fixed disk transfer rate and a constant-bit-rate media type. Assuming a constant service time, we quantify the impact of waiting time variance attributed to random placement across disks on the distribution of retrieval time. As shown in Table 4.1, some block requests experience retrieval time (ω) longer

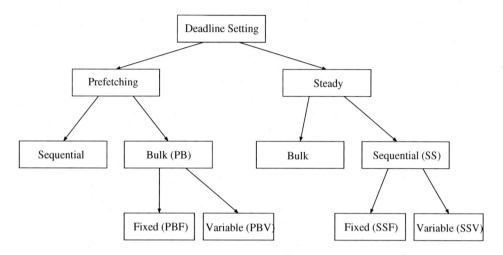

Figure 4.2. A taxonomy of deadline setting techniques.

than the average. For example, when the system load is 0.8, 1.9% of requests experience hiccups, i.e., longer retrieval time than a time period (1.0 sec), with 0.045 % of these requests experiencing delays longer than two time periods (2.0 sec). With double buffering, a display does not incur a hiccup when retrieval time of a block is either equal to or less than a time period long. However, whenever ω is greater than a time period, a display incurs a hiccup. Thus, the probability of hiccup is $p[\omega > T_p]$.

4.2   A Taxonomy of Deadline-driven Approaches

We can straightforwardly generalize double buffering to N *buffering* by using N buffers and prefetching N-1 blocks before initiating a display. The system can continue to request a block every time a memory buffer becomes empty. This reduces the probability of hiccup to $p[\omega > (N-1)T_p]$ because the retrieval time of a block must now exceed (N-1) time periods in order for a display to incur a hiccup. Assuming a clip consists of n blocks which have display sequence from B_0 to B_{n-1} and N buffers, at most N-1 blocks[1] (from B_0 to B_{N-2}) can be prefetched and accumulated before starting display to provide a more tolerable variance in block retrieval time for the consecutive blocks (B_{N-1} to B_{n-1}). This is because prefetching provides a time gap of $(N-1)T_p$ between the currently displayed block and the recently requested block. When the display of the first block B_0 is initiated and N-1 blocks

---

[1]One buffer should be available to receive a block while one of N-1 blocks is being displayed.

are prefetched, a request for B_{N-1} is issued. Then, unless its retrieval time is longer than $(N-1)T_p$, B_{N-1} will be available when it is needed (after $(N-1)T_p$). Therefore, hiccup probability of this approach can be defined as $p[\omega > (N-1)T_p]$. As demonstrated by Table 4.1, when the system load is 0.85, the hiccup probability decreases by a factor of 20 when N is increased from 2 to 3. With N buffering, there exist alternative ways of prefetching data and scheduling block retrievals. Figure 4.2 outlines a taxonomy of different techniques using a deadline -driven servicing policy where each block is tagged with a deadline and the disk services requests using an earliest-deadline-first (EDF) policy.

This taxonomy differentiates between two stages of block retrieval on behalf of a display: (1) *prefetching* stage that requests the first N-1 blocks and (2) *steady* stage that requests the remaining blocks. The system may employ a different policy to tag blocks that constitute each stage. Furthermore, blocks can be issued either in a *Bulk* or *Sequential* manner. With Bulk, all requests are issued at the same time, while with Sequential, requests are issued one at a time whenever a buffer in client's memory becomes free. Note that Bulk is irrelevant during a steady stage because it is very expensive to prefetch the entire clip at the client. This explains why the Bulk branch is shown as a leaf node of steady in Figure 4.2. Similarly, Sequential is irrelevant during prefetching because N buffers are available and our objective is to minimize startup latency[2]. The remaining leaves of the taxonomy are categorized based on how they assign deadline to each block: either *Fixed* or *Variable*. With Fixed, all block requests have identical deadlines while, with Variable, requests might have different deadlines.

During the steady stage (SS), a client issues a block request when a buffer in its memory becomes free. Typically, a memory buffer becomes free every time period because the display time of a block is one time period long. With SSF, a fixed deadline, $(N-1)T_p$[3], is assigned to all steady requests to maximize the tolerable variance of block retrieval time to prevent hiccups. However, with SSV, deadlines are determined by the number of blocks in the buffer. If the number of undisplayed blocks in the buffer is k when a block request is issued, then its deadline is set to $k \times T_p$. SSV strives to maintain the maximum data in the buffer by making the buffer full as soon as possible, while SSF strives to prevent data starvation in the buffer. The results demonstrate that both techniques provide an almost identical

---

[2]If block requests are issued sequentially during the prefetching stage, the startup latency would increase as a linear function of N, see Figure 4.3.

[3]We are using this notation for simplicity but the real deadline is $t_{issue} + (N-1)T_p$, where t_{issue} is the time that this request is issued.

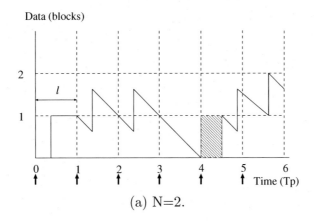

(a) N=2.

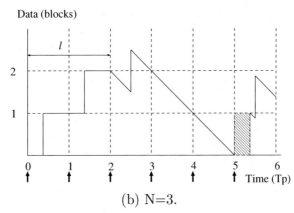

(b) N=3.

Figure 4.3. N buffering technique: prefetching with Sequential.

performance.

4.2.1   PB: Bulk Dispatching of Blocks During Prefetching Stage

While the prefetching with Sequential approach minimizes the hiccup probability, startup latency increases linearly as a function of the number of buffers, N. Thus, even though we can guarantee a desired hiccup probability, the increased startup latency may not fit to some applications; especially latency sensitive applications such as interactive applications. This linear increase is caused by the sequential request scheduling because it issues one request per time period. By changing the request scheduling, we can avoid this problem: when a clip is referenced, $N - 1$ requests for the first $N - 1$ blocks are concurrently issued as in Figure 4.4.a (bulk prefetching). This

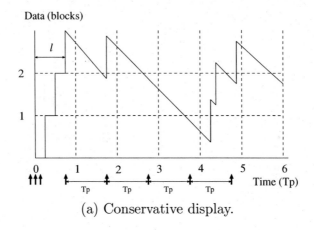

(a) Conservative display.

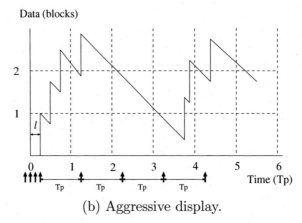

(b) Aggressive display.

Figure 4.4. N buffering with prefetching bulk requests (N=4).

section describes alternative strategies for (a) when to initiate the display of
A relative to the arrival of the N-1 blocks and (b) how to set the deadline for
these bulk requests. Consider each in turn. A client may initiate display in
two alternative ways: either 1) once all N-1 blocks have arrived, termed conservative display (CD), or 2) upon the arrival of block A_0, termed aggressive
display (AD). In Section 4.3, we compare these two alternatives. The results
demonstrate that AD is superior to CD.

The deadline assigned to the first N-1 blocks can be either fixed (termed
Fixed, PBF) or variable (termed *Variable*, PBV). With PBV, block B_i is
tagged with $(i+1)T_p$ as its deadline. Assuming that a client initiates display
when all N-1 blocks have arrived (i.e., CD), the startup latency is determined
by the longest retrieval time of the first N-1 requests, $max(\omega_0, ..., \omega_{N-2})$

	Schemes	Number of buffers (N)					
		2	3	4	5	6	7
Avg. retrieval	Skip	296.4	338.7	348.8	351.9	352.6	352.6
time (msec)	Wait	336.6	347.9	351.9	352.4	352.6	352.6
Hiccup	Skip	0.011388	0.001195	0.000238	0.000043	0.000002	$< 10^{-6}$
probability	Wait	0.042648	0.005485	0.001064	0.000212	0.000005	$< 10^{-6}$

Table 4.2. Skip vs. Wait (utilization = 0.807).

where ω_i is the retrieval time of block B_i. PBF is more aggressive because it can set the deadline for all N-1 requests to T_p in order to minimize startup latency. These requests might compete with block requests issued by other clients that are in their steady stage, increasing the probability of hiccups. However, this increase is negligible because the number of clients that are in their prefetching stage is typically small.

Section 4.3 compares these four alternatives, namely, PBF-CD, PBF-AD, PBV-CD, PBV-AD. The results demonstrate that PBV-AD provides a performance almost identical to PBF-AD. These two techniques are superior to the other alternatives.

4.2.2 Two Approaches to Handle Hiccups

While our proposed techniques strive to minimize hiccups, they cannot eliminate them all together. Moreover, the policy used at the client to respond to a hiccup impacts the server. To elaborate, a client may respond in two alternative ways to a hiccup: either wait for the missing data indefinitely (termed *Wait*) or skip the missing data and continue the display with remaining blocks (termed *Skip*). In the first case, the display is resumed upon the arrival of the missing data. This means that the server must service all block requests, even those whose deadline has been violated. With Skip, the server may discard these block requests because a client no longer needs them, minimizing the server load.

With Skip, in addition to skipping content with hiccups, every time that a display incurs a hiccup, the probability of it incurring another hiccup increases exponentially. This is because the waiting tolerance of N buffering decreases to that of N-1 buffering since the number of buffers in the client's memory is reduced to N-2. One may extend Skip to delay the display of remaining blocks by one T_p in order to prevent this undesirable situation. (There is no advantage to making Skip delay for multiple time periods.)

As detailed in Section 4.3, Skip results in a lower hiccup probability when compared with Wait because it reduces the server load. In passing, it is important to note that a client should not issue block requests while waiting because its buffers may overflow.

4.3   Evaluation

We evaluated techniques described in this chapter using a trace driven simulation study. This trace was generated synthetically using a Poisson arrival pattern. In all simulations presented in this section, we assumed a single media type with 4 Mb/s bandwidth requirement and a block size of 0.5 Mbytes ($T_p = 1$ second). Blocks were distributed across twenty Quantum Atlas XP32150 disks using random. Our repository consisted of 50 different edited clips each consisting of 120 logical blocks. We assumed a uniform distribution of access to the clips.

All results demonstrate that the hiccup probability decreases as a function of N (number of buffers), see Figure 4.5.a. For example, we can reduce the hiccup probability to less than one in a million by increasing the number of buffers to seven. This probability implies that a hiccup only happens approximately once in 12 days when a client continuously displays clips. This would be satisfactory for almost all applications.

Table 4.2 shows that Skip provides a lower hiccup probability than Wait because Skip minimizes the server's load by not servicing requests that have already violated their deadlines. This difference becomes more profound when the number of buffers decreases because this increases the percentage of requests that miss their deadline. It is important to note that the use of Skip is application dependent. Those applications that can tolerate skipping content should use Skip in order to minimize the probability of hiccup. For the rest of this evaluation, we assume the Wait scheme.

Table 4.3 shows a comparison of alternative PB techniques. The results demonstrate that the probability of hiccups is almost identical with all techniques. Figure 4.5.b and 4.6.a show the average startup latency and the worst latency as a function of N. Figure 4.6.b shows the distribution of startup latency when N=7. The y-axis of this figure is the probability of a display incurring a certain startup latency. For example, the peak point of PBV-AD illustrates that 44% of displays experience startup latency between 100 msec and 150 msec. The aggressive display (AD) approach provides a better startup latency distribution, i.e., a smaller average and variance, than the conservative display (CD) approach. Overall, both PBV-AD and PBF-AD are superior to the other alternatives when it comes to startup latency. Hence, this can satisfy almost all the latency-sensitive applications, even in the worst case scenario (336 msec), with a hiccup probability that is less than one in a million ($< 10^{-6}$).

	Schemes	Number of buffers (N)					
		2	3	4	5	6	7
Avg. retrieval time (msec)	PBF-CD	335.7	348.1	352.5	354.3	355.4	357.1
	PBF-AD	335.7	348.1	352.7	354.3	355.6	357.0
	PBV-CD	335.7	347.8	350.5	352.7	353.8	355.9
	PBV-AD	335.7	347.7	352.5	354.0	355.6	357.3
Hiccup probability	PBF-CD	0.041364	0.005574	0.001200	0.000293	0.000027	$< 10^{-6}$
	PBF-AD	0.041364	0.005598	0.001213	0.000281	0.000023	$< 10^{-6}$
	PBV-CD	0.041364	0.005276	0.001115	0.000196	0.000021	$< 10^{-6}$
	PBV-AD	0.041364	0.005681	0.001212	0.000261	0.000021	$< 10^{-6}$
Avg. startup latency (msec)	PBF-CD	318.2	186.6	148.7	195.9	206.2	224.2
	PBF-AD	318.2	148.5	140.5	139.4	138.5	139.8
	PBV-CD	318.2	369.0	375.3	380.1	389.8	420.6
	PBV-AD	318.3	147.7	139.5	138.1	136.9	137.7
Standard deviation of startup latency	PBF-CD	281.7	144.2	93.5	59.5	48.3	52.9
	PBF-AD	281.7	106.0	65.4	48.0	40.7	40.5
	PBV-CD	281.7	325.3	334.6	326.2	327.5	351.2
	PBV-AD	281.7	103.8	64.4	45.1	38.1	37.4
Worst startup latency (msec)	PBF-CD	2861	2708	2430	1883	881	512
	PBF-AD	2861	2534	2524	1999	980	406
	PBV-CD	2861	3515	4409	4373	4740	5087
	PBV-AD	2861	2552	2437	1828	891	336

Table 4.3. PB techniques (utilization = 0.807).

Similar trends were observed with our other simulation studies that utilized different parameters such as utilization, block size, number of disks. Generally, a lower hiccup probability is observed when the system utilization is lower. The number of disks does not impact the performance unless it is too small (less than 6). With a smaller block size, the portion of disk seek time becomes greater and it results in a higher disk utilization and a lower hiccup probability.

4.4 Conclusions

This chapter presented and evaluated bulk prefetching techniques with deadline-driven scheduling for the design of streaming media servers. The simulation results demonstrated the proposed techniques can minimize the average and worst startup latency while maintaining the hiccup probability at less than 10^{-6} with the expense of extra memory space. This study can be extended to determine the optimal number of buffers and their size using an analytic model for the configuration of various SM applications.

Hiccup probability

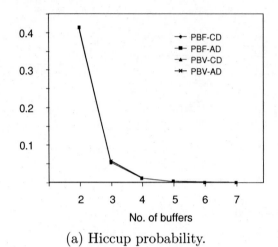

(a) Hiccup probability.

Avg. startup latency (msec)

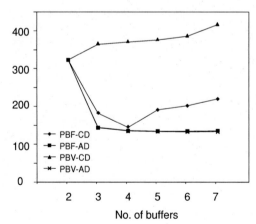

(b) Avg. startup latency (N=7).

Figure 4.5. Startup latency distribution of PB techniques.

Worst startup latency (sec)

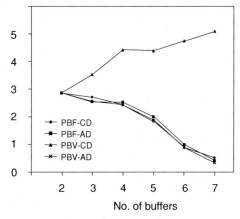

(a) Worst startup latency.

Probability

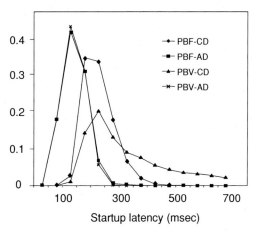

(b) Startup latency distribution (N=7).

Figure 4.6. Startup latency distribution of PB techniques.

Chapter 5

HETEROGENEOUS DISK PLATFORM SM SERVERS

To achieve the high bandwidth and massive storage required for multi-user SM servers, disk drives are commonly combined into disk arrays to support many simultaneous I/O requests, as described in Chapter 3. Such large-scale systems potentially suffer from two limitations that might introduce heterogeneity into their storage disk array. First, disks are mechanical devices that can fail. We will introduce the fault characteristics of disks and fault-tolerance techniques in Chapter 6. Because the technological trend for magnetic disk drives is one of annual increase in both performance and storage capacity (40% to 60% annually), it is usually more cost-effective to replace a failed disk with a new model (there is a 50% to 100% annual decrease in storage cost [83]). Moreover, in the fast-paced disk industry, the corresponding model may be unavailable because it was discontinued by the manufacturer. Second, if the disk array needs to be expanded due to increased demand of either bandwidth or storage capacity, it is usually most cost-effective to add current-generation disks. Hence, a SM server should manage these heterogeneous resources intelligently in order to maximize their utilization and to guarantee jitter-free video and audio rendering.

Only a few multi-disk designs and implementations to display SM from heterogeneous storage systems have been reported in the research literature. They can broadly be classified into two groups: (1) designs that partition a heterogeneous storage system into multiple, homogeneous subservers, and (2) designs that present a *virtual* homogeneous view on top of the physical heterogeneity. We will consider the two groups in turn, in Section 5.1 and in Section 5.2. Subsequently, in Section 5.3, we extend our discussion to multizone disk drives. Finally, in Section 5.4, we present our conclusions.

5.1   Partitioning Techniques

By partitioning a heterogeneous storage system into a set of homogeneous subservers, each can be configured optimally, thus improving the disk bandwidth utilization [73], [35], [179]. With this approach, each video object is striped only across the storage devices of the subserver on which it is placed. The access pattern to a collection of video or movie clips is usually quite skewed; for example, 20% of the clips may account for 80% of all retrieval requests [70]. Because the subservers will have different performances (bandwidth and/or storage capacity) it becomes important to select the appropriate server on which to place each video clip so that the imposed load is balanced evenly. One metric that can be employed to make such placement decisions is the *bandwidth to space ratio* (BSR) of each subserver [35]. However, should the access pattern change over time, then a subserver may become a bottleneck. Load-balancing algorithms are commonly used to counter such effects but they have the disadvantage of being detective rather than preventive, i.e., they can only try to remedy the situation after an imbalance or a bottleneck has already occurred. Data replication can help to reduce the likelihood of bottlenecks but the replicated objects are wasting valuable storage space and hence the resulting system may not be very cost-effective.

5.2   Non-Partitioning Techniques

Non-partitioning techniques place data over all the disk drives present in a heterogeneous, multi-node storage system [174], [27], [188]. Therefore, they are generally not affected by changing access patterns. A number of techniques exist and they differ in how they place data and schedule disk bandwidth. In its simplest form, such a technique is based on the bandwidth ratio of the different devices leading to the concept of *virtual disks*. A variation of the technique considers capacity constraints as well. These decisions impact how much of the available disk bandwidth and storage space can be utilized, and how much memory is required to support a fixed number of simultaneous displays.

We first assume a disk platter that is location independent, with a constant average transfer rate for each disk model, even though most modern disks are multi-zone drives that provide a different transfer rate for each zone. Section 5.3 describes one technique by which this simplifying assumption can be eliminated.

5.2.1 Optimization Criterion: Bandwidth, Space, or Both

The two most important parameters of disk drives for SM storage applications are their bandwidth and storage space. If a storage array is homogeneous, then a data placement technique such as round-robin or random will result in an equal amount of data on each disk drive. Consequently, each disk drive will be accessed equally frequently during data retrievals. As a result, load balancing is achieved for both storage space and disk bandwidth.

If a storage system is heterogeneous, load balancing is not guaranteed any longer. Consider a simple example of two disk drives. Assume that disk 2 has f_s-times more storage space and f_b-times more bandwidth than disk 1. If we are lucky and $f_s = f_b = f$ then we can adjust our data placement technique to place f-times more data on disk 2 compared with disk 1. This will result in an equal utilization of space and bandwidth on both disks. However, most of the time $f_s \neq f_b$ (among commercial disk generations storage space tends to increase faster than disk bandwidth). Therefore, the algorithm designer is now faced with a choice: should she aim for balanced space or bandwidth utilization? If a configuration fully utilizes the disk bandwidth some storage space may be unused, or conversely, if all of the space is allocated then some bandwidth may be idle. The techniques described in the next few sections will address this question in different ways.

One might wonder if there is a way to fully utilize all the disk space *and* all the disk bandwidth. The only possible solution is to consider a third parameter when designing the data placement algorithm: the access frequency to each stored object. Since generally some objects are more popular than others, the amount of bandwidth that they require per stored data byte is also different. This factor can be used to optimize both the disk space and retrieval bandwidth at the same time.

However, there are several disadvantages to such an approach. Objects can no longer be declustered across the complete disk array; they must be stored on a subset of the available disks. Therefore, load balancing can only be guaranteed if the access pattern to all objects does not significantly change over time. Any such change will result in hotspots on some disks and underutilization of others. This unpredictability may result in a worse overall bandwidth utilization than a technique that does not consider the object access frequency, and hence will waste a fixed portion of the overall bandwidth.

5.2.2 Bandwidth-to-Space-Ratio (BSR) Media Placement

As mentioned in the previous section, one way to fully utilize the bandwidth and space of all devices in a heterogeneous storage system is to consider the access frequency as well. One such technique considers the Bandwidth-to-Space-Ratio (BSR) of the disks plus the expected bandwidth of every object (based on its access frequency) [35]. The BSR policy may or may not utilize striping. In its simplest case it places each object completely on a single device. As a result, it may create several replicas of an object on different disks to meet the client demands.

The BSR technique works as follows. First it computes the bandwidth to space ratio of all the storage devices. Second, it computes the expected load of every video object. The expected load is based on the popularity of the object, i.e., how often it will be accessed. The policy then decides *how many* replicas of each object need to be created and *on which* devices these copies must be placed. The BSR policy is an *online* policy, hence it will periodically re-compute both the number of replicas and their placement. As a result of this step additional replicas may be created or existing ones deleted.

Recall that the performance of a technique such as BSR depends on how quickly access frequency changes can be detected and remedied. If such changes are infrequent then very good performance can be achieved. With frequently changing loads, however, hotspots and bottlenecks may develop and the creation and movement of replicas will waste part of the available bandwidth.

5.2.3 Grouping Techniques

Grouping techniques combine physical disks into logical ones and assume a uniform characteristic for all logical disks. To illustrate, the six physical disks of Figure 5.1 are grouped to construct two homogeneous logical disks. In this figure, a larger physical disk denotes a newer disk model that provides both a higher storage capacity and a higher performance[1]. The blocks of a movie X are assigned to the logical disks in a round-robin manner to distribute the load of a display evenly across the available resources [142], [12], [126], [52]. A logical block is declustered across the participating physical disks [71]. Each piece is termed a *fragment* (e.g., $X_{0.0}$ in Figure 5.1). The size of each

---

[1]While this conceptualization might appear reasonable at first glance, it is in contradiction with the current trend in the physical dimensions of disks that calls for a comparable volume when the same form factor is used (e.g., 3.5" platters). For example, the ST31200WD holding 1 GByte of data has approximately the same physical dimensions as compared with the ST34501WD holding 4 GBytes.

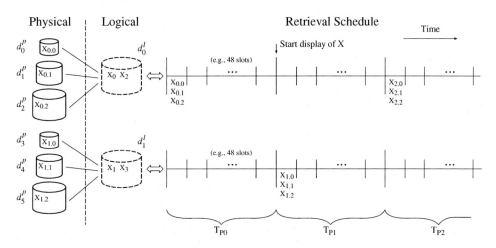

Figure 5.1. Disk Grouping.

fragment is determined such that the service time ($T_{Service}$) of all physical disks is identical (for an example see Figure 5.2).

Disk Model (Seagate)	Fragment Size [KByte]	Fragment display time [sec]	Number of Fragments	% Space for traditional data
		$N_D = 48, D^l = 2, N = 96$		
ST34501WD	$\mathcal{B}_2 = 687.4$	1.534	6,463	17.4%
ST32171WD	$\mathcal{B}_1 = 395.4$	0.883	5,338	0%
ST31200WD	$\mathcal{B}_0 = 166.0$	0.371	6,206	14.0%

Table 5.1. Disk Grouping configuration example with three disk drive types ($2 \times$ ST31200WD, $2 \times$ ST32171WD, and $2 \times$ ST34501WD). It supports $N = 96$ displays of MPEG-2 ($R_C = 3.5$ Mb/s) with a logical block size of $\mathcal{B}^l = \mathcal{B}_0 + \mathcal{B}_1 + \mathcal{B}_2 = 1,248.7$ KByte.

With a fixed storage capacity and pre-specified fragment sizes for each physical disk, one can compute the number of fragments that can be assigned to a physical disk. The storage capacity of a logical disk (number of logical blocks it can store) is constrained by the physical disk that can store the fewest fragments. For example, to support 96 simultaneous displays of MPEG-2 streams ($R_C = 3.5$ Mb/s), the fragment sizes[2] of Table 5.1 enable each logical disk to store only 5,338 blocks. The remaining space of the other two physical disks (ST34501WD and ST31200WD) can be used to store tra-

---

[2]Fragment sizes are approximately proportional to the disk transfer rates, i.e., the smallest fragment is assigned to the disk with the slowest transfer rate. However, different seek overheads may result in significant variations (see Figure 5.2).

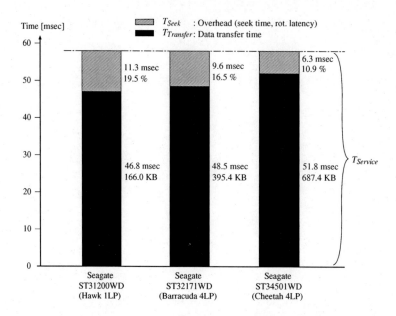

Figure 5.2. Mechanical positioning overhead (seek and rotational delay) when retrieving blocks from three different disk models. (The block sizes correspond to the configuration shown in Table 5.1.) The retrieval block sizes have been adjusted to equalize the overall service time. Note that at first the Cheetah 4LP model seems to have the lowest overhead. However, this is only the case because of its large block size. For small block sizes the Cheetah wastes a higher percentage of its bandwidth. This trend is common for newer disk models.

ditional data types, e.g., text, records, etc. During off-peak times, the idle bandwidth of these disks can be employed to service requests that reference this data type.

When the system retrieves a logical block into memory on behalf of a client, all physical disks are activated simultaneously to stage their fragments into memory. This block is then transmitted to the client.

To guarantee a hiccup-free display, the display time of a logical block must be equivalent to the duration of a time period:

$$T_P = \frac{\mathcal{B}^l}{R_C} \qquad (5.1)$$

Moreover, if a logical disk services N_D simultaneous displays then the duration of each time period must be greater or equal to the service time to read N_D fragments from every physical disk i (d_i^p) that is part of the logical disk.

Thus, the following constraint must be satisfied:

$$T_P \geq N_D \times T_{Service_i} = N_D \times \left(\frac{\mathcal{B}_i}{R_{D_i}} + T_{Seek_i}\left(\frac{\#cyl_i}{N_D} \right) \right) \tag{5.2}$$

Substituting T_P with its definition from Equation 5.1 plus the additional constraint that the time period must be equal for all physical disk drives (K) that constitute a logical disk, we obtain the following system of equations (note: we use T_{Seek_i} as an abbreviated notation for $T_{Seek_i}\left(\frac{\#cyl_i}{N_D} \right)$):

$$\mathcal{B}^l = \sum_{i=0}^{K-1} \mathcal{B}_i$$

$$\frac{\mathcal{B}^l}{R_C \times N_D} = \frac{\mathcal{B}_0}{R_{D_0}} + T_{Seek_0}$$

$$\frac{\mathcal{B}^l}{R_C \times N_D} = \frac{\mathcal{B}_1}{R_{D_1}} + T_{Seek_1} \tag{5.3}$$

$$\vdots \quad = \quad \vdots$$

$$\frac{\mathcal{B}^l}{R_C \times N_D} = \frac{\mathcal{B}_{K-1}}{R_{D_{K-1}}} + T_{Seek_{K-1}}$$

Solving Equation Array 5.3 yields the individual fragment sizes $\mathcal{B}_i$ for each physical disk type i:

$$\mathcal{B}_i = \frac{R_{D_i} \times [T_{Seek_i} \times N_D \times R_C - \sum_A + \sum_B]}{\sum_{z=0}^{K-1} R_{D_z} - R_C \times N_D} \tag{5.4}$$

with

$$\sum_A \stackrel{\text{def}}{=} \sum_{x=0}^{i-1} R_{D_x} \times T_{Seek_i} + \sum_{x=i+1}^{K-1} R_{D_x} \times T_{Seek_i}$$

$$\sum_B \stackrel{\text{def}}{=} \sum_{y=0}^{i-1} R_{D_y} \times T_{Seek_y} + \sum_{y=i+1}^{K-1} R_{D_y} \times T_{Seek_y}$$

Figure 5.2 illustrates how each fragment size $\mathcal{B}_i$ computed with Equation 5.4 accounts for the difference in seek times and data transfer rates to obtain identical service times for all disk models.

To support N_D displays with one logical disk, the system requires $2 \times N_D$ memory buffers. Each display necessitates two buffers for the following reason: One buffer contains the block whose data is being displayed (it was staged in memory during the previous time period), while a second is employed by the read command that is retrieving the subsequent logical block into memory[3]. The system toggles between these two frames as a function of time until a display completes. As one increases the number of logical disks (D^l), the total number of displays supported by the system $(N = D^l \times N_D)$ increases, resulting in a higher amount of required memory by the system:

$$M = 2 \times N \times \mathcal{B}^l \tag{5.5}$$

In the presence of $N - 1$ active displays, we can compute the maximum latency that a new request might observe. In the worst case scenario, a new request might arrive referencing a clip X whose first fragment resides on logical disk 0 (d_0^l) and miss the opportunity to utilize the time period with sufficient bandwidth to service this request. Due to a round-robin placement of data and activation of logical disks, this request must wait for D^l time periods before it can retrieve X_0. Thus, the maximum latency $T_{Latency}^{Max}$ is:

$$T_{Latency}^{Max} = D^l \times T_P = D^l \times \frac{\mathcal{B}^l}{R_C} \tag{5.6}$$

These equations quantify the resource requirement of a system with this technique.

Staggered Grouping

Staggered Grouping is an extension of the grouping technique of the previous section. It minimizes the amount of memory required to support N_D simultaneous displays based on the following observation: Not all fragments of a logical block are needed at the same time. Assuming that each fragment is a contiguous chunk of a logical block, fragment $X_{0.0}$ is required first, followed by $X_{0.1}$ and $X_{0.2}$ during a single time period. Staggered Grouping orders the physical disks that constitute a logical disk based on their transfer rate and assumes that the first portion of a block is assigned to the fastest disk, second portion to the second fastest disk, etc. The explanation for this constrained placement is detailed in the next paragraph. Next, it staggers the retrieval of each fragment as a function of time. Once a physical disk is activated, it retrieves N_D fragments of blocks referenced by the N_D active displays that require these fragments (Figure 5.3). The duration of this time

---

[3]With off-the-shelf disk drives and device drivers, a disk read command requires the address of a memory frame to store the referenced block. The amount of required memory can be reduced by employing specialized device drivers that allocate and free memory on demand [120].

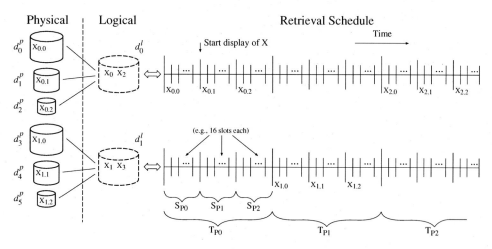

Figure 5.3. Staggered Grouping.

is termed a sub-period (S_{P_i}) and is identical in length for all physical disks. Due to the staggered retrieval of the fragments, the display of an object X can start at the end of the first sub-period (as opposed to the end of a time period as is the case for Disk Grouping).

The ordered assignment and activation of the fastest disk first is to guarantee a hiccup-free display. In particular, for slower disks, the duration of a sub-period might exceed the display time of the fragment retrieved from those disks. Fragment sizes are approximately proportional to a disk's bandwidth and consequently slower disks are assigned smaller fragments. As an example, with the parameters of Table 5.1, each sub-period is 0.929 seconds long (a time period is 2.787 seconds long) with each disk retrieving 16 fragments: the two logical disks read 48 blocks each to support $N = 96$ displays, hence, 16 fragments are retrieved during each of the three sub-periods. The third column of this table shows the display time of each fragment. More data is retrieved from the Seagate ST34501WD (with a display time of 1.534 seconds) to compensate for the bandwidth of the slower ST31200WD disk (with a display time of 0.371 seconds). If the first portion of block X ($X_{0.0}$) is assigned to the ST31200WD then the client referencing X might suffer from a display hiccup after 0.371 seconds because the second fragment is not available until 0.929 seconds (one sub-period) later. By assigning the first portion to the fastest disk (Seagate ST34501WD), the system prevents the possibility of hiccups. Recall that an amount of data equal to a logical block is consumed during a time period.

The amount of memory required by the system is a linear function of

```
procedure PeakMem (S_P, R_C, B_0, ... , B_{K-1}, K);
    /* Consumed: the amount of data displayed during a sub-period S_P.
    /* K: the number of physical disks that form a logical disk.
    Consumed = S_P × R_C;
    Memory = Peak = B_0 + Consumed;
    for i = 1 to K - 1 do
        Memory = Memory + B_i - Consumed;
        if (Memory > Peak) then Peak = Memory;
    if ((B_0 + B_1) > Peak) then Peak = B_0 + B_1;
    return(Peak);
end PeakMem;
```

Figure 5.4. Algorithm for the dynamic computation of the peak memory requirement with Staggered Grouping.

the peak memory (denoted $\widehat{P}$) required by a display during a time period, N_D and D^l, i.e., $M = N_D \times D^l \times \widehat{P}$. For a display, the peak memory can be determined using the simple dynamic computation of Figure 5.4. This figure computes the maximum of two possible peaks: one that corresponds to when a display is first started (labeled (1) in Figure 5.5) and a second that is reached while a display is in progress (labeled (2) in Figure 5.5). Consider each in turn. When a retrieval is first initiated, its memory requirement is the sum of the first and second fragment of a block, because the display of the first fragment is started at the end of the first sub-period (and the block for the next fragment must be allocated in order to issue the disk read for B_1). For an in-progress display, during each time period that consists of K sub-periods, the display is one sub-period behind the retrieval. Thus, at the end of each time period, one sub-period worth of data remains in memory (the variable termed $Consumed$ in Figure 5.4). At the beginning of the next time period, the fragment size of B_0 is added to the memory requirement because this much memory must be allocated to activate the disk. Next, from the combined value, the amount of data displayed during a sub-period is subtracted and the size of the next fragment is added, i.e., the amount of memory required to read this fragment. This process is repeated for all sub-periods that constitute a time period. Figure 5.5 illustrates the resulting memory demand as a function of time. Either of the two peaks could be highest depending on the characteristics and the number of physical disks involved. In the example of Figure 5.3 (and with the fragment sizes of Table 5.1), the peak amount of required memory is 1,103.7 KBytes, which is reached at the beginning of each time period (peak (2) in Figure 5.5). This compares favorably with the two logical blocks (2.44 MBytes) per display re-

Peak Memory [KB]

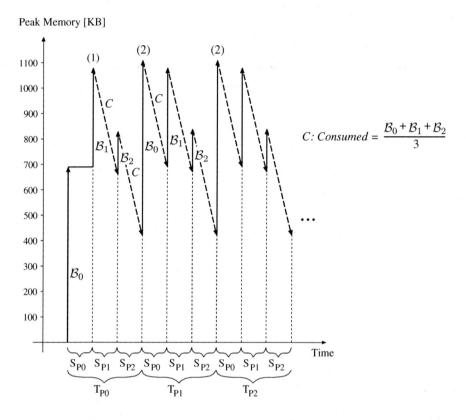

Figure 5.5. Memory requirement for Staggered Grouping.

quired by Disk Grouping. If no additional optimization is employed then the maximum value of either peak (1) or (2) specifies the memory requirement of a display.

We now derive equations that specify the remaining parameters of this technique analogous to the manner in the previous section. The sub-periods S_{P_i} are all equal in length and collectively add up to the duration of a time period T_P:

$$S_P \equiv S_{P_0} = S_{P_1} = \ldots = S_{P_{K-1}} \tag{5.7}$$

$$T_P = \sum_{i=1}^{K-1} S_{P_i} = K \times S_P \tag{5.8}$$

The total block size $\mathcal{B}^l$ retrieved must be equal to the consumed amount of data:

$$\mathcal{B}^l = \sum_{i=0}^{K-1} \mathcal{B}_i = R_C \times \sum_{i=0}^{K-1} S_{P_i} = R_C \times T_P \tag{5.9}$$

Furthermore, the sub-period S_P must satisfy the following inequality for the block size $\mathcal{B}_i$ on each physical disk i (d_i^p):

$$S_P \geq N_D \times \left(\frac{\mathcal{B}_i}{R_{D_i}} + T_{Seek_i}\left(\tfrac{\#cyl_i}{N_D} \right) \right) \tag{5.10}$$

Solving for the fragment sizes $\mathcal{B}_i$ leads to a system of equations similar to the Disk Grouping Equation 5.3 with the difference of T_P being substituted by S_P. As a consequence of the equality $T_P = K \times S_P$, every instance of N_D is replaced by the expression $N_D \times K$. The solution (i.e., the individual block size $\mathcal{B}_i$) of this system of equations is analogous to Equation 5.4 where, again, every instance of N_D is replaced with $N_D \times K$ (and hence, for notational convenience, T_{Seek_x} is equivalent to $T_{Seek_x}\left(\frac{\#cyl_x}{N_D \times K} \right)$).

The total number of streams for a Staggered Grouping system can be quantified as:

$$N = K \times D^l \times N_D \tag{5.11}$$

Recall that there is no closed form solution for the peak memory requirement (see Figure 5.4 for its algorithmic computation). Hence, the total amount of memory M at the server side is derived as:

$$M = N \times \text{PeakMem}(S_P, R_C, \mathcal{B}_0, \dots, \mathcal{B}_{K-1}, K) \tag{5.12}$$

Consequently, the maximum latency $T_{Latency}^{Max}$ is given by:

$$T_{Latency}^{Max} = K \times D^l \times S_P = D^l \times T_P = D^l \times \frac{\mathcal{B}^l}{R_C} \tag{5.13}$$

5.2.4   Logical/Virtual Homogeneous Disks

This technique separates the concept of logical disk from the physical disks all together. Moreover, it forms logical disks from fractions of the available bandwidth and storage capacity of the physical disks, as shown in Figure 5.6. These two concepts are powerful abstractions that enable this technique to utilize an arbitrary mix of physical disk models. To illustrate, consider the following example.

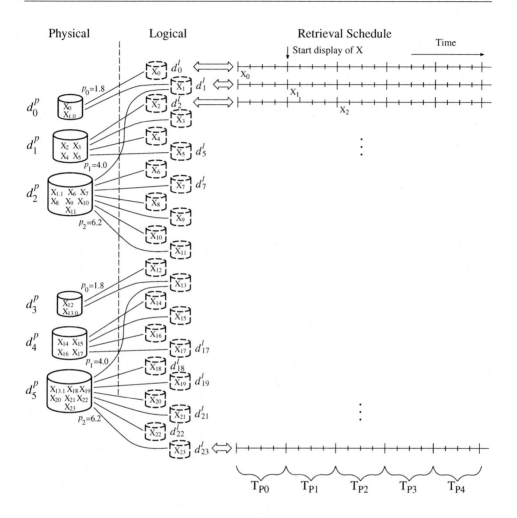

Figure 5.6. Disk Merging.

Example 5.2.1: Assume the six physical disk drives of the storage system shown in Figure 5.6 are configured such that the physical disks d_0 and d_3 each support 1.8 logical disks. Because of their higher performance the physical disks d_1 and d_4 realize 4.0 logical disks each and the disks d_2 and d_5 support 6.2 logical disks. Note that fractions of logical disks (e.g., 0.8 of d_0 and 0.2 of d_2) are combined to form additional logical disks (e.g., d_1^l, which contains block X_1 in Figure 5.6). The total number of logical disks adds up to $\lfloor (2 \times 1.8) + (2 \times 4.0) + (2 \times 6.2) \rfloor = 24$. Assuming $N_D = 4$ MPEG-2 (3.5 Mb/s) streams per logical disk, the maximum throughput that can be achieved is 96 streams ($24 \times N_D = 24 \times 4$). □

The detailed design of this technique is as follows. First, it selects how many logical disks should be mapped to each of the *slowest physical* disks and denotes this factor with p_0[4]. With no loss of generality, for Disk Merging we index the types in order of increasing performance. For example, in Figure 5.6, the two slowest disks d_0 and d_3 each represent 1.8 logical disks, i.e., $p_0 = 1.8$.

Next, the time period T_P and the block size necessary to support p_0 logical disks on a physical disk can be established by extending Equation 5.2 and solving for the block size $\mathcal{B}^l$ as follows:

$$T_P \geq N_D \times \left(\frac{\mathcal{B}^l \times p_0}{R_{D_0}} + \#seek(p_0) \times T_{Seek_0}\left(\frac{\#cyl_0}{\#seek(p_0) \times N_D} \right) \right) \tag{5.14}$$

$$\mathcal{B}^l = \frac{N_D \times R_C \times R_{D_0} \times \#seek(p_0) \times T_{Seek_0}\left(\frac{\#cyl_0}{\#seek(p_0) \times N_D} \right)}{R_{D_0} - p_0 \times N_D \times R_C} \tag{5.15}$$

As with the two previously introduced techniques, all the logical disks in the system must be identical. Therefore, T_P and $\mathcal{B}^l$ obtained from Equations 5.14 and 5.15 determine how many logical disks map to the other, faster disk types i, $1 \leq i \leq m - 1$, in the system. Each factor p_i must satisfy the following constraint:

$$\frac{\mathcal{B}^l}{N_D \times R_C} = \frac{\mathcal{B}^l \times p_i}{R_{D_i}} + \#seek(p_i) \times T_{Seek_i}\left(\frac{\#cyl_i}{\#seek(p_i) \times N_D} \right) \tag{5.16}$$

To compute the values for p_i from the above equation we must first quantify the number of seeks, i.e., $\#seek(p_i)$, that are induced by p_i. The function $\#seek(p_i)$ consists of two components:

$$\#seek(p_i) = \lceil p_i \rceil + (y_i - 1) \tag{5.17}$$

Because the disk heads need to seek to a new location prior to retrieving a chunk of data of any size—be that a full logical block or just a fraction thereof—the number of seeks is at least $\lceil p_i \rceil$. However, there are scenarios when $\lceil p_i \rceil$ underestimates the necessary number of seeks because the fractions of logical disks no longer add up to 1.0. In these cases, the expression $(n_i - 1)$ denotes any additional seeks that are needed. To illustrate, if there are two physical disk types with $p_0 = 3.7$ and $p_1 = 5.6$ then it is impossible to directly combine the fractions 0.7 and 0.6 to form complete

---

[4]Subscripts for p_i indicate a particular disk type, not individual disks.

logical disks. Hence, one of the fractions must be further divided into parts, termed *sub-fragments*. However, this process will introduce additional seeks. Specifically, assuming y_i sub-fragments, the number of seeks will increase by $(y_i - 1)$. Because seeks are wasteful operations, it is important to find an optimal division for the fragments to minimize the total overhead. For large storage systems employing a number of different disk types, it is a challenge to find the best way to divide the fragments.

Further complicating matters is the phenomenon that the value of y_i depends on p_i and, conversely, p_i might change once y_i is calculated. Because of this mutual dependency, there is no closed form solution for p_i from Equation 5.16. However, numerical solutions can easily be found. An initial estimate for p_i may be obtained from the bandwidth ratio of disk types i and 0 ($p_i \approx \frac{R_{D_i}}{R_{D_0}} \times p_0$). When iteratively refined, this value converges rapidly towards the correct ratio.

Based on the number of physical disks in the storage system and the p_i values for all the types the total number of logical disks (D^l) is defined as[5]:

$$D^l = \left\lfloor \sum_{i=0}^{m-1} (p_i \times \#disk_i) \right\rfloor \tag{5.18}$$

Once a homogeneous collection of logical disks is formed, the block size $\mathcal{B}^l$ of a logical disk determines the total amount of required memory, the maximum startup latency, and the total number of simultaneous displays. The equations for these are an extension of those described in Chapter 2 and are as follows.

The total number of streams a system can sustain is proportional to the number of logical disks and the number of streams each logical disk can support:

$$N = N_D \times D^l \tag{5.19}$$

The total amount of memory M is:

$$M = 2 \times N \times \mathcal{B}^l \tag{5.20}$$

While the maximum latency $T_{Latency}^{Max}$ is no more than:

$$T_{Latency}^{Max} = D^l \times T_P = D^l \times \frac{\mathcal{B}^l}{R_C} \tag{5.21}$$

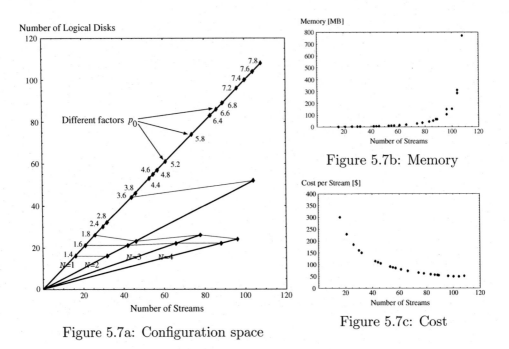

Figure 5.7a: Configuration space

Figure 5.7b: Memory

Figure 5.7c: Cost

Figure 5.7. Sample system configurations for different values of p_0 ($1.0 \leq p_0 \leq 7.8$ with increments of $\Delta p_0 = 0.2$) for a storage system with 6 physical disks ($2 \times$ ST31200WD, $2 \times$ ST32171WD, and $2 \times$ ST34501WD). Streams are of type MPEG-2 with a consumption rate of $R_C = 3.5$ Mb/s.

The overall number of logical disks constructed primarily depends on the initial choice of p_0. Figure 5.7 shows possible configurations for a system based on two disks each of type ST31200WD, ST32171WD, and ST34501WD. In Figure 5.7a, system throughputs are marked in the configuration space for various sample values of p_0. A lower value of p_0 results in fewer, higher performing logical disks. For example, if $p_0 = 1.8$ then each logical disk can support up to 4 displays simultaneously ($N_D \leq 4$). Figure 5.7a shows the total throughput for each $N_D = 1, 2, 3, 4$. As p_0 is increased, more logical disks are mapped to a physical disk and the number of concurrent streams supported by each logical disk decreases. For example, a value of $p_0 = 7.2$ results in logical disks that can support just one stream ($N_D = 1$). However, the overall number of logical disks also roughly increases four-fold. The maximum value

---

[5]m denotes the number of different disk types employed and $\#disk_i$ refers to the number of disks of type i.

Disk Model (Seagate)	p_i	Logical block size [MB]	Logical disk size [MB]	% Space for traditional data	Overall % space for traditional data
			$N_D = 1, D^l = 96, N = 96$		
ST34501WD	$p_2 = 25.2$		172.2	23.3%	
ST32171WD	$p_1 = 15.6$	0.554	132.1	0%	$W = 14.4\%$
ST31200WD	$p_0 = 7.2$		139.7	5.4%	
			$N_D = 4, D^l = 24, N = 96$		
ST34501WD	$p_2 = 6.2$		699.8	26.4%	
ST32171WD	$p_1 = 4.0$	0.780	515.3	0%	$W = 16.5\%$
ST31200WD	$p_0 = 1.8$		558.9	7.8%	

Table 5.2. Two configurations from Figure 5.7 shown in more detail. Their points in the configuration space of Figure 5.7a are $\langle p_0 = 1.8, N_D = 4 \rangle$ and $\langle p_0 = 7.2, N_D = 1 \rangle$, respectively. Streams are of type MPEG-2 ($R_C = 3.5$ Mb/s).

of the product $p_0 \times N_D$ is approximately constant[6] and limited by the upper bound of $\frac{R_{D_0}}{R_C}$. In Figure 5.7 this limit is $\frac{R_{D_0}=3.47MB/s=27.74Mb/s}{R_C=3.5Mb/s} = 7.93$. Pairs of $\langle p_0, N_D \rangle$ that approach this maximum will yield the highest possible throughput (e.g., $\langle 3.6, 2 \rangle$ or $\langle 7.8, 1 \rangle$). However, the amount of memory required to support such a high number of streams increases exponentially (see Figure 5.7b). The most economical configurations with the lowest cost per stream (i.e., approximately 100 streams, see Figure 5.7c) can be achieved with a number of different, but functionally equivalent, $\langle p_0, N_D \rangle$ combinations.

The number of possible configurations in the two-dimensional space shown in Figure 5.7a will increase if p_0 is incremented in smaller deltas than $\Delta p_0 = 0.2$. Hence, for any but the smallest storage configurations it becomes difficult to manually find the minimum cost system based on the performance objectives for a given application. However, an automated configuration planner can rapidly search through all configurations to find the ones that are most appropriate.

5.2.5   Random Disk Merging (RDM)

The previously introduced technique of Disk Merging can also be termed Deterministic Disk Merging (DDM) because it is based on round-robin data placement which results in an orderly, round-robin retrieval of blocks for

---

[6]It is interesting to note that, as a result of this property, the perceived improvement of the average seek distance d from $d = \frac{\#cyl_i}{N_D}$ (Equation 5.2) to $d = \frac{\#cyl_i}{\#seek(p_i) \times N_D}$ (Equation 5.16) does not lower the seek overhead.

streams that are sequentially viewed. This allows the DDM technique to provide a guaranteed retrieval time for each block.

For applications that require non-sequential access to data blocks such as interactive games or visualization applications the round-robin data placement does not provide any benefits. In those cases *random* data placement has been found to be more suitable [116], [144].

If data blocks are randomly placed on different disk drives, then the number of block retrievals per disk drive is not known exactly in advance but only statistically. Hence, such a scheme cannot make any absolute guarantees for the retrieval time of a specific block. However, *on average* the block retrieval times will be comparable to the round-robin case.

Because Equations 5.1 through 5.16 essentially partition the effective bandwidth of all the physical disk drives into equal amounts (equaling the data rate of one logical disk drive) we can still use those equations in our configuration planner for systems that employ random data placement in conjunction with cycle-based scheduling (for example RIO [116]).

One aspect that does not apply to systems with random block placement is the declustering of data blocks into fragments. For example, if a logical disk drive was allocated across three physical disk drives according to 20%, 30%, and 50% of its bandwidth, one would randomly distribute 20%, 30%, and 50% of the data blocks assigned to this logical disk to the respective physical drives. There would be no need to decluster each block into 20%, 30%, and 50% fragments because no absolute guarantees about the retrieval time are made. Without declustering the number of seek operations is reduced and more of the disk transfer rate can be harnessed.

Furthermore, allocating blocks randomly on the surface of a multi-zoned physical disk drive will automatically result in a disk data rate that approximates the average transfer rate of all the zones. Hence, no explicit technique to address multi-zoning needs to be employed.

Because of the previous observations we can simplify the seek function for systems that use random placement to

$$\#seek(p_i) = p_i \qquad (5.22)$$

In essence we count fractional numbers of seek operations because, for example, 20% of a logical disk will result in one complete seek operation approximately 20% of the time this logical disk is accessed.

The description of RDM in this section is based on an *average case* analysis. A worst case analysis requires statistical and queuing models and will not be presented here. Note that because the worst and average case are identical for DDM it will outperform RDM in a worst case scenario. Hence, it is application-dependent whether RDM or DDM are most appropriate under any specific circumstances.

Storage Space Considerations

Utilizing most of the bandwidth of the physical disks in a system to lower the cost per stream CPS is certainly a compelling aspect of selecting a system configuration (i.e., p_0 and N_D). However, for some applications the available storage space is equally or more important. With Disk Merging the storage space associated with each logical disk may vary depending onto which physical disk it is mapped. For example, Table 5.2 shows two possible configurations for the same physical disk subsystem. Because the round-robin assignment of blocks results in a close to uniform space allocation of data, the system will be limited by the size of the smallest logical disk. Both sample configurations support $N = 96$ streams. While one fails to utilize 14.4% of the total disk space in support of SM ($W = 14.4\%$), the other wastes 16.5% of the disk space ($W = 16.5\%$). In general, this space is not wasted and can be used to store traditional data, such as text and records, that are accessed during off-peak hours. Alternatively, the number of logical disks can purposely be decreased on some of the physical disks to equalize the disk space. The latter approach will waste some fraction of the disk bandwidth. In summary, if a configuration fully utilizes the disk bandwidth some storage space may be unused, or conversely, if all of the space is allocated then some bandwidth may be idle.

5.2.6   Configuration Planner for Disk Merging

The same hardware configuration can support a number of different system parameter sets depending on the selected value of p_0, see Figure 5.7. The total number of streams, the memory requirement, the amount of storage space unusable for continuous media, and ultimately the cost per stream are all affected by the choice of a value of p_0. The value of p_0 is a real number and theoretically an infinite number of values are possible. But even if p_0 is incremented by some discrete step size to bound the search space, it would be very time consuming and tedious to manually find the best value of p_0 for a given application. To alleviate this problem this section presents a configuration planner that can select an optimal value for p_0 based on the performance objective desired by a system.

Figure 5.8 shows the schematic structure of the configuration planner. It works in two stages: *Stage 1* enumerates all the possible configuration tuples based on a user supplied description of the disk subsystem under consider-

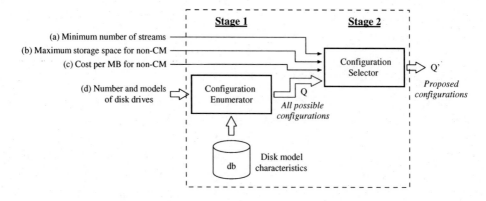

Figure 5.8. Configuration planner structure.

ation. The calculations are based on the analytical models of Section 5.2.4. *Stage 2* filters the output of Stage 1 by removing configurations that do not meet the performance objectives of the application. The result of Stage 2 is a set of proposed configurations that provide the lowest cost per stream or, alternatively, make the best use of the available storage space.

Operation

The detailed operation of the planner is as follows. The number and type of disks of the storage system under consideration are input parameters into Stage 1. This information is used to access a database that stores the detailed physical characteristics of these disk types. (This information is similar to the examples listed in Tables 2.1 and 2.2.) With $p_0 = 1$ the analytical models of Section 5.2.4 are then employed (by iterating over values of N_D) to produce output values for the total number of streams N, the number of logical disks D^l, the memory required M, the maximum latency $T_{Latency}^{Max}$ (for DDM), the logical block size $\mathcal{B}^l$, the storage space usable for continuous media S, the percentage of the total storage space that cannot be used for continuous media W, the cost per stream CPS (see Equation 5.23), and $p_i, i \geq 1$. When N_D reaches its upper bound of $\frac{R_{D_0}}{R_C \times p_0}$, the value of p_0 is incremented by a user specified amount, termed the step size and denoted with Δp_0, and the iteration over N_D is restarted. This process is repeated until the search space is exhausted. Stage 1 produces all possible tuples $Q = \langle N, D^l, M, T_{Latency}^{Max}, \mathcal{B}^l, S, W, CPS, p_0, p_1, \ldots, p_{m-1} \rangle$ for a given storage subsystem. Table 5.3 shows an example output for a 30 disk storage system with three disk types.

To provide the output of Stage 1 in a more useful and abbreviated form

N	D^l	M [MB]	L_{max} [s]	$\mathcal{B}$ [MB]	S [MB]	W [%]	CPS [$]	p_0	p_1	p_2
49	49	1.916	2.190	0.0196	49294.0	33.440	306.16	1.0	1.5^a	2.4
98	49	7.416	4.238	0.0378	49294.0	33.440	153.14	1.0	1.5^a	2.4
156	52	19.827	7.553	0.0635	52312.0	29.365	96.28	1.0	1.6^b	2.6
220	55	45.037	12.868	0.1024	55330.0	25.290	68.39	1.0	1.7^c	2.8
285	57	94.771	21.662	0.1663	57342.0	22.574	52.96	1.0	1.8^d	2.9
366	61	215.046	40.961	0.2938	61366.0	17.140	41.57	1.0	1.9^e	3.2
455	65	632.904	103.331	0.6955	63792.9	13.863	34.36	1.0	2.1	3.4
104	104	10.132	11.579	0.0487	63042.4	14.877	144.33	1.5^a	3.4	5.5
214	107	49.150	28.085	0.1148	63007.7	14.923	70.32	1.5^a	3.5	5.7
318	106	147.867	56.330	0.2325	62418.9	15.719	47.63	1.5^f	3.5	5.6
420	105	445.231	127.209	0.5300	63648.5	14.058	36.77	1.5	3.4	5.6
525	105	3062.561	700.014	2.9167	63648.5	14.058	34.40	1.5	3.4	5.6
90	90	6.811	7.784	0.0378	45270.0	38.874	166.74	2.0	2.7	4.3
210	105	42.989	24.565	0.1024	52815.0	28.686	71.63	2.0	3.2	5.3
366	122	215.046	81.922	0.2938	61366.0	17.140	41.57	2.0	3.9^g	6.3
173	173	31.018	35.449	0.0896	63670.2	14.029	86.88	2.5^f	5.6	9.2
332	166	163.513	93.436	0.2463	63356.7	14.452	45.67	2.5^a	5.4	8.7
519	173	2490.946	948.932	2.3998	63670.2	14.029	33.70	2.5^f	5.6	9.2
146	146	18.556	21.207	0.0635	48958.7	33.893	102.87	3.0	4.4	7.2
364	182	213.871	122.212	0.2938	61030.7	17.593	41.80	3.0	5.8^h	9.4
212	212	46.600	53.258	0.1099	60934.9	17.722	70.97	3.5^a	6.7	11.0
478	239	900.622	514.641	0.9421	63151.2	14.730	33.26	3.5	7.8^i	12.6
210	210	42.989	49.131	0.1024	52815.0	28.686	71.63	4.0	6.5	10.5
292	292	109.716	125.389	0.1879	64022.6	13.553	51.75	4.5^a	9.4	15.3
281	281	93.441	106.789	0.1663	56537.2	23.660	53.71	5.0	8.8^j	14.3
360	360	213.857	244.408	0.2970	63415.4	14.373	42.26	5.5^a	11.7	18.8
364	364	213.871	244.424	0.2938	61030.7	17.593	41.80	6.0	11.6^k	18.8
425	425	429.615	490.988	0.5054	63936.1	13.670	36.30	6.5	13.7	22.3
456	456	634.295	724.909	0.6955	63933.1	13.674	34.29	7.0	14.7^l	23.9
515	515	1928.946	2204.510	1.8728	63557.8	14.181	32.87	7.5	16.7	27.3

^{a} through ^{l}: Fractions of logical disks must sometimes be divided into parts such that complete logical disks can be assembled (see Section 5.2.4 and [185]). For example, fractions of 10×0.7 and 10×0.8 become useful when 0.7 is divided into two parts, 0.2+0.5. Then 5 disks can be assembled from the 0.5 fragments and 10 disks from a combination of the 0.2+0.8 fragments. The configuration planner automatically finds such divisions if they are necessary. For this table the complete list is as follows: ^{a}0.2+0.3, ^{b}0.2+0.4, ^{c}0.2+0.5, ^{d}0.1+0.2+0.5, ^{e}0.1+0.8, ^{f}0.1+0.4, ^{g}0.2+0.7, ^{h}0.2+0.6, ^{i}0.2+0.2+0.4, ^{j}0.1+0.7, ^{k}0.2+0.2+0.2 and ^{l}0.1+0.1+0.5.

Table 5.3. Planner *Stage 1* output for 30 disks with Deterministic Disk Merging, DDM ($10 \times$ ST31200WD, $10 \times$ ST32171WD, and $10 \times$ ST34501WD). The raw tuples Q are sorted by increasing values of p_0 (third column from the right). Streams are of type MPEG-2 with $R_C = 3.5$ Mb/s and the value of p_0 was iterated over a range of $1.0 \leq p_0 \leq 7.5$ ($\frac{R_{D_{ST31200WD}}}{R_C} = 7.93$) with increments of $\Delta p_0 = 0.5$. $\$MB = \$1$ per MB of memory and $\$Disk_i = \$500$ per disk drive.

the data is fed into Stage 2, which eliminates tuples Q that fail to satisfy the performance objectives of the application, and then orders the remaining tuples according to an optimization criterion. The input parameters to Stage 2 are (a) the minimum desired number of streams, (b) the maximum desired cost per stream, and (c) a hypothetical cost per MByte of disk space that cannot be utilized by CM applications. Stage 2 operates as follows. First, all tuples Q that do not provide the minimum number of streams desired are discarded. Second, an optimization component calculates the logical cost CPS' per stream for each remaining tuple Q' according to Equation 5.24,

$$CPS = \frac{(M \times \$MB) + \sum_{i=0}^{D-1} \$Disk_i}{N_D \times D^l} \tag{5.23}$$

$$CPS' = CPS + \left(\frac{W}{100\% - W} \times S \times \$W \right) \tag{5.24}$$

where $\$MB$ is the cost per MByte of memory, $\$Disk_i$ is the cost of each physical disk drive i, and $\$W$ is the cost per MByte of unused storage space.

By including a hypothetical cost for each MByte of storage that cannot be utilized for CM (input parameter (c)), Stage 2 can optionally optimize for (i) the minimum cost per stream, or (ii) the maximum storage space for continuous media, or (iii) a weighted combination of (i) and (ii).

The final, qualifying tuples Q' will be output by Stage 2 in sorted order of ascending total cost per stream C' as shown in Table 5.4. For many applications the lowest cost configuration will also be the best one. In some special cases, however, a slightly more expensive solution might be advantageous. For example, the startup latency $T_{Latency}^{Max}$ of a stream is a linear function of the number of logical disk drives in the system (see Equation 5.21). Hence, a designer might choose, for example, the second configuration from the top in Table 5.4 at a slightly higher cost per stream but for a drastically reduced initial latency (103 seconds versus 491 seconds, maximum). Alas, in some cases the final decision for a certain configuration is application-specific, but the planner will always provide a comprehensive list of candidates to select from.

Search Space Size

It is helpful to quantify the search space that the planner must traverse. If it is small then an exhaustive search is possible. Otherwise, heuristics must be employed to reduce the time or space complexity. The search space is defined by Stage 1 of the planner. The algorithm consists of two nested loops, as illustrated in Figure 5.9.

N	D^l	M [MB]	L_{max} [s]	$\mathcal{B}$ [MB]	W [%]	CPS [\$]	CPS' [\$]	p_0	p_1	p_2
425	425	429.615	490.988	0.5054	13.670	36.30	542.50	6.5	13.7	22.3
455	65	632.904	103.331	0.6955	13.863	34.36	547.72	1.0	2.1	3.4
519	173	2490.946	948.932	2.3998	14.029	33.70	553.19	2.5	5.6	9.2
525	105	3062.561	700.014	2.9167	14.058	34.40	554.98	1.5	3.4	5.6
420	105	445.231	127.209	0.5300	14.058	36.77	557.35	1.5	3.4	5.6
515	515	1928.946	2204.510	1.8728	14.181	32.87	557.98	7.5	16.7	27.3
360	360	213.857	244.408	0.2970	14.373	42.26	574.49	5.5	11.7	18.8
478	239	900.622	514.641	0.9421	14.730	33.26	578.71	3.5	7.8	12.6
332	166	163.513	93.436	0.2463	14.452	45.67	580.84	2.5	5.4	8.7
318	106	147.867	56.330	0.2325	15.719	47.63	629.69	1.5	3.5	5.6
366	61	215.046	40.961	0.2938	17.140	41.57	676.27	1.0	1.9	3.2
364	182	213.871	122.212	0.2938	17.593	41.80	693.26	3.0	5.8	9.4

Table 5.4. Planner *Stage 2* output for 30 disks with Deterministic Disk Merging, DDM (10 × ST31200WD, 10 × ST32171WD, and 10 × ST34501WD). The input data for this example was as listed in Table 5.3. These final tuples Q' are sorted according to their increasing, adjusted cost CPS' (fourth column from the right) which includes a hypothetical cost for each MByte that cannot be utilized for continuous media (\$$W$ = \$0.05 per MByte in this example, \$$MB$ = \$1 per MB of memory and \$$Disk_i$ = \$500 per disk drive). The minimum number of streams desired was 300.

> **for** $p_0 = 1$ **to** $\frac{R_{D_0}}{R_C}$ **step** Δp_0 **do**
>   **for** $N_D = 1$ **to** $\frac{R_{D_0}}{R_C \times p_0}$ **step** 1 **do**
>   /* Computation of the tuples Q. */
>   **end for;**
> **end for;**

Figure 5.9. The two nested loops that define the configuration planner search space.

Figure 5.9 also shows the three parameters that define the size of the search space: the disk bandwidth of the slowest disk drive R_{D_0}, the bandwidth requirements of the media R_C, and the step size Δp_0. The outer loop, iterating over p_0, is terminated by its upper bound of $\frac{R_{D_0}}{R_C}$. This upper bound describes into how many logical disks the slowest physical disk can be divided. The inner loop iterates over N_D to describe the number of streams supported by one logical disk. Hence, the upper bound of the inner loop is dependent on the outer loop. A low value of p_0 results in fewer, higher performing logical disks. As p_0 is increased, more logical disks are mapped to a physical disk and the number of concurrent streams N_D supported by each logical disk decreases. Therefore the maximum value of the product

$p_0 \times N_D$ is approximately constant at a value of $\frac{R_{D_0}}{R_C}$. The total number of executions of the program block enclosed by the two loops can be estimated at:

$$E_{\#loops} = \frac{1}{\Delta p_0} \times \left(\left\lfloor \frac{R_{D_0}}{R_C} \right\rfloor + \left\lfloor \frac{R_{D_0}}{2 \times R_C} \right\rfloor + \left\lfloor \frac{R_{D_0}}{3 \times R_C} \right\rfloor + \ldots + 1 \right) \quad (5.25)$$

$$\approx \frac{1}{\Delta p_0} \times \frac{R_{D_0}}{R_C} \times \left(1 + \frac{1}{2} + \frac{1}{3} + \frac{1}{4} + \ldots + \frac{1}{\frac{R_{D_0}}{R_C}} \right) \quad (5.26)$$

There is no closed form solution of Equation 5.25. However, since the number of terms enclosed by the parenthesis is $\left\lfloor \frac{R_{D_0}}{R_C} \right\rfloor$, a loose upper bound for $E_{\#loops}$ is:

$$E_{\#loops} < \frac{1}{\Delta p_0} \times \left(\frac{R_{D_0}}{R_C} \right)^2 \quad (5.27)$$

Table 5.5 shows two sample disk configurations, both with approximately the same aggregate disk bandwidth (200 MB/s) but with a different number of disk types. In Configuration A the slowest disk is of type ST31200WD which limits the ratio $\frac{R_{D_0}}{R_C}$ to 7.93. The actual number of configuration points produced by the planner is 139 (see Table 5.6). The estimated search space size (using Equation 5.25) is $E_{\#loops} = 160$, or about 15.1% higher. The same calculation for Configuration B yields $E_{\#loops} = 580$ (with 469 actual points, an over-estimation of 23.7%, see Table 5.6). Note that the number of possible configuration points depends on the ratio between the media bandwidth requirements and the disk drive production rate for the slowest disk in the system. It does not depend on the overall number of disk drives. Table 5.6 shows a few additional configuration spaces and estimates. Note that at each configuration point, for all disk types other than the slowest one, p_i must be iteratively computed.

The upper bound illustrated in Equation 5.27 is less than quadratic and the examples of Table 5.6 indicate that the exponent is approximately 1.4 (a five-fold increase in the disk bandwidth results in a ten-fold increase in the search space, i.e., $\left(\frac{18 \text{ MB/s}}{3.47 \text{ MB/s}} \right)^{1.4} = 5.2^{1.4} = 10 = \frac{1381}{139}$).

However, to find an optimal configuration if DDM is employed the planner must consider that fragments sometimes need to be divided into subfragments (see Section 5.2.4). Unfortunately, the runtime of the algorithm

Configuration	A	B
Aggregate disk bandwidth	200 MB/s	200 MB/s
Number of different disk types	3	2
Total number of disks	33	20
Number of disks of type ST34501WD	5	8
Number of disks of type ST32171WD	8	12
Number of disks of type ST31200WD	20	–

Table 5.5. Two sample storage subsystem configurations: A with three disk types, and B with two disk types. Both provide the same aggregate disk bandwidth of 200 MB/s.

Slowest disk type		Search space	
Model	Bandwidth	Planner	Estimate ($E_{\#loops}$)
ST31200WD	(3.47 MB/s)	139	160 (+15.1%)
ST32171WD	(7.96 MB/s)	469	580 (+23.7%)
ST34501WD	(12.97 MB/s)	898	1030 (+14.7%)

Table 5.6. Sample configuration planner search space sizes ($R_C = 3.5$ Mb/s). All configurations used three disk types with 10 disks per type.

to find the best fragment division is exponential [185]. For storage systems up to approximately 1,000 disk drives it is still feasible to execute an exhaustive search. (This was done for all our experiments, see see Section 5.2.6 and Figure 5.11.) However, for much larger configurations heuristics need to be employed. For example, the exponential algorithm might be modified to take advantage of a polynomial-time approximation [94].

Step Size Δp_0

Ideally the value for p_0 is a real number and infinitely many values are possible within a specified interval. The smaller the step size Δp_0, the closer this ideal is approximated. However, a disk subsystem can only support a finite number of streams N and the number of streams is a discrete value. Because of this constraint, many values of p_0 will map to very similar final configurations. Since the step size Δp_0 is directly related to the execution time of the planner, it should be chosen sufficiently small to generate all possible configurations but large enough to avoid excessive and wasteful computations. Figure 5.10a shows the optimal number of streams supported for a storage system with $5 \times$ ST31200WD, $10 \times$ ST32171WD, and $15 \times$ ST34501WD disk drives for a decreasing value of Δp_0. For $\Delta p_0 < 0.2$ the throughput improvement is marginal while the planner execution time becomes very long as illustrated in Figure 5.10b. As mentioned earlier, the fragment combi-

Number of Streams

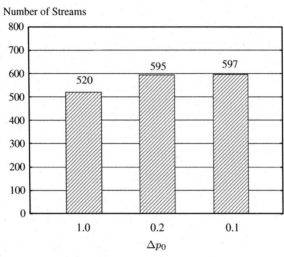

Fig. 5.10a. Number of streams at the lowest cost.

Planner Execution Time [sec]

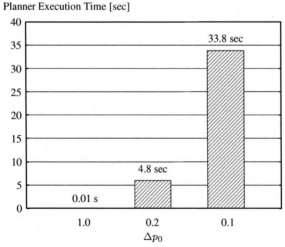

Fig. 5.10b. Runtime of the configuration planner.

Figure 5.10. Decreasing the step size Δp_0 results in an improved, more optimal system throughput (i.e., number of streams, Fig. 5.10a) with the same hardware configuration. However, the configuration planner execution time increases dramatically, see Fig. 5.10b. For $\Delta p_0 < 0.2$ the throughput improvement is marginal while the planner execution time becomes very long. (Configuration of $5 \times$ ST31200WD, $10 \times$ ST32171WD, and $15 \times$ ST34501WD disk drives.)

nation that is performed by the planner is a very expensive algorithm (see Section 5.2.6 for a discussion). Furthermore, the smaller Δp_0, the more possible fragment combinations there are (e.g., for $\Delta p_0 = 0.5$ there is only one fragment combination "0.5+0.5" while for $\Delta p_0 = 0.1$ there are 29 [185]). From these results we conclude that values of 0.1 or 0.2 result in an efficient and complete search space traversal.

Planner Execution Runtime

The runtime of the planner increases with both the number of physical disk drives D and the number of different disk drive types m. The time complexity for Deterministic Disk Merging (DDM) is much larger than for the random version (RDM) due to the fragment combination problem as described in Section 5.2.4.

The runtime of the DDM planner is exponential in the number of different disk drive types m for an exhaustive search. However, m is typically small (less than twenty) even for large storage system. If too many generations of disk drives with extraordinary performance differences are employed it becomes increasingly difficult to utilize all the resources (bandwidth and storage space). Still, because the runtime of the fragment combination algorithm is of the order $\mathcal{O}(f^3)$ execution times become large with an increasing number of fragments f. Figure 5.11 shows some sample execution times for the algorithm compiled with a gcc v2.8.1 C compiler on a Sun Enterprise Server 450 with two Sparc II (300 MHz) processors running Solaris 2.6.

Further illustrated in Figure 5.11 are the results for Random Disk Merging (RDM) which requires very little computation time and is only dependent on the number of disk types m and not the total number of physical disk drives D. For most of our measurements the execution time of the planner for RDM was below 100 milliseconds. Hence, the configuration planner can be used to configure very large RDM storage systems.

Planner Extension to $p_0 < 1$

Thus far, an implicit assumption for Disk Merging has been that a physical disk would be divided into one or more logical disks, i.e., $p_0 \geq 1$. However, we have already investigated how to combine block fragments into complete logical disks. By generalizing this idea we can extend the Disk Merging technique to include p_0 values that are finer in granularity, i.e., $0 < p_0 \leq 1$. Consider Example 5.2.6.

Proof: Figure 5.12 shows a storage subsystem with five physical disk drives. Assume that disks d_3 and d_4 are approximately 62% faster than the disks

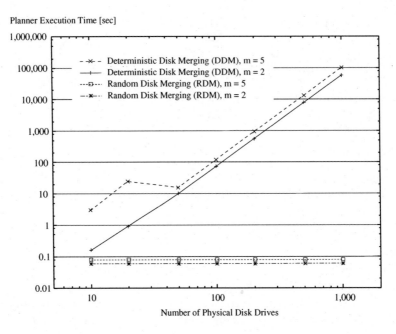

Figure 5.11. Planner execution times with $m = 2$ and $m = 5$ disk types for Deterministic Disk Merging (DDM) and Random Disk Merging (RDM). Measurements were obtained from a Sun Enterprise Server 450 with two Sparc II (300 MHz) processors, Solaris 2.6 and the gcc v2.8.1 C compiler, $\Delta p_0 = 0.2$.

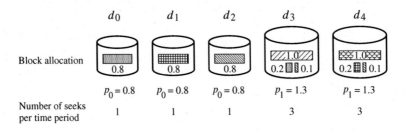

Figure 5.12. A Disk Merging example configuration where the smallest physical drives support less than one full logical disk, i.e., $p_0 < 1$.

d_0, d_1, and d_2. Then a possible configuration is $p_0 = 0.8$ and $p_1 = 1.3$ for a total of five logical drives. The physical disks d_0, d_1, and d_2 each store data in chunks of 0.8 logical blocks $\mathcal{B}^l$. This necessitates that either d_3 or d_4 must be activated whenever a complete logical block is retrieved from either d_0, d_1, or d_2. These fragment should be $0.2 \times \mathcal{B}^l$ in size. However, because disks d_3 and d_4 are partitioned into 1.3 logical disks, the fragment for one of the slower disks is divided in half (0.1+0.1) and assigned to both d_3 and d_4. □

N	D^l	M [MB]	L_{max} [s]	$\mathcal{B}$ [MB]	S [MB]	W [%]	CPS [\$]	p_0	p_1	p_2
130	65	16.515	9.437	0.0635	60893.2	17.779	115.51	0.8	2.2	3.5
195	65	39.922	15.209	0.1024	60893.2	17.779	77.13	0.8	2.2	3.5
252	63	76.834	21.953	0.1524	61830.0	16.514	59.83	0.8	2.1	3.4
310	62	137.677	31.469	0.2221	60848.6	17.839	48.83	0.8	2.1	3.3
360	60	235.093	44.780	0.3265	61830.0	16.514	42.32	0.8	2.0	3.2
441	63	634.061	103.520	0.7189	61830.0	16.514	35.45	0.8	2.1	3.4

Table 5.7. Planner *Stage 1* output with 30 disks (10 × ST31200WD, 10 × ST32171WD, and 10 × ST34501WD) for a single value of $p_0 = 0.8$.

There are tradeoffs associated with choosing p_0 to be less than 1.0. First, such a configuration can reduce the number of logical disks in the system which in turn reduces the worst case latency. The number of logical disks in Example 5.2.6 is five whereas, with values of $p_0 \geq 1$, the planner would have constructed at least six logical disks. Second, when the value of p_0 is set to very small values, then several physical disks are combined to support just one logical disk. A logical block $\mathcal{B}^l$ is effectively declustered across several disk drives. In its purest form, all physical disks form just one logical disk. Each block retrieval then activates all the disks in parallel. This is a special case of the Disk Grouping technique with all the disks forming just one group. Small values of p_0 can result in large logical block sizes $\mathcal{B}^l$. And in turn, large blocks translate into a large memory requirement and potentially increased cost.

These tradeoffs can automatically be quantified by extending the configuration planner to operate for values of p_0 less than 1.0. Table 5.7 shows a sample output of the planner for a 30 disk storage system. The table has been abbreviated to show just the configurations for $p_0 = 0.8$.

DDM and RDM Comparison

Table 5.8 compares the most cost-effective solutions produced by the configuration planner for a fixed, 30-disk storage system with the use of either DDM or RDM (the results are representative of other storage system sizes as well). As expected, DDM results in a lower number of streams at a slightly higher price as compared with RDM. Recall that DDM uses track-pairing to achieve the average transfer rate of a multi-zone disk drive. Track-pairing results in two seeks for each data transfer and hence the overhead is increased. RDM places blocks in random locations on each disk platter and thus achieves the average disk transfer rate. For comparison purposes Ta-

Storage Configuration	Parameter	DDM	DDM without Track-Pairing	RDM
10 × {ST31200WD, ST32171WD, ST34501WD}	Streams N	497 @ $32.73	515 @ $31.22	520 @ $31.21
	Blocksize $\mathcal{B}^l$	1.2737 MB	1.0491 MB	1.1808 MB
	Throughput %	89.5%	92.8%	93.7%
	Total Cost	$16,266.81	$16,078.30	$16,229.20
	Total Memory	1,266.1 MB	1,080.6 MB	1,227.9 MB
10 × {ST32171WD, ST34501WD, ST39102LW}	Streams N	772 @ $21.78	800 @ $20.52	805 @ $20.50
	Blocksize $\mathcal{B}^l$	1.1727 MB	0.8846 MB	0.9312 MB
	Throughput %	86.8%	90.0%	90.6%
	Total Cost	$16,814.16	$16,416.00	$16,502.50
	Total Memory	1,810.6 MB	1,415.4 MB	1,499.2 MB

Table 5.8. Configuration planner output for Deterministic Disk Merging (DDM) and Random Disk Merging (RDM) for two 30 disk storage systems. DDM performs worse because track-pairing is increasing the number of seeks by approximately a factor of two. Streams are of type MPEG-2 with $R_C = 3.5$ Mb/s. $MB = 1 per MB of memory and $Disk_i = 500 per disk drive.

ble 5.8 shows an additional column that illustrates the performance of DDM without track-pairing (i.e., hypothetically assuming single-zone disks). Under such circumstances the performance of DDM and RDM would be very similar. The remaining difference is due to block fragmentation. No block fragmentation is required for RDM: if a disk supports 0.2 of a logical disk then simply 20% of the blocks assigned to that logical disk are stored there. These differences result in a slightly better resource utilization for RDM than for DDM with the same physical storage system. All configurations achieve a high overall throughput utilization of 86.8% to 93.7% (the bandwidth of the total number of streams supported relative to the maximum bandwidth of the storage system, assuming no overhead: $\frac{N \times R_C}{\sum_{i=0}^{D-1} R_{D_i}}$). Note that only statistical guarantees can be made for the disruption-free stream playback based on RDM. Hence, it depends on the application whether RDM or DDM is more appropriate.

5.2.7   Validation

The purpose of a configuration planner is to guide a system designer when implementing a real system. To be useful, it must reveal and predict the behavior of a real system reliably and accurately. With the configuration planner, as with any non-trivial software program, the determination of correctness is a challenge. We have employed multiple strategies to provide compelling evidence that the planner indeed closely models an actual system.

The analytical models developed in Section 5.2.4 were the first tool used to evaluate the performance of the planner. These models have the advantage of being relatively simple to compute and easy to verify. However, they cannot encompass the full complexity of a storage system and hence are based on some arguable simplifying assumptions. Used in conjunction with other methods, however, their point estimates provide compelling evidence and reinforcement for the accuracy and correctness of our methodology.

The primary tool to evaluate the planner was a simulator. It included a detailed disk model that was calibrated with parameters extracted from commercially available disk drives. The file system was the same as the one used in the implementation of our prototype, Mitra [76]. The benefits of this approach were two-fold. First, any influence that the file system might have on the results gathered from the simulation would be identical to the effects experienced in a real implementation. Second, there was high confidence that the code was largely bug-free because it had been tested with a number of different storage system configurations on platforms running for extended periods of time on either UNIX (HP-UX^{TM} and OSF/1^{TM}) or Windows NT^{TM}. To model user behavior, the simulator included a module to generate synthetic workloads based on various Poisson and Zipf distributions [190]. A detailed description of the simulation infrastructure is provided in Section 5.2.7. The main advantage of the simulator was that it allowed us to model some of the complexities in more detail than in the analytical models.

The third tool to substantiate the results provided by both, the simulator and the analytical models, was an actual *implementation* of the storage subsystem. The strategy of using the file system of Mitra in the simulator made it possible to replace the modules that encapsulated a model of the magnetic disk drives with actual physical devices. Hence, results such as individual disk utilizations could be measured directly from real devices and the resulting statistics can be expected to match the real performance of a heterogeneous CM storage system closely. Two disadvantages limited the use of this implementation. First, the workload had to execute in real-time. This was substantially (i.e., several orders of magnitude) slower than using simulated disk drives. Second, we were restricted to the actual hardware devices at hand. Hence, we used the implementation primarily to validate the simulator with some specific configurations.

Simulation Infrastructure

The simulator was implemented using the C programming language on an HP 9000 735/125 workstation running HP-UX™. Other than the standard libraries no external support is needed. The simulator was implemented in a modular fashion and consists of the following components:

1. The *disk emulation* module imitates the response and behavior of a magnetic disk drive. The level of detail of such a model depends largely upon the desired accuracy of the results. Several studies have investigated disk modeling in great detail [139], [180], [75]. Our model includes mechanical positioning delays (seeks and rotational latency) as well as variable transfer rates due to the common zone-bit-recording technique. The disk emulation module can be simultaneously instantiated multiple times for various parameter sets describing different disk models in a disk array.

2. The logical-to-physical *disk mapping layer* translates read and write requests directed to logical disks into corresponding operations mapped onto the actual, physical devices.

3. The *file system* module provides the abstraction of files on top of the disk models and is responsible for the allocation of blocks and the maintenance of the free space. We selected the *Everest* file system (developed for our CM server prototype Mitra) for its ability to minimize disk space fragmentation, to provide guaranteed contiguous block allocation, and for its real-time performance [63].

4. The *loader* module generates a synthetic set of continuous media objects that are stored in the file system as part of the initialization phase of the simulator. The parameters for this module include the total number of objects produced, their display time length, and their consumption rate (for example 1.5 Mb/s as required by MPEG-1 or 3.5 Mb/s for MPEG-2).

5. The *scheduler* module translates a user request into a sequence of real-time block retrievals. It implements the concept of a time period and enables the round-robin migration of consecutive block reads on behalf of each stream. Furthermore, it ensures that all real-time deadlines are met. To simulate a partitioned system consisting of multiple sub-servers, the scheduler can be instantiated multiple times, each copy configured with its own parameter set (time period, block size, etc.).

6. Finally, the *workload* generator models user behavior and produces a synthetic trace of access requests to be executed against the stored objects. Both the distribution of the request arrivals as well as the distribution of the object access frequency can be individually specified. For the purpose of our simulations, the request inter-arrival times were Poisson distributed while the object access frequency was modeled according to Zipf's law [190].

Metric	Value
Number of video clips	50
Bandwidth of video clips	3.5 Mb/s (e.g., MPEG-2)
Length of a video clip	5 minutes $\equiv$ 131.25 MB
User request distribution	Poisson
Access frequency distribution	Zipf with parameter 0.271
Disk drives	4$\times$ ST31200WD, 1$\times$ ST34501WD
Total CM storage capacity	7,377.3 MB
Allocated CM storage capacity	6,562.5 MB (89%)
Simulation time	12 hours
Target disk utilization	20%, 40%, 60%, and 80%

Table 5.9. Parameters for the simulation and implementation used to validate the configuration planner.

The storage system chosen for the comparison was influenced by the physical devices at hand. The setup included two different disk models and a total of five disks: four Seagate Hawk 1LP (ST31200WD) drives and one Seagate Cheetah 4LP (ST34501WD) drive. Based on this storage system the configuration planner computed a total of 78 possible configurations. We selected one of them—its parameters are shown in Table 5.10—to illustrate the results.

N	D^l	M [MB]	L_{max} [sec]	$\mathcal{B}$ [MB]	S [MB]	W [%]	CPS [\$]	p_0	p_1
55	11	125.814	28.758	1.1438	7377.3	11.786	75.01	1.5	5.4

Table 5.10. Planner output with 5 disks (4 $\times$ ST31200WD and 1 $\times$ ST34501WD) used for validation purposes.

For this particular configuration, the planner predicted the maximum number of concurrent streams to be $N = 55$. The metric used to compare the analytical, simulation, and implementation results was the *disk utilization* as defined by:

$$\varrho = \frac{\text{Total disk service time}}{\text{Total execution time}} = \frac{\sum T_{Service}}{\sum T_P} \qquad (5.28)$$

In the case of servicing the maximum number of streams, i.e., N, the disk utilization ϱ was expected to reach 100%. The continuous media clips selected for our experiments were 5 minutes long. (Table 5.9 summarizes the parameters used for the comparison.) Hence, a request arrival rate $\lambda_{5min} = 55$ per five minutes or equivalently $\lambda = 660$ per hour, would achieve 100% utilization. We tested and compared the analytical, simulation, and implementation results for a projected load of 20% ($\lambda = 132$), 40% ($\lambda = 264$), 60% ($\lambda = 396$), and 80% ($\lambda = 528$).

Figures 5.13, 5.14, 5.15, and 5.16 show the resulting disk utilization for the analytical model (i.e., predicted by the planner), the simulation, and the implementation. Overall the results are a very close match. Specifically, for the four ST31200WD disks $d_0^p, \ldots, d_3^p$ the simulated utilization is slightly less than predicted, but the difference is no more than -1.5% in the worst case. Equally impressive are the implementation results which are within +3.0% of the predicted values.

To interpret the results for the d_4^p physical disk (ST34501WD), we must describe an additional consideration. Note that the analytical disk utilization for this disk has been adjusted by a factor of $\frac{5}{5.4} = 0.926$. The reason for this is as follows. The number of logical disks in Table 5.10 is $D^l = 11$. However, the planner actually computed the possible number of disks to be $(4 \times p_0) + (1 \times p_1) = (4 \times 1.5) + 5.4 = 11.4$, i.e., 5.4 logical disks were mapped to the ST34501WD disk drive. Naturally, only complete disks can be used in the final storage assignment and hence, 0.4 logical disks must be left idle on one of the physical disks. We chose this to be the case on disk d_4^p. Therefore, the predicted load on this disk is reduced by a factor of $\frac{5}{5.4}$. Note that this rounding adjustment only plays a perceptible role in this case because of the small number of logical disks involved. For all but the smallest storage configurations this effect quickly becomes negligible as the number of logical disks increases. With the described adjustment, the difference in disk utilization for d_4^p is less than -6.9% for the simulation and less than -1.9% for the implementation.

Finally we evaluated our techniques by comparing the non-partitioning Disk Merging scheme with a simple partitioning technique, i.e., a configuration where identical physical disks were grouped into independent subservers. In this case each subserver was configured optimally for a given load.

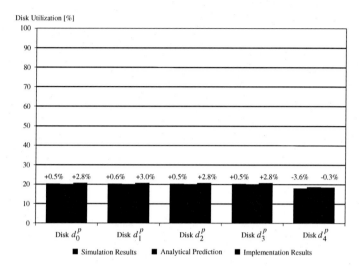

Figure 5.13. Verification results for 20% load ($\lambda = 132$).

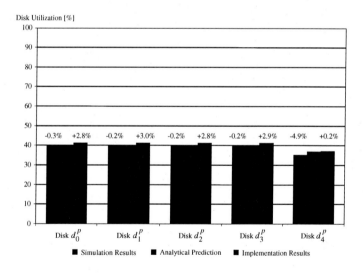

Figure 5.14. Verification results for 40% load ($\lambda = 264$).

Similar to a real system, our simulation modeled shifts in the user access pattern as a function of time. Tests were conducted at system loads of 70%, 80%, and 90%, respectively. The 90% load revealed the fundamental problem of any partitioned system: some resources became a bottleneck while others remained idle, waiting for work. As expected in the case of the partitioned CM server, one of the three subservers became overloaded and as a result

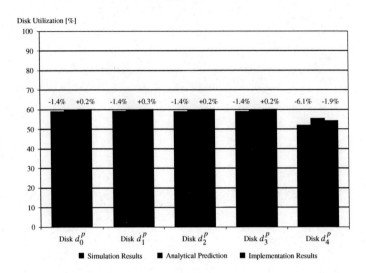

Figure 5.15. Verification results for 60% load ($\lambda = 396$).

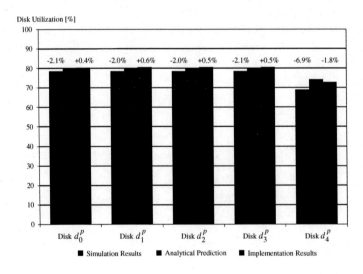

Figure 5.16. Verification results for 80% load ($\lambda = 528$).

the average startup latency for all users increased. Disk Merging was able to handle the situation more gracefully. It was sensitive to neither the frequency of access to clips nor their assignment. As a result the latency increase was much less dramatic. (See [185] for a detailed description of these results.)

One disadvantage of non-partitioning schemes as compared with partitioned systems is their increased vulnerability to disk failures. However, as part of our research we have investigated fault-tolerance and high availability techniques and their application to heterogeneous storage systems. Our investigation resulted in the design of a non-overlapping parity technique termed *HERA* that can be successfully applied to Disk Merging [189].

5.3   Techniques with Multi-zone Disk Drives

Numerous studies [12], [126], [176], [13], [67], [76] have described techniques in support of a hiccup-free display assuming a homogeneous collection of single-zone disk drives. Single-zone disk drives provide a constant transfer rate. A multi-zone disk drive consists of a number of regions (termed zones) that provide a different storage capacity and transfer rate. For example, the Seagate Barracuda 18 provides 18 gigabyte of storage and consists of 9 zones (see Table 5.11). To the best of our knowledge, there are only four techniques in support of hiccup-free display with multi-zone disk drives: IBM's Logical Track [166], Hewlett Packard's Track Pairing [15] and USC's FIXB [65] and deadline driven [68] techniques. Studies that investigate stochastic analytical models in support of admission control with multi-zone disks, e.g., [119], are orthogonal because they investigate only admission control (while the above four techniques describe how the disk bandwidth should be scheduled, the block size for each object, and admission control). Moreover, we are not aware of a single study that investigates hiccup-free display using a heterogeneous collection of multi-zone disk drives.

This section extends the four techniques to a heterogeneous disk subsystem. While these extensions are novel and a contribution in their own right, we believe that the primary contribution of this study is the performance comparison of these techniques and quantification of their tradeoff. This is because three of the described techniques assume certain characteristics about the target platform. Our performance results enable a system designer to evaluate the appropriateness of a technique in order to decide whether it is worthwhile to refine its design by eliminating its assumptions.

The rest of this section is organized as follows. Section 5.3.1 introduces five hiccup-free display techniques for a heterogeneous disk subsystem: two for IBM's Logical Track (termed OLT1 and OLT2), and one for each of the other techniques. Section 5.3.8 quantifies the performance tradeoff associated with these techniques. Our results demonstrate tradeoffs between cost per simultaneous stream supported by a technique, its startup latency, throughput, and the amount of disk space that it wastes. For example, while USC's FIXB results in the best cost/performance ratio, the potential maximum latency incurred by each user is significantly larger than the other techniques.

	Seagate Barracuda 4LP Introduced in 1994, 2 GBytes capacity, with a $1,200 price tag			
Zone id	Size (MB)	Track Size (MB)	No. of Tracks	Rate (MB/s)
0	506.7	0.0908	5579	10.90
1	518.3	0.0903	5737	10.84
2	164.1	0.0864	1898	10.37
3	134.5	0.0830	1620	9.96
4	116.4	0.0796	1461	9.55
5	121.1	0.0767	1579	9.20
6	119.8	0.0723	1657	8.67
7	103.2	0.0688	1498	8.26
8	101.3	0.0659	1536	7.91
9	92.0	0.0615	1495	7.38
10	84.6	0.0581	1455	6.97
	Seagate Cheetah 4LP Introduced in 1996, 4 GBytes capacity, with a $1,100 price tag			
Zone id	Size (MB)	Track Size (MB)	No. of Tracks	Rate (MB/s)
0	1017.8	0.0876	11617	14.65
1	801.6	0.0840	9540	14.05
2	745.9	0.0791	9429	13.23
3	552.6	0.0745	7410	12.47
4	490.5	0.0697	7040	11.65
5	411.4	0.0651	6317	10.89
6	319.6	0.0589	5431	9.84
	Seagate Barracuda 18 Introduced in 1998, 18 GBytes capacity, with a $900 price tag			
Zone id	Size (MB)	Track Size (MB)	No. of Tracks	Rate (MB/s)
0	5762	0.1268	45429	15.22
1	1743	0.1214	14355	14.57
2	1658	0.1157	14334	13.88
3	1598	0.1108	14418	13.30
4	1489	0.1042	14294	12.50
5	1421	0.0990	14353	11.88
6	1300	0.0923	14092	11.07
7	1268	0.0867	14630	10.40
8	1126	0.0807	13958	9.68

Table 5.11. Three different Seagate disk models and their zone characteristics.

The choice of a technique is application dependent: one must analyze the requirements of an application and choose a technique accordingly. For example, with nonlinear editing systems, the deadline driven technique is more desirable than the others because it minimizes the latency incurred by each users [97].

5.3.1   Five Techniques

In order to describe the alternative techniques, we assume a configuration consisting of K disk models: D_0, D_1, ..., D_{K-1}. There are q_i disks belonging to disk model i: d_0^i, d_1^i, ..., $d_{q_i-1}^i$. A disk drive of model D_i consists of m_i zones. To illustrate, Figure 5.17 shows a configuration consisting of two disk models D_0 and D_1 ($K=2$). There are two disks of each model ($q_0=q_1=2$), numbered d_0^0, d_1^0, d_0^1, d_1^1. Disks of model 0 consist of 2 zones ($m_0=2$) while those of model 1 consist of 3 zones ($m_1=3$). Zone j of a disk (say d_0^0) is denoted as $Z_j(d_0^0)$. Figure 5.17 shows a total of 10 zones for the 4 disk drives and their unique indexes. The kth physical track of a specific zone is indexed as $PT_k(Z_j(d_0^i))$.

We use the set notation, { : }, to refer to a collection of tracks from different zones of several disk drives. This notation specifies a variable before the colon and, the properties that each instance of the variable must satisfy after the colon. For example, to refer to the first track from all zones of the disk drives that belong to disk model 0, we write: $\{PT_0(Z_j(d_i^0)) \ : \ \forall i,j \ where \ 0 \leq j < m_0 \ and \ 0 \leq i < q_0\}$. With the configuration of Figure 5.17, this would expand to:
$\{PT_0(Z_0(d_0^0)), PT_0(Z_0(d_1^0)), PT_0(Z_1(d_0^0)), PT_0(Z_1(d_1^0))\}$.

5.3.2   IBM's Logical Track [166]

This section starts with a description of this technique for a single multizone disk drive. Subsequently, we introduce two variations of this technique, OLT1 and OLT2, for a heterogeneous collection of disk drives. While OLT1 constructs several logical disks from the physical disks, OLT2 provides the abstraction of only one logical disk. We describe each in turn.

With a single multi-zone disk drive, this technique constructs a logical track from each distinct zone provided by the disk drive. Conceptually, this approach provides equi-sized logical tracks with a single data transfer rate such that one can apply traditional continuous display techniques [12], [176], [13], [67], [116], [126]. With K different disk models, a naive approach would construct a logical track LT_k by utilizing one track from each zone:

$$LT_k \ = \ \{ \ PT_k(Z_j(d_p^i)) \ : \ \forall i,j,p \ \}$$

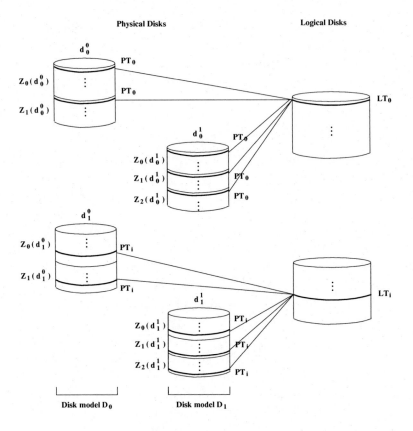

Figure 5.17. OLT1

where $0 \leq j < m_i$ and $0 \leq i < K$ and $0 \leq p < q_i$. With this technique, the value of k is bounded by the zone with the fewest physical tracks, i.e., $0 \leq k < Min[NT(Z_j(d^i_{q_i}))]$, where $NT(Z_j(d^i_{q_i}))$ is the number of physical tracks in zone j of disk model D_i. Large logical tracks result in a significant amount of memory per simultaneous display, rendering a continuous media server economically unavailable. In the next section, we describe two optimized versions of this technique that render its memory requirements reasonable.

5.3.3 Optimized Logical Track 1 (OLT1)

Assuming that a configuration consists of the same number of disks for each model[7], OLT1 constructs logical disks by grouping one disk from each disk

---

[7]This technique is applicable as long as the number of disks for each model is a multiple of the model with the fewest disk drives: if $min(q_i)$ $(0 \leq i < K)$ denotes the model with fewest disks, then q_j is a multiple of $min(q_i)$.

model (q logical disks). For each logical disk, it constructs a logical track consisting of one track from each physical zone of a disk drive. To illustrate, in Figure 5.17, we pair one disk from each model to form a logical disk drive. The two disks that constitute the first logical disk in Figure 5.17, i.e., disks d_0^0 and d_0^1, consist of a different number of zones (d_0^0 has 2 zones while d_0^1 has 3 zones). Thus, a logical track consists of 5 physical tracks, one from each zone.

Logical disks appear as a homogeneous collection of disk drives with the same bandwidth. There are a number of well known techniques that can guarantee hiccup-free display given such an abstraction, see [12], [176], [13], [67], [116], [126], [175]. Briefly, given a video clip X, these techniques partition X into equi-sized blocks that are striped across the available logical disks [13], [175], [176]: one block per logical disk in a round-robin manner. A block consists of either one or several logical tracks.

Let T_i denote the time to retrieve m_i tracks from a single disk of model D_i consisting of m_i zones: $T_i = m_i \times (a\ revolution\ time + seek\ time)$. Then, the transfer rate of a logical track (R_{LT}) is: $R_{LT} = \frac{size\ of\ a\ logical\ track}{Max[T_i]}$ $\forall i,\ 0 \leq i < K$.

In Figure 5.17, to retrieve a LT from the first logical disk, d_0^0 incurs 2 revolution times and 2 seeks to retrieve two physical tracks, while disk d_0^1 incurs 3 revolutions and 3 seeks to retrieve three physical tracks. Assuming a revolution time of 8.33 milliseconds (7200 rpm) and the average seek time of 10 milliseconds for both disk models, d_0^0 requires 36.66 milliseconds ($T_0 = 36.66$) while d_0^1 requires 54.99 ($T_1 = 54.99$) milliseconds to retrieve a LT. Thus, the transfer rate of the LT is determined by disk model D_1. Assuming that a LT is 1 megabyte in size, its transfer rate is $\frac{size\ of\ a\ logical\ track}{Max[T_0,T_1]} = \frac{1\ megabyte}{54.99\ milliseconds} = 18.19$ megabytes per second.

This example demonstrates that OLT1 wastes disk bandwidth by requiring one disk to wait for another to complete its physical track retrievals. In our example, this technique wastes 33.3% of D_0's bandwidth. In addition, this technique wastes disk space because the zone with the fewest physical tracks determines the total number of logical tracks. In particular, this technique eliminates the physical tracks of those zones that have more than NT_{min} tracks, $NT_{min} = Min[NT(Z_j(d_{q_i}^i))]$, i.e., it eliminates $PT_k(Z_j(d_{q_i}^i))$ that have $NT_{min} \leq k < NT(Z_j(d_{q_i}^i))$, for all i and j, $0 \leq i < K$ and $0 \leq j < m_i$.

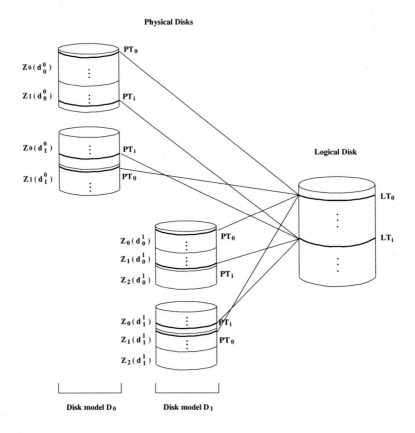

Figure 5.18. OLT2

5.3.4   Optimized Logical Track 2 (OLT2)

OLT2 extends OLT1 with the following additional assumption: each disk model has the same number of zones, i.e., m_i is identical for all disk models, $0 \leq i < K$. Using this assumption, it constructs logical tracks by pairing physical tracks from zones that belong to different disk drives. This is advantageous for two reasons. First, it eliminates the seeks required per disk drive to retrieve the physical tracks. Second, assuming an identical revolution rate of all heterogeneous disks, it prevents one disk drive from waiting for another.

The details of OLT2 are as follows. First, it reduces the number of zones of each disk to that of the disk with fewest zones: $m_{min} = Min[m_i]$ for all i, $0 \leq i < K$. Hence, we are considering only zones, $Z_j(d_k^i)$ for all i, j, and k ($0 \leq i < K$, $0 \leq j < m_{min}$, and $0 \leq k < q$). For example, in Figure 5.18, the slowest zone of disks of d_0^1 and d_1^1 (Z_2) are eliminated such that all disks utilize only two zones. This technique requires m_{min} disks of each disk model (totally $m_{min} \times K$ disks). Next, it constructs LTs such that no two physical tracks (from two different zones) in a LT belong to one physical disk drive. A logical track LT_k consists of a set of physical tracks:

$$LT_k = \{PT_{k \bmod NT_{min}}(Z_{l \bmod m_{min}}(d^i_{j \bmod m_{min}})) :$$
$$\forall i, j \ where \ 0 \le i < K \ and \ 0 \le j < m_{min}\} \qquad (5.29)$$

where $l = \lfloor \frac{k}{NT_{min}} \rfloor + j$. The total number of LTs is $m_{min} \times NT_{min}$, thus $0 \le k < m_{min} \times NT_{min}$.

OLT2 may use several possible techniques to force all disks to have the same number of zones. For each disk with δ_z zones more than m_{min}, it can either (a) merge two of its physically adjacent zones into one, repeatedly, until its number of logical zones is reduced to m_{min}, (b) eliminate its innermost δ_z zones, or (c) a combination of (a) and (b). With (a), the number of simultaneous displays is reduced because the bandwidth of two merged zones is reduced to the bandwidth of the slower participating zone. With (b), OLT2 wastes disk space while increasing the average transfer rate of the disk drive, i.e., number of simultaneous displays. In [65], we describe a configuration planner that empowers a system administrator to strike a compromise between these two factors for one of the techniques described in this study (HetFIXB). The extensions of this planner in support of OLT2 is a part of our future research direction.

5.3.5   Heterogeneous Track Pairing (HTP)

We describe this technique in two steps. First, we describe how it works for a single multi-zone disk drive. Next, we extend the discussion to a heterogeneous collection of disk drive. Finally, we discuss the tradeoff associated with this technique.

Assuming a single disk drive (say d^0_0) with $\#TR_0$ tracks, Track Pairing [15] pairs the innermost track ($TR_{\#TR_0-1}(d^0_0)$) with the outermost track ($TR_0(d^0_0)$), working itself towards the center of the disk drive. The result is a logical disk drive that consists of $\frac{\#TR_0}{2}$ logical tracks that have approximately the same storage capacity and transfer rate. This is based on the (realistic) assumption that the storage capacity of tracks increases linearly as one moves from the innermost track to the outermost track. Using this logical disk drive, the system may utilize one of the traditional continuous display techniques in support of hiccup-free display.

Assuming a heterogeneous configuration consisting of K disk models, HTP utilizes Track Pairing to construct track pairs for each disk. If the number of disks for each disk model is identical ($q_0 = q_1 = ... = q_{K-1}$), HTP constructs q_i groups of disk drives consisting of one disk from each of the K disk models. Next, it realizes a logical track that consists of K track pairs, one track pair from each disk drive in the group. These logical tracks constitute a logical disk. Obviously, the disk with the fewest number of tracks determines the total number of logical tracks for each logical disk. With such a collection of homogeneous logical disks, one can use one of the popular hiccup-free display techniques. For example, similar to both OLT1 and OLT2, one can stripe a video clip into blocks and assign the blocks to the logical disks in a round-robin manner.

HTP wastes disk space in two ways. First, the number of tracks in a logical disk is determined by the physical disk drive with fewest track pairs. For example, if a configuration consists of two heterogeneous disks, one with 20,000 track pairs and the other with 15,000 track pairs, then the resulting logical disk will consist of 15,000 track pairs. In essence, this technique eliminates 5,000 track pairs from the first disk drive. Second, while it is realistic to assume that the storage capacity of each track increases linearly from the innermost track to the outermost one, it is not 100% accurate [15]. Once the logical tracks are realized, the storage capacity of each logical track is determined by the track with the lowest storage capacity.

5.3.6   Heterogeneous FIXB (HetFIXB)

Please refer to Section 2.5.3 for a description of how the system guarantees a continuous display with a single multi-zone disk drive and the FIXB technique. Here, we describe the extensions of this technique to heterogeneous disk drives.

With a heterogeneous collection of disks, we continue to maintain a T_{scan} per disk drive. While the duration of a T_{scan} is identical for all disk drives, the amount of data produced by each T_{scan} is different. We compute the block size for each disk model (recall that blocks are equi-sized for all zones of a disk) such that the faster disks compensate for the slower disks by producing more data during their T_{scan} period. HetFIXB aligns the T_{scan} of each individual disk drive with one another such that they all start and end in a T_{scan}.

To support N simultaneous displays, HetFIXB must satisfy the following equations.

$$M = \sum_{i=0}^{K-1} M_i, \text{ where } M_i = m_i \times B_i \qquad (5.30)$$

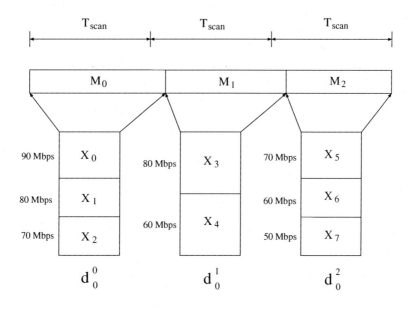

Figure 5.19. HetFIXB

$$AvgR_i \ : \ AvgR_j \ = \ M_i \ : \ M_j, \ \ 0 \leq i,j < K \tag{5.31}$$

$$T_{scan} = Tp/K, \text{ where } Tp = \frac{M}{R_C} \tag{5.32}$$

$$T_{scan_i} = T_{cseek} + \sum_{j=0}^{m_i-1} N(\frac{B_i}{R(Z_j(D_i))} + seek_i) \leq T_{scan} \tag{5.33}$$

where $0 \leq i < K$.

To illustrate, assume a configuration consisting of 3 disks, see Figure 5.19. Assume the average transfer rates of disks, $AvgR_0 = 80$ Mbps, $AvgR_1 = 70$ Mbps, and $AvgR_2 = 60$ Mbps respectively. When $R_C = 4$ Mbps, 1.5 Mbytes of data ($M = 1.5$ MB) is required every 3 seconds ($T_P = 3$ sec) to support a hiccup-free display. Based on the ratio among the average transfer rates of disk models, $M_0 = 0.5715$ MB, $M_1 = 0.5$ MB, and $M_2 = 0.4285$ MB. Thus, $B_0 = M_0/m_0 = 0.19$ MB, $B_1 = M_1/m_1 = 0.25$ MB, $B_2 = M_2/m_2 = 0.14$ MB. An object X is partitioned into blocks and blocks are assigned into zones in a round-robin manner. When a request for X arrives, the system retrieves X_0, X_1, and X_2 ($M_0 = 3 \times B_0$ amount of data) from D_0 during the first T_{scan}. A third of M (0.5 MB) is consumed during the same T_{scan}.

Hence, some amount of data, 0.0715 MB, remains unconsumed in the buffer. In the next T_{scan}, the system retrieves X_3 and X_4 ($M_1 = 2 \times B_1$ amount of data) from D_1. While the same amount of data (0.5 MB) is retrieved and consumed during this T_{scan}, the accumulated data (0.0715 MB) still remains in the buffer. Finally, during the last T_{scan}, the system retrieves X_5, X_6, and X_7 ($M_2 = 3 \times B_2$ amount of data) from D_2. Even though the amount of data retrieved in this T_{scan} (0.4285 MB) is smaller than the amount of data displayed during a T_{scan} (0.5 MB), there is no starvation because 0.0715 megabytes of data is available in the buffer. This process is repeated until the end of display.

5.3.7   Heterogeneous Deadline Driven (HDD)

With this technique, blocks are assigned across disk drives in a random manner. A client issues block requests, each tagged with a deadline. Each disk services block requests with the Earliest Deadline First policy. In [68], we showed that the assignment of blocks to the zones in a disk should be independent of the frequency of access to the blocks. Thus, blocks are assigned to the zones in a random manner. The size of the blocks assigned to each disk model is different. They are determined based on the average weighted transfer rate of each disk model. Let WR_i denote the weighted average transfer rate of disk model i:

$$WR_i = \sum_{j=0}^{m_i-1} [S(Z_j(D_i)) \times R(Z_j(D_i)) / \sum_{k=0}^{m_i-1} S(Z_k(D_i))] \qquad (5.34)$$

$$WR_i : WR_j = B_i : B_j, \ 0 \le i, j < K \qquad (5.35)$$

Assuming $B_i \ge B_j$ where $i < j$ and $0 \le i, j < K$, an object X is divided into blocks such that the size of each block X_i is $B_{i \bmod K}$. Blocks with the size of B_i are randomly assigned to disks belonging to model i. A random placement may incur hiccups that are attributed to the statistical variation of the number of block requests per disk drive, resulting in varying block retrieval time. Traditionally, double buffering has been widely used to absorb the variance of block retrieval time: while a block in a buffer is being consumed, the system fills up another buffer with data. However, we generalize double buffering to N buffering and prefetching N-1 buffers before initiating a display. This minimizes the hiccup probability by absorbing a wider variance of block retrieval time, because data retrieval is N-1 blocks ahead of data consumption.

We assume that, upon a request for a video clip X, a client: (1) concurrently issues requests for the first N-1 blocks of X (to prefetch data), (2) tags the first N-1 block requests, X_i $(0 \leq i < N)$, with a deadline, $\sum_{j=0}^{i} \frac{sizeof(X_j)}{R_C}$, (3) starts display as soon as the first block (X_0) arrives. For example, when $N = 4$, first three blocks are requested at the beginning. Then, the next block request is issued immediately after the display is initiated. Obviously, there are other ways of deciding both the deadline of the prefetched blocks and when to initiate display blocks. In [68], we analyzed the impact of these alternative decisions and demonstrated that the combination of the above two choices enhances system performance.

5.3.8   Performance Evaluation

In this section, we quantify the performance tradeoffs associated with alternative techniques. While OLT1, OLT2, HTP and HetFIXB were quantified using analytic models, HDD was quantified using a simulation study. We conducted numerous experiments analyzing different configurations with different disk models from Quantum and Seagate. Here, we report on a subset of our results in order to highlight the tradeoffs associated with different techniques. In all results presented here, we used the three disk models shown in Table 5.11. Both Barracuda 4LP and 18 provide a 7,200 rpm while the Cheetah provides a 10,000 rpm. Moreover, we assumed that all objects in the database require a 4 Mbps bandwidth for their continuous display.

Figure 5.20 shows the cost per stream as a function of the number of simultaneous displays supported by the system (throughput) for three different configurations. Figure 5.20a shows a system that is installed in 1994 and consists of 10 Barracuda 4LP disks. Figure 5.20b shows the same system two years later when it is extended with 10 Cheetah disks. Finally, Figure 5.20c shows this system in 1998 when it is extended with 10 Barracuda 18 disks. To estimate system cost, we assumed: a) the cost of each disk at the time when they were purchased with no depreciation cost, and b) the system is configured with sufficient memory to support the number of simultaneous displays shown on the x-axis. We assumed that the cost of memory is $7/MB, $5/MB, and $3/MB in 1994, 1996, and 1998, respectively. Additional memory is purchased at the time of disk purchases in order to support additional users. (Once again, we assume no depreciation of memory.) While one might disagree with our assumptions for computing the cost of the system, note that the focus of this study is to compare the different techniques. As long as the assumptions are kept constant, we can make observations about the proposed techniques and their performance tradeoff.

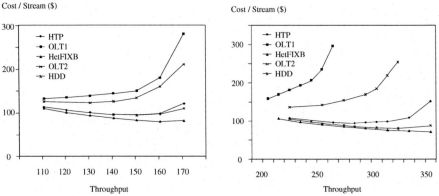

Figure 5.20a: One disk model (homogeneous).

Figure 5.20b: Two disk models (heterogeneous).

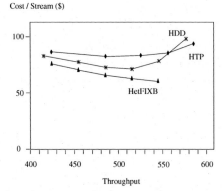

Figure 5.20c: Three disk models (heterogeneous).

Figure 5.20. Throughput and cost per stream.

In these experiments, OLT2 constructed logical zones in order to force all disk models to consist of the same number of zones. This meant that OLT2 eliminated the innermost zone (zone 10) of Barracuda 4LP, splitting the fastest three zones of Cheetah into six zones, and splitting the outermost zone of Barracuda 18 into two. Figure 5.20c does not show OLT1 and OLT2 because: a) their cost per stream is almost identical to that shown in Figure 5.20b, and b) we wanted to show the difference between HetFIXB, HDD, and HTP.

Figure 5.20 shows that HetFIXB is the most cost effective technique, however, it supports fewer simultaneous displays as a function of heterogeneity. For example, with one disk model, it provides a throughput similar to the other techniques. However, with 3 disk models, its maximum through-

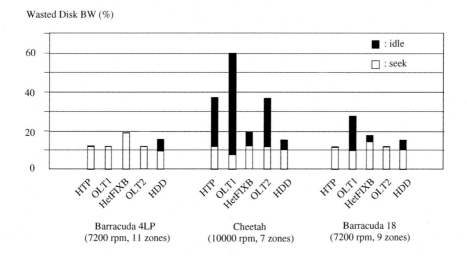

Figure 5.21. Wasted disk bandwidth.

put is lower than those provided by HDD and HTP. This is dependent on the physical characteristics of the zones because HetFIXB requires the duration of T_{scan} to be identical for all disk models. This requirement results in fragmentation of the disk bandwidth which in turn limits the maximum throughput of the system. Generally speaking, the greater the heterogeneity, the greater the degree of fragmentation. However, the zone characteristics ultimately decide the degree of fragmentation. One may construct logical zones in order to minimize this fragmentation, see [65]. This optimization is not reported because of strict space limitations imposed by the call for paper. It raises many interesting issues that are not presented here. Regardless, the comparison shown here is fair because our optimizations are applicable to all techniques.

OLT1 provides inferior performance as compared to the other techniques because it wastes a significant percentage of the available disk bandwidth. To illustrate, Figure 5.21 shows the percentage of wasted disk bandwidth for each disk model with each technique when the system is fully utilized (the trend holds true for less than 100% utilization). OLT1 wastes 60% of the bandwidth provided by Cheetah and approximately 30% of Barracuda 18. Most of the wasted bandwidth is attributed to these disks sitting idle. Cheetahs sit idle because they provide 10,000 rpm as compared to 7,200 rpm provided by the Barracudas. Barracuda 4LP and 18 disks sit idle because of their zone characteristics. In passing, while different techniques provide approximately similar cost per performance ratios, each wastes bandwidth in a different manner. For example, both HTP and HetFIXB provide approximately similar cost per performance ratios, HTP wastes 40% of Cheetah's bandwidth while HetFIXB wastes only 20% of the bandwidth provided by this disk model. HTP makes up for this limitation by harnessing a greater percentage of the bandwidth provided by Barracuda 4LP and 18.

Figure 5.22 shows the maximum latency incurred by each technique as a function of the load imposed on the system. In this figure, we have eliminated OLT1 because of its prohibitively high latency. (One conclusion of this study is that OLT1 is not a competitive strategy.) The results show that HetFIXB provides the worst latency while HDD's maximum latency is below 1 second. This is because HetFIXB forces a rigid schedule with a disk zone being activated in an orderly manner (across all disk drives). If a request arrives and the zone containing its referenced block is not active then it must wait until the disk head visits that zone (even if idle bandwidth is available). With HDD, there is no such a rigid schedule in place. A request is serviced as soon as there is available bandwidth. Of course, this is at the risk of some requests missing their deadlines. This happens when many requests collide on a single disk drive due to random nature of requests to the disks. In these experiments, we ensured that such occurrences impacted one in a million requests, i.e., a hiccup probability is less than one in a million block requests.

OLT2 and HTP provide a better latency as compared to HetFIXB because they construct fewer logical disks [12], [70]. While OLT2 constructs a single logical disk, HTP constructs 10 logical disks, and HetFIXB constructs 30 logical disks. In the worst case scenario (assumed by Figure 5.22), with both HTP and HetFIXB, all active requests collide on a single logical disk (say $d_{bottleneck}$). A small fraction of them are activated while the rest wait for this group of requests to shift to the next logical disk (in the case of HetFIXB, they wait for one T_{scan}). Subsequently, another small fraction is activated on $d_{bottleneck}$. This process is repeated until all requests are activated. Figure 5.22 shows the incurred latency by the last activated request.

With three disk models (Figure 5.20c), OLT1 and OLT2 waste more than 80% of disk space, HTP and HDD waste approximately 70% of disk space, while HetFIXB wastes 44% of the available disk space. However, this does NOT mean that HetFIXB is more space efficient than other techniques. This is because the percentage of wasted disk space is dependent on the physical characteristics of the participating disk drives: number of disk models, number of zones per disk, track size of each zone, storage capacity of individual zones and disk drives. For example, with two disk models (Figure 5.20b), HetFIXB wastes more disk space when compared with the other techniques.

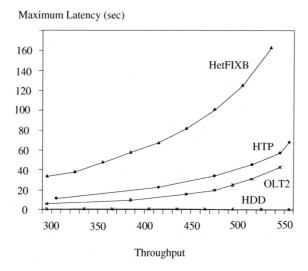

Figure 5.22. Maximum startup latency.

Modern magnetic disk drives feature variable transfer rates due to a technique called zone-bit recording (ZBR) which increases the amount of data being stored on a track as a function of its distance from the disk spindle. Several techniques have been proposed to harness the average transfer rate of the zones [85], [15], [65], [62], [171], as described in Chapter 2. These techniques are orthogonal to Disk Grouping, Staggered Grouping, and Disk Merging. In this section, we will only describe one approach and why its design is orthogonal. Assuming that the number of tracks in every zone is a multiple of some fixed number, [85] constructs Logical Tracks (LT) from the same numbered physical track of the different zones. The order of tracks in a LT is by zone number. When displaying a clip, the system reads a LT on its behalf per time period. This forces the disk to retrieve data from the constituting physical tracks in immediate succession by zone order. An application observes a constant disk transfer rate for each LT retrieval. LTs are employed in Disk Grouping and Staggered Grouping as follows. With these techniques, each logical block is declustered across the participating physical disks into fragments. These fragments are then assigned to one or more LTs on each of the physical disks according to the size of the fragments. Similarly, for Disk Merging the logical blocks are assigned to LTs.

5.4 Conclusions

This chapter introduced and evaluated three different non-partitioning techniques that guarantee a continuous display of objects with a heterogeneous disk subsystem. These techniques are Disk Grouping, Staggered Grouping, and Disk Merging. All three techniques conceptualize the physical heterogeneity of a storage system as a set of logical disks with homogeneous characteristics. As a result of this abstraction, conventional scheduling and data placement algorithms can be applied.

Disk Grouping aggregates sets of physical disks into groups to form logical disks. Each set must consist of the same mix of physical disk drives. Data is placed on and retrieved from the logical disks. Each data block is in effect declustered across the physical disks that form a logical disk. Partial blocks are called fragments and their size is adjusted according to the performance of the physical disk they reside on. If a block retrieval is initiated from a logical disk, all appropriate fragments are read in parallel, recombined in memory, and transmitted to the client display station.

Staggered Grouping operates similar to Disk Grouping, except that it reduces the amount of main memory that is required for the same throughput. The improvement is based on the observation that not all fragments of a logical block are needed at the same time at the client side. By staggering the retrieval of the fragments in time, memory frames can be reused.

Finally, Disk Merging separates the concept of logical disks from the physical disks altogether. Logical disks are formed from fractions of the bandwidth and storage capacity of the available physical devices. Consequently, Disk Merging is more flexible than the other two techniques because it can support an arbitrary mix of physical disks. This flexibility increases the effort necessary to find the best possible configuration for a specific application. Consequently, a configuration planner was introduced to automate and speed up the configuration process.

To evaluate our techniques we first compared the three non-partitioning schemes with one another using analytical methods. All utilize system resources well, with Staggered Grouping and Disk Merging providing a slightly lower cost per stream (by minimizing the amount of required memory). While Disk Merging provides the lowest minimum startup time, the variance in startup latency is reduced with both Disk Grouping and Staggered Grouping.

Next, we chose Disk Merging as the representative of the non-partitioning schemes to compare it with a simple partitioning technique. Similar to a real system, our simulation modeled shifts in the user access pattern as a function of time. Tests were conducted at system loads of 70%, 80%, and 90%, respectively. The 90% load revealed the fundamental problem of any partitioned system: some resources become a bottleneck while others remain idle, waiting for work. In the specific case of the partitioned CM server, one of the three sub-servers became overloaded and as a result the average startup latency for all users increased. Disk Merging was able to handle the situation more gracefully (both Disk and Staggered Grouping are expected to provide a performance similar to Disk Merging). It was sensitive to neither the frequency of access to clips nor their assignment. As a result the latency increase was much less dramatic.

One disadvantage of non-partitioning schemes as compared with partitioned systems is their increased vulnerability to disk failures. Hence, we investigated fault-tolerance and high availability techniques and their application to non-partitioned storage systems. The result was a framework of broad observations that can build the foundation for a more in-depth investigation. We provided one example, where we analyzed how the class of non-overlapping parity techniques can be applied to a non-partitioning storage approach (i.e., Disk Merging). We specified both design rules and algorithms that provide the necessary independence of logical disks among each other and across parity groups for the successful application of parity-based data redundancy.

Chapter 6

FAULT TOLERANCE ISSUES IN SM SERVERS

To provide sufficiently large storage and bandwidth capacities, SM servers typically contain a large number of disks. Current disk drives are fairly reliable with a mean time to failure (MTTF; for a definition of terms see Table 6.2) of up to 1,200,000 hours (see Table 6.1 for the reliability of some commercially available disk drives). However, if a large number of disks is aggregated into a SM storage system, the chance of one disk failing increases. For example, assuming that all the disks are independent and that their lifetimes are exponentially distributed [128], [78], then a disk array consisting of 1,000 disks experiences a mean time to service loss (MTTSL) of $1/1,000^{th}$ of a single disk, equivalent to 1,000 hours or approximately 42 days. The

Disk Series	Storage Capacity	MTBF (power-on hours)
Hawk 1LP	1 GB	500,000
Barracuda 4LP	2 GB	1,000,000
Cheetah 4LP	4 GB	1,000,000
Cheetah 18	18 GB	1,000,000
Cheetah 36	36 GB	1,000,000
Cheetah 73LP	73 GB	1,200,000

Table 6.1. Single disk reliability for six commercial disk drives. In these examples, the manufacturer, Seagate Technology LLC, reported the reliability as mean-time-between-failures (MTBF). A simple approximation for MTBF is $MTBF = MTTF + MTTR$ where MTTR denotes the mean-time-to-repair [161, pp 206]. Because failed disks are typically replaced and not repaired, the notion of MTTR refers to replacing the disk and rebuilding its content onto a new disk. This process can usually be completed within several hours. Hence $MTTR \ll MTTF$ and thus for most practical purposes the following approximation can be used: $MTTF \approx MTBF$.

Term	Definition
MTTF $= \frac{1}{\lambda}$	Mean time to failure; mean lifetime of an individual, physical disk with failure rate λ
MTTDL	Mean time to data loss
MTTSL	Mean time to service loss
MTTR $= \frac{1}{\mu}$	Mean time to repair of a physical disk with repair rate μ
d_i^p	Physical disk drive i
d_i^l	Logical disk drive i
D	Number of physical disk drives
D^l	Number of logical disk drives
G	Parity group size
$\mathcal{G}_i$	Parity group i; a data stripe allocated across a set of G disks and protected by a parity code
L	Disk utilization (also referred to as *load*) of an individual, logical disk
$R(t)$	Reliability function

Table 6.2. List of terms used repeatedly in this chapter and their respective definitions.

problem is further complicated by the common practice of striping media data to distribute the load of a display evenly across the disks. As a result, a single disk failure will affect all of the stored clips. Hence, display disruptions might be experienced by many of the display stations.

Frequently, SM servers store all their data on tertiary devices (e.g., tape libraries) as well as on disk arrays. In such a case, a disk failure does not lead to loss of data. However, because of the real-time requirements of SM streams it will likely result in service disruptions. Furthermore, rebuilding the data on a new disk from the tertiary may require considerable time and occupy substantial amounts of resources (bandwidth) because of the typically large size of SM objects.

Consequently, it is important for many applications that a SM server employ fault tolerance techniques to guarantee high availability. This chapter investigates techniques to increase the reliability and availability of large-scale SM servers that are based on heterogeneous storage subsystems. First, in Section 6.1, we briefly describe the target architecture. Second, in Section 6.2, we present an overview of high availability techniques for homogenous storage systems. Third, in Section 6.3, we extend the high availability techniques for heterogenous storage systems. Forth, in Section 6.4, we present a basic reliability model to evaluate these techniques. Fifth, in Section 6.5, we present our simulation results. Finally, in Section 6.6, we present our concluding remarks.

6.1   Target Architecture

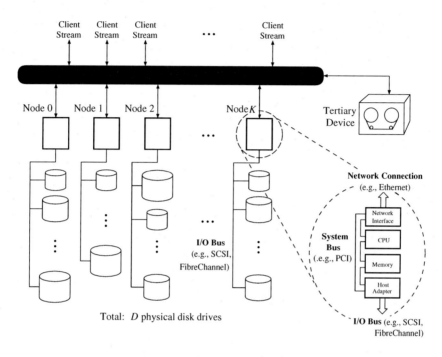

Figure 6.1. Multi-node SM server architecture.

To support a large number of clients we assume a *multi-node* SM server architecture as illustrated in Figure 6.1[1]. Each node has a set of local disks attached to it. The nodes are linked with each other through a high-speed interconnection network to which the clients are connected as well. Furthermore, a tertiary device, which stores the complete database of all multimedia objects in the system, is attached to the network.

In this chapter we investigate two types of failures in the context of the presented architecture: (1) disk failures and (2) node failures[2]. Without techniques to improve reliability and availability, any such failure might result in display disruptions for some or all of the client display stations. Because the complete database is stored on a tertiary device, repaired disks or nodes can be rebuilt and data will not be permanently lost. In summary, a server can be said to operate in one of three modes: (a) *normal* mode, i.e., all nodes are fully operational, (b) *degraded* mode, i.e., some disk or node has failed, and (c) *rebuild* mode, i.e., the data on a repaired disk or node is

---

[1]Figure 6.1 extends the architecture presented in [76] with heterogeneous storage.

[2]The reliability aspects of the interconnection network is beyond the scope of this book.

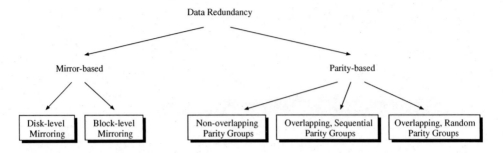

Figure 6.2. Classification of improved reliability techniques for SM servers.

being restored. All of these three modes must be considered when designing reliability techniques to mask disk and node failures, such that the service to the client stations can be continued.

6.2   Overview of High Availability Techniques

To increase the reliability and availability of multi-disk storage systems it is common to add data redundancy in order to survive the data loss of a failed disk. Two basic paradigms for data redundancy exist: (1) mirroring the original data [17] and (2) constructing parity information [128]. Both of these basic techniques can be refined into several alternative variations, each with its own specific characteristics. Figure 6.2 shows five broad classes of reliability techniques. Most algorithms that have been reported in the literature can be subsumed under one of these five classes. We now proceed to give an overview of each class in turn.

6.2.1   Mirror-based Techniques

With mirroring, each data block is replicated within the storage system[3]. Mirroring can be applied at either the *disk-level* or at the *block-level*. When applied at the disk-level, the complete contents of each disk are duplicated onto a mirror disk. In case of a disk failure, all data accesses that reference the failed disk are redirected to its mirror. Mirroring results in a 100% storage space overhead, but there is little processing overhead when a system runs in degraded mode. There is no need to reconstruct data as is the case with parity-based techniques.

With disk-level mirroring, half of the storage system bandwidth is idle

---

[3]We will assume that data is replicated once because this is the most common practice. However, data could be replicated several times.

during normal operation. Block-level mirroring addresses this inefficiency by distributing the mirrored blocks among all remaining disks in the storage system. Hence, if a disk fails, the resulting accesses to the block copies are spread among all remaining disks. Random data placement can be used to balance the load (e.g., *doubly striping* [115]) and only a small fraction of the bandwidth must be reserved on each disk for failure mode. However, because in degraded mode all disks are accessed to reconstruct the unavailable data, only a single disk failure can be tolerated. Alternatively, if the mirrored blocks are distributed only across a subset of all disks, then a higher number of disk failures may be tolerated (the *Tiger* project implements this concept [19]). Furthermore, because of the random data placement with block-level mirroring, additional storage overhead is incurred through metadata that is needed to remember the location of each block mirror pair.

With random placement some mirror data may be placed on different disks within the same node and hence node failures cannot be tolerated. By slightly altering the allocation strategy to place mirror data only on mutually exclusive nodes this problem can be eliminated.

6.2.2   Parity-based Techniques

In parity-based systems the disks are partitioned into *parity groups*. The parity information is computed across the blocks in a parity group, most commonly with an XOR function. The large storage space overhead of mirroring is reduced since for a parity group size of G only one $1/G^{th}$ of the space is dedicated to data redundancy. The increased storage efficiency is traded for a lower fault tolerance. Failures are tolerated at a rate of one per parity group[4]. Furthermore, processing power to perform XOR calculations is needed when a parity group operates in degraded mode. The products based on *Streaming RAID*™ implement this technique [175].

In the basic case, the disks of a storage system are partitioned into multiple *non-overlapping* parity groups. This scheme can tolerate one disk failure per parity group. However, when a parity group operates in degraded mode, each access to the failed disk triggers the retrieval of blocks from all of the disks within this parity group to reconstruct the lost data. Thus, the load on all operational disks increases by 100% under failure, making this parity group a hot spot for the entire system. To distribute the additional load more evenly, parity groups may be rotated such that they overlap [171]. In this scenario, termed *overlapping, sequential* (see Figure 6.2), a failure will

---

[4]Reed-Solomon codes as used in *P+Q redundancy* (RAID Level 6) can reduce this vulnerability [30].

generate additional load on the $(G-1)$ disks before and $(G-1)$ disks after the failed disk drive. The improved load-balance is traded for a reduced fault tolerance of approximately one failure per two parity groups.

A third approach alleviates the possibility of bottlenecks even further by assigning blocks randomly to parity groups. By abandoning the requirement for data blocks on sequential disk drives to be placed in the same parity group, the additional load generated by a failed disk can be almost uniformly distributed across all operational disks. Hence, in this *overlapping, random* (see Figure 6.2) approach, very little bandwidth must be reserved on each disk during normal operation. Mappings of parity group members to disks that exhibit the desired property are, for example, a complete combinatorial approach which results in $\binom{D}{G}$ different parity groups, or a design devised using the theory of *balanced incomplete block designs* [84], [125]. Such a system is somewhat less reliable because only one disk failure can be tolerated in the whole storage system. Furthermore, additional meta data is needed for each block to associate it with a parity group.

With the pseudo-random placement algorithms described above, some parity data may be placed within the same node and hence node failures cannot be tolerated. By extending the allocation strategy such that parity data is placed only on mutually exclusive nodes this problem can be eliminated.

6.2.3   Issues Specific to Streaming Media Servers

The requirements for SM servers in regards to high availability techniques are somewhat different than the requirements for more traditional, transaction-oriented database applications.

Write Performance

For example, the write performance of some parity-based techniques is lacking, which would be a concern for many database applications. For SM servers, however, reading of data is much more prevalent than writing. Only when objects are replaced or a repaired disk is rebuilt are data writes necessary. Furthermore, because data blocks are static, parity information needs to be generated only once.

Bandwidth Requirements

SM objects require a continuous stream of data at a pre-specified rate to be displayed without disruptions. If the server bandwidth becomes overcommitted due to either disk or node failures, service is not just slowed down, as would be the case for traditional, transaction oriented applications, but the client stations will experience hiccups and discontinuities in their display, which might not be tolerable.

Parity Reconstruction

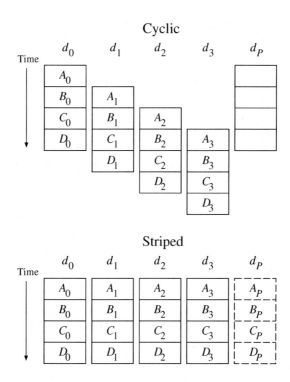

Figure 6.3. Cyclic and striped retrieval. With cyclic retrieval scheduling block reads are staggered in time for each stripe, e.g., $A_0 \ldots A_3$, and less buffer memory is required. But parity computations (which require concurrent access to a complete stripe) are not supported without buffering.

Many SM servers use a *cyclic retrieval* approach to service SM streams. With such a scheme one block of data is read from a disk on behalf of each stream during every time period. The block is temporarily stored in memory and transmitted to the client display station during the next time period. An alternative scheduling technique is *striped retrieval* [175], [60] with which D data blocks are read in parallel from all D disk drives in the system on behalf of each client stream during every time period. Figure 6.3 shows the difference of cyclic versus striped retrieval as a function of time. The main advantage of the cycle-based approach over the striped technique is a substantial reduction of the memory requirement in the server.

However, in a system that uses parity information to protect against data loss, all data blocks on the remaining disks in a parity group must be retrieved to reconstruct any missing data. For example, if disk d_2 in Figure 6.3 were to fail, then the block A_2 can be reconstructed by reading all the other blocks of stripe A as well as the parity block A_p and performing the following exclusive or operation: $A_2 = A_0 \oplus A_1 \oplus A_3 \oplus A_p$. With cyclic retrieval scheduling, consecutive data blocks of a stripe are read in subsequent rounds. The data belonging to one parity group is therefore not usually available at the same time. A partial solution is to keep the data blocks in memory that are retrieved before the failed disk is accessed (A_0 and A_1 in our example). However, the blocks of the parity group normally retrieved in the rounds after the failed disk is referenced (e.g., A_3) must still be read ahead of schedule when a retrieval reference to block A_2 on the failed disk necessitates the data reconstruction. Therefore, additional memory and bandwidth is necessary in the server when a parity group is running in degraded mode.

A number of alternative approaches with different tradeoffs have been proposed in the literature to address this problem. Streaming RAID™ [175] uses striped retrieval during both normal and degraded mode, therefore avoiding any scheduling changes during fail-over, but at a very high memory cost. The Staggered-group scheme [13] reads a parity group in a single cycle but plays it out over n cycles, roughly halving the memory requirement of Streaming RAID. The Non-clustered and Improved-bandwidth schemes proposed by the same authors read an entire parity group only in degraded mode, essentially switching from cyclic to striped retrieval during a fail-over. However, because of the scheduling conflicts that arise during such a switch some data blocks are lost and display disruptions are expected at the client station.

Software RAID [171] further refines the above techniques by using cyclic and striped retrieval concurrently in a parity group that runs in degraded mode. Block accesses that reference operational disks are serviced with cyclic scheduling while block accesses to failed disks result in parallel data retrievals on all intact disks. As an optimization, blocks that are known to be required for a parity computation in the future will be kept in memory until needed once they are retrieved. Enough memory and bandwidth must be reserved to allow this mode of operation.

6.3 High Availability for Heterogeneous Storage Systems

Being able to utilize a heterogeneous storage system can be very beneficial for a large scale SM server implementation. The service operator is free to choose the most cost-effective disk drives on the market. However, the implementation of high availability techniques is somewhat more complex for heterogeneous systems because additional constraints are placed on the location of mirror and parity data.

In the following sections we will investigate these aspects and illustrate solutions.

6.3.1 Mirror-based Techniques

The three non-partitioning techniques introduced earlier in Chapter 5, Disk Grouping, Staggered Grouping, and Disk Merging all mask the heterogeneity of the physical storage system and present the abstraction of a set of homogeneous, logical disk drives to the target application. This abstraction can be used to implement traditional mirroring techniques on top of the logical storage layout.

Disk Grouping and Staggered Grouping

From the viewpoint of fault tolerance, Disk Grouping and Staggered Grouping behave similarly, and hence, are described together. When implementing disk-level mirroring, each *logical* disk is replicated on another logical disk. Duplicating a logical disk is straightforward because by definition all logical disks consist of an equivalent set of physical disks. Hence, copying the underlying physical disks will effectively replicate the logical disk. If it is desired that the system should survive node failures then both logical disks of a mirror pair should be attached to mutually exclusive nodes, i.e., none of the physical disks constituting one logical disk should be connected to the same node as any physical disk of its mirror pair.

If block-level mirroring is desired, a technique such as doubly striping [115] can be employed. Figure 6.4 shows an example of the block placement layout of doubly striping if the system must survive node failures. The algorithm distributes mirror blocks of a logical disk evenly among all the remaining logical disks in the system but excluding disks that are attached to the same node. Therefore, in case of a disk or node failure the resulting load will be uniformly spread across the operational disks or nodes.

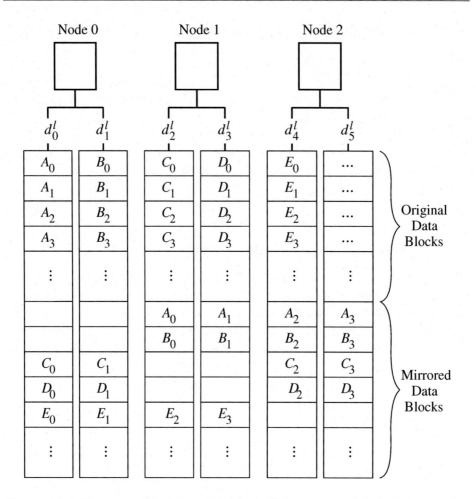

Figure 6.4. Example of doubly striping for a SM server consisting of three nodes, each connected to two logical disk drives. Every logical disk itself consists of several physical drives. The block labels are only used to illustrate the placement of the data and do not imply association with a specific SM object. Furthermore, unlabelled blocks do *not* represent unused blocks, i.e, they have only been inserted to clarify the pattern.

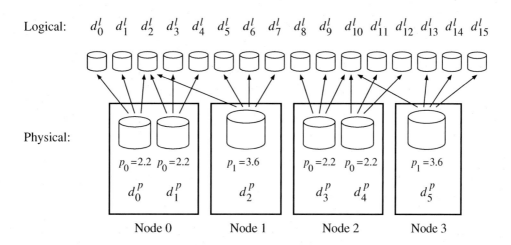

Figure 6.5. Example of Disk Merging with a four node, six disk server. The six physical disks are mapped to sixteen logical disks.

Disk Merging

The basic approach of replicating logical disks can also be used for Disk Merging. However, additional constraints must be considered because of the potential dependencies among the logical disks. With Disk Grouping or Staggered Grouping the failure of a physical disk can only affect one logical disk. With Disk Merging a failure at the physical level may cause multiple logical disks to become unavailable simultaneously.

Consider disk-level mirroring first. When choosing a mirror for a logical disk d_i^l then a candidate disk d_j^l *cannot* be a suitable target if either or both of the following two statements apply:

(a) The logical disk d_j^l or any fraction thereof is attached to the same node as d_i^l.

(b) The logical disk d_j^l maps completely or partially to the same physical disk as d_i^l.

Design rule (a) simply ensures that node failures are survivable. Hence, mirror data should not be stored on the same node. This rule may be ignored if only disk failures need to be handled. Rule (b) captures the fact that, if a logical disk and its mirror copy were to map to the same physical disk, then a failure of that disk would render both copies unavailable.

Example 6.3.1: Figure 6.5 shows a SM server configuration consisting of four nodes. Nodes 0 and 2 are connected to two physical disk drives each

(d_0^p and d_1^p, respectively, d_3^p and d_4^p), while one disk is attached to each of the nodes 1 and 3 (d_2^p, respectively, d_5^p). In this example, a suitable mirror target for the logical disk d_0^l would be any one of the disks $d_5^l, \ldots, d_{15}^l$ (rule (a)). However, for d_2^l only disks $d_8^l, \ldots, d_{15}^l$ would be mirror candidates because both rules (a) and (b) apply. □

A server employing disk-level mirroring and operating under failure conditions might experience hot spots because the load is typically not evenly spread across the operational nodes. Block-level mirroring can improve this situation. However, the constraints imposed by rules (a) and (b) on the selection of mirror locations also apply to block-level mirroring and these restrictions may put some lower bound on how well the system load can be balanced.

With doubly-striping the contents of a logical disk is replicated in equal fractional parts on all eligible mirror candidates of the system. For example, $1/11^{th}$ of the data of logical disk d_0^l in Figure 6.5 can be copied onto each of the disks d_5^l through d_{15}^l (3 nodes). For disk d_2^l, however, $1/8^{th}$ of the contents must be copied onto each of the disks d_8^l through d_{15}^l (just 2 nodes), because d_5^l, d_6^l, and d_7^l share the same physical disk with d_2^l and must be excluded. If we assume that the normal load of a single logical disk is L, then we can quantify the load distribution after a failure. For example, if node 0 were to fail, the *load increase* on node 2 and 3 would be higher ($\Delta L = +48.9\%$) than on node 1 ($\Delta L = +13.6\%$) as illustrated in Figure 6.6. The load increase would be smaller if only a disk were to fail. Figure 6.7 shows the load distribution in a scenario were disk d_0^p is no longer available.

Note that these load imbalances are most pronounced for small systems with few nodes like the one on which the previous examples are based. For large-scale installations with many nodes, the additional load can likely be spread more evenly. However, the magnitude of these effects is dependent on the ratio of logical to physical disks. The interaction between this ratio and subsequent load imbalances is currently not well understood and deserves further investigation.

6.3.2   Parity-based Techniques

Parity-based techniques are popular because they require less storage overhead than mirroring schemes. For example with a parity group size of $G = 4$ the storage space equivalent to only one out of four disks (i.e., 25%) needs to be reserved for redundant information. This compares favorably with the 50% space overhead required by mirror-based techniques. However, parity-based schemes must perform additional XOR computations while operating

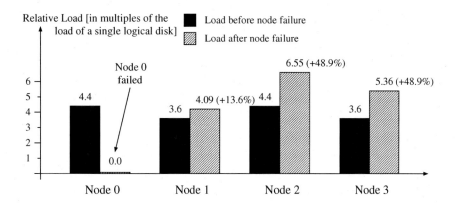

Figure 6.6. Load distribution example with mirroring after a *node* failure in a four node server. The load is represented as multiples of the load of a single logical disk, denoted L. If operation with a failed node must continue with no service degradation then each logical disk can only be loaded with approximately $2/3^{rd}$ of its maximum bandwidth in normal mode.

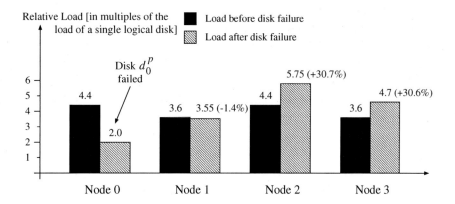

Figure 6.7. Load distribution example with mirroring after a *disk* failure in a four node server. The load is represented as multiples of the load of a single logical disk, denoted L. If operation with a failed node must continue with no service degradation then each logical disk can only be loaded with approximately $3/4^{th}$ of its maximum bandwidth in normal mode.

with a failed disk. This need for additional CPU resources must be considered when designing a system, otherwise the real-time performance required by a SM server might suffer. Tewari et al. [171] report XOR execution times of 10 to 15 msecs for a parity group size of 5, based on a data block size of 256 KB and a CPU clock rate of 100 MHz. In a high performance server special hardware accelerators might be employed to speed up such computations.

Disk Grouping and Staggered Grouping

Analogous to the techniques described with mirroring, parity-based schemes can readily be applied on top of the logical disk storage layer that is provided by both Disk Grouping and Staggered Grouping. No special considerations are required.

Disk Merging

For Disk Merging we must again, as with mirroring, consider that several logical disks may map to the same physical disk. Alas, in a parity-based scheme all the disks that form a parity group must be independent. This is necessary because with a traditional XOR-based parity computation exactly *one* data block of a stripe can be reconstructed as long as *all the other* blocks of that particular stripe are available.

Because all logical disks that map to a single physical disk will fail simultaneously they must be assigned to independent parity groups. Hence, the number of independent parity groups D^l/G needs to be larger than any number of logical disks that map to a single physical disk.

This can be formally expressed with the following parity group size constraint:

$$G \leq \left\lfloor \frac{D^l}{p_i} \right\rfloor \qquad \text{for } 0 \leq i < m \qquad (6.1)$$

Figure 6.8 shows an example storage system with six physical disk drives that map to sixteen logical disks. The parity group size G is required to be either less than or equal to both $\left\lfloor \frac{D^l}{p_0} \right\rfloor = \left\lfloor \frac{16}{2.2} \right\rfloor = 7$ and $\left\lfloor \frac{D^l}{p_1} \right\rfloor = \left\lfloor \frac{16}{3.6} \right\rfloor = 4$. Hence, the maximum parity group size equals 4 which can be accommodated by creating $16/4 = 4$ or more parity groups $\mathcal{G}_i$. For illustration purposes, we will use a simple, non-overlapping parity group scheme. One possible assignment of the sixteen logical disks $d_0^l, \ldots, d_{15}^l$ to four parity groups $\mathcal{G}_1, \ldots, \mathcal{G}_4$ is as follows (also illustrated in Figure 6.8): $\mathcal{G}_1 = \{d_2^l, d_9^l, d_{12}^l, d_{15}^l\}$, $\mathcal{G}_2 = \{d_0^l, d_3^l, d_5^l, d_{10}^l\}$, $\mathcal{G}_3 = \{d_1^l, d_4^l, d_6^l, d_{13}^l\}$, and $\mathcal{G}_4 = \{d_7^l, d_8^l, d_{11}^l, d_{14}^l\}$.

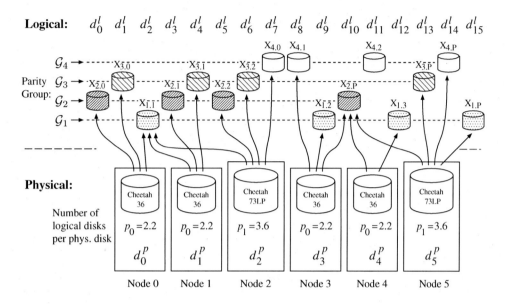

Figure 6.8.   Example of parity group assignment for Disk Merging (non-overlapping parity groups). Six physical disks are mapped to sixteen logical disks. All of a physical disk's logical disks must map to different parity groups ($G = 4$).

From the above parity group assignment we can further determine which physical disks participate in each of the parity groups ("$\mapsto$" denotes "maps to"):

$$\mathcal{G}_1 \mapsto d_0^p, d_1^p, d_2^p, d_3^p, d_4^p, d_5^p \tag{6.2}$$

$$\mathcal{G}_2 \mapsto d_0^p, d_1^p, d_2^p, d_3^p, d_4^p, d_5^p \tag{6.3}$$

$$\mathcal{G}_3 \mapsto d_0^p, d_1^p, d_2^p, d_5^p \tag{6.4}$$

$$\mathcal{G}_4 \mapsto d_2^p, d_3^p, d_4^p, d_5^p \tag{6.5}$$

Note that the parity groups $\mathcal{G}_1$ and $\mathcal{G}_2$ map to all six physical disks in the system. Recall that a parity group can only tolerate a single physical disk failure and still continue operation. Hence, if it is required for the configuration of Figure 6.8 that node failures must be survivable then each physical disk needs to be attached to an independent node.

Figure 6.9 illustrates the additional load on each of the six nodes for the case of a node 0 (or equivalently disk d_0^p) failure. Loads are measured in multiples of the load of a single logical disk which is denoted with L. For non-overlapping parity groups the load increases 100% on each operational disk within a parity group running in degraded mode (i.e., $2 \times L$). The failure of disk d_0^p will effectively render d_0^l, d_1^l, and d_2^l unavailable. Hence, the parity

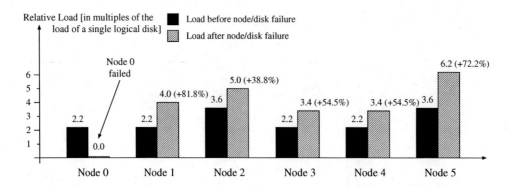

Figure 6.9. Load distribution example with parity-based fault tolerance after a *node* failure in a six node server ($D^l = 16, G = 4$). The load is represented as multiples of the load of a single logical disk, denoted L. If operation with a failed node must continue with no service degradation then each logical disk can only be loaded with approximately 50% of its maximum bandwidth in normal mode.

groups $\mathcal{G}_1$, $\mathcal{G}_2$, and $\mathcal{G}_3$ will run in degraded mode, doubling the load on each of its remaining logical disks. Therefore, the normalized load in the system will be $((3+3)+(3+3)+(3+3)+4) \times L = 22 \times L$, a 37.5% increase over the nominal load of $16 \times L$.

In the previous example we have assumed one specific assignment of logical to physical disks. In general (and this is especially the case for larger storage configurations) a number of different alternative assignments are possible. Unquestionably, some will have better load distributions than others. Finding the best mapping function that minimizes any load imbalance is an open research problem, and its investigation is part of our future research directions.

6.4   Basic Reliability Modelling

With the help of data replication or parity coding, the reliability of a disk array can be greatly improved. Several studies have quantified the reliability of homogeneous disk arrays with mirroring [115], parity coding schemes (RAID) [128], [117], [78], [30], [13], or a combination of both [14].

The primary metric to compare the reliability of multi-disk storage systems is typically the *mean time to failure* (MTTF) of a system. The most critical failures typically involve loss of data. Hence in such situations, the MTTF is rephrased as the *mean time to data loss* (MTTDL). This would be a metric for traditional transaction oriented applications, for example. For a SM server, however, we assume an architecture with an integrated tertiary device that holds a copy of the complete database (see Figure 6.1). Consequently, data cannot be permanently lost even when the secondary storage (i.e., disk storage) fails. However, such a failure may lead to service disruptions for the client community. Therefore, we will use the term *mean time to service loss* (MTTSL) to denote a failure that results in some data being unavailable from the secondary storage.

6.4.1 Non-overlapping Parity

The concepts of reliability or fault tolerance involve a large number of issues concerning software, hardware (mechanics and electronics), and environmental (e.g., power) failures. A large body of work already exists in the field of reliable computer design and it will provide the foundation for this section. Because it is beyond the scope of this chapter to cover all the relevant issues in depth, we will restrict our presentation to the class of non-overlapping, parity-based schemes to illustrate the techniques involved.

Analytical Model for Reliability

The *reliability function* of a system, denoted $R(t)$, is defined as the probability that the system will perform satisfactorily from time zero to time t, given that it is initially operational [161, pp 7]. When the higher failure rates of a component at the beginning (infant mortality or burn-in period) and at the end (wear-out period) of its lifetime are excluded, then there is strong empirical evidence that the *failure rate* λ during its normal lifetime is approximately constant [161], [78]. This is equivalent to an exponential distribution of the product's lifetime and gives rise to a reliability function $R(t)$ that can be expressed as:

$$R(t) = e^{-\lambda t} \tag{6.6}$$

Perhaps the most commonly encountered measure of a system's reliability is its mean time to failure (MTTF) which is defined as follows:

$$MTTF = \int_0^\infty R(t)\,dt \tag{6.7}$$

With the exponential lifetime distribution $R(t)$ substituted from Equation 6.6 the MTTF simply becomes:

$$MTTF = \int_0^\infty e^{-\lambda t}\,dt = \frac{1}{\lambda} \tag{6.8}$$

In a heterogeneous storage environment it is possible for each physical disk drive d_i^p to have its own $MTTF_i$ and hence its own failure rate λ_i (see Table 6.1).

The mean lifetime of a system can be greatly prolonged if it contains redundant components that can be repaired when a failure occurs. This is the case for parity based storage systems, where a parity group can run in degraded mode until a failed disk is replaced and its data is recovered. Note that the repair time for a non-redundant system, such as a single disk drive, is by definition unimportant for its analytically computed reliability[5] because if any of its components fail, then the whole system has failed (see the discussion of Table 6.1). A redundant system can continue to operate while being repaired (albeit with less performance) and its mean time to repair ($MTTR$) is crucial because it provides a "window of opportunity" during which a second component failure may result in a system failure. Hence a short $MTTR$ is very desirable. The mean time to repair (MTTR)[6] is often difficult to model analytically, and it must usually be estimated or measured [161, pp 205]. If the operating environment of the system is known, then an estimate can be given. For example, if spare disks are at hand and service personnel is part of the staff at the site location, a MTTR of a few hours should be realistic. If spare parts must be ordered or service personnel called in then the MTTR may be in the order of days or even weeks. For the analytical models the repair rate is commonly denoted μ and for exponential distributions $MTTR = \frac{1}{\mu}$. In a heterogeneous environment the repair rate may not be the same for all disk types, hence in the most general case each disk has its own repair rate μ_i.

Markov Model for a Single Parity Group

Markov models provide a powerful tool for basic reliability modelling if the system is composed of several processes (such as a failure process and a repair process) [161], [78]. Each component of such a model is assumed to be independent. However, logical disks produced by the Disk Merging technique are not a priori independent, because several of them may map to the same physical disk drive. Conversely, fractions of a single logical disk may be spread across several physical disks. In Section 6.4.1 we will derive the mean time to failure for each individual logical disk in a storage system, such that the necessary independence for the Markov model is preserved.

---

[5]However important it may be to the owner of the device.

[6]It is customary to refer to MTTR as the mean-time-to-repair even though for most practical purposes a failed magnetic disk will be replaced and not repaired. In such a case the MTTR should include the time to rebuild the lost data on the new disk.

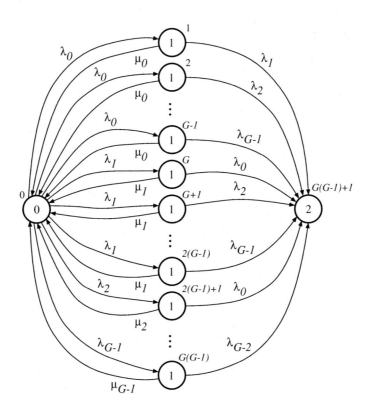

Figure 6.10. Markov model for a heterogeneous disk array (one parity group). The labels in each state denote the number of disk failures encountered by the array. State 2 is a trapping state, i.e., the probability of exiting is zero, meaning the array has failed.

Figure 6.10 shows the Markov model for a single parity group comprised of G independent disk drives. The states are labelled with the number of disk failures that the array is experiencing. For example, in state 0 all disks are operational. Then with probability λ_i disk d_i becomes unavailable and a transition is made to one of the states labelled 1. With probability μ_i repairs are completed and the array is again in state 0. Or, with probability λ_j a second disk fails and the transition to state 2 indicates that an unrecoverable failure has occurred. As illustrated, the number of states of the model is $(G \times (G-1)) + 2$, i.e., it is quadratic with respect to the parity group size G. The evaluation of such a model is computationally quite complex, especially for larger values of G. Hence we propose the following two simplifying assumptions:

1. The repair rate μ_i is the same for all the disks in the system, i.e., $\mu =$

$\mu_0 = \mu_1 = \ldots = \mu_{G-1}$. It is likely that the time to notify the service personnel is independent of the disk type and will dominate the actual repair time. Furthermore, disks with a higher storage capacity are likely to exhibit a higher disk bandwidth, leading to an approximately constant rebuild time.

2. The probability of a transition from any of the states 1 to state 2 (a second failure) is the sum of the failure rates of all the remaining disks of the group (i.e., $\sum_{j=0}^{G-1} \lambda_j$ minus the one failure rate that led to the transition from state 0 to 1). In the most general case (because of heterogeneity) each disk can have a different failure rate. Hence, the correct solution requires $G - 1$ cases in each of which we subtract the failure rate of the first disk that failed from the sum. To establish a single case and a closed form solution we propose to always subtract the smallest failure rate λ_{min} *of any disk in the group*. Hence, by ordering (without loss of generality) λ_j according to decreasing values, $\lambda_0 \geq \lambda_1 \geq \ldots \geq \lambda_{G-1}$, we can express the probability of a transition from state 1 to 2 to be $\sum_{i=0}^{G-2} \lambda_j$. This approximation will lead to the highest value for the sum $\sum_{i=0}^{G-2} \lambda_j$ and therefore the shortest parity group lifetime. In other words, our estimate for the lifetime is conservative and in most cases the real lifetime will be slightly longer.

With these assumptions the Markov model is simplified to three states as shown in Figure 6.11 and can be solved using a set of linear equations and Laplace transforms [161], [78].

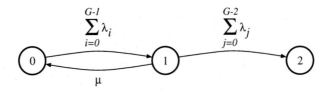

Figure 6.11. Simplified Markov model for a heterogeneous disk array.

If the expression of $R(t)$ does not need to be obtained, $MTTDL$ can be found more easily [78]. Beginning in a given state i, the expected time until the first transition into a different state j can be expressed as

$$\text{E[state } i \text{ to state } j] = \text{E[time in state } i \text{ per visit]}$$
$$+ \sum_{k \neq i} \text{P(transition state } i \text{ to state } k) \times \text{E[state } k \text{ to state } j] \quad (6.9)$$

where

$$E[\text{time in state } i \text{ per visit}] = \frac{1}{\sum \text{rates out of state } i} \qquad (6.10)$$

and

$$P(\text{transition state } i \text{ to state } k) = \frac{\text{rate of transition to state } k}{\sum \text{rates out of state } i} . \qquad (6.11)$$

The solution to this system of linear equations includes an expression for the expected time beginning in state 0 and ending on the transition into state 2, that is, for $MTTDL$. For the Markov model in Figure 6.11, this system of equations is

$$E[\text{state 0 to state 2}] = \frac{1}{\sum_{i=0}^{G-1} \lambda_i} + \frac{\sum_{i=0}^{G-1} \lambda_i}{\sum_{i=0}^{G-1} \lambda_i} E[\text{state 1 to state 2}] \qquad (6.12)$$

$$E[\text{state 1 to state 2}] = \frac{1}{\mu + \sum_{j=0}^{G-2} \lambda_j} + \frac{\mu}{\mu + \sum_{j=0}^{G-2} \lambda_j} E[\text{state 0 to state 2}]$$

$$+ \frac{\sum_{j=0}^{G-2} \lambda_j}{\mu + \sum_{j=0}^{G-2} \lambda_j} E[\text{state 2 to state 2}] \qquad (6.13)$$

$$E[\text{state 2 to state 2}] = 0 . \qquad (6.14)$$

The resulting mean lifetime of a single parity group of independent heterogeneous disks is shown in Equation 6.15.

$$MTTDL = \frac{\mu + \sum_{i=0}^{G-1} \lambda_i + \sum_{j=0}^{G-2} \lambda_j}{\sum_{i=0}^{G-1} \lambda_i \times \sum_{j=0}^{G-2} \lambda_j} \qquad (6.15)$$

Under most circumstances the repair rate $\mu = 1/MTTR_{disk}$ will be much larger than any of the failure rates $\lambda_i = 1/MTTF_{disk_i}$. Thus, the numerator of Equation 6.15 can be simplified as presented in Equation 6.16 (see [78, pp 141]).

$$MTTDL = \frac{\mu}{\sum_{i=0}^{G-1} \lambda_i \times \sum_{j=0}^{G-2} \lambda_j}$$

$$= \frac{1}{\sum_{i=0}^{G-1} \frac{1}{MTTF_{disk_i}} \times \sum_{j=0}^{G-2} \frac{1}{MTTF_{disk_j}} \times MTTR_{disk}} \qquad (6.16)$$

If all the failure rates λ_i are equal, i.e., $\lambda = \lambda_0 = \lambda_1 = \ldots = \lambda_{G-1}$ then Equation 6.16 will, as expected, correspond to the derivation for a homogeneous disk RAID level 5 array with group size G [128]

$$MTTDL_{homogeneous} = \frac{MTTF_{disk}^2}{G \times (G-1) \times MTTR_{disk}} \qquad (6.17)$$

where $MTTF_{disk}$ is the mean-time-to-failure of an individual disk, $MTTR_{disk}$ is the mean-time-to-repair of a single disk, and G is the parity group size. Both Equations 6.16 and 6.17 capture the probability of an independent double disk failure within a single parity group.

Example 6.4.1: Consider a parity coded disk array consisting of 3 Cheetah 36 and 2 Cheetah 73LP disk drives ($G = 5$). The mean-time-to-failure of the Cheetah 36 series is $MTTF_{disk} = 1,000,000$ hours and for the Cheetah 73LP series it is $MTTF_{disk} = 1,200,000$ hours (see Table 6.1). If we assume an average repair time of $MTTR_{disk} = 6$ hours, then the $MTTSL$ approaches a stunning 1,064,000 years. $\square$

By comparison, a homogeneous array consisting of five Cheetah 36 disks would have a similar mean lifetime of $MTTSL = 951,000$ years.

Logical Disk Independence

The Markov model of the previous section assumes that all the disks are independent. This assumption is not guaranteed for Disk Merging, because one logical disk may map to several physical disks and vice versa. Hence, to be able to apply Equation 6.16 at the logical level of a Disk Merging storage system, we will need to derive the mean time to failure of each individual, logical disk. Two cases are possible: (1) a logical disk maps to exactly one physical disk and (2) a logical disk maps to multiple physical disks. Consider each case in turn.

If a logical disk maps completely to one physical disk then its life expectancy is equivalent to the lifetime of that physical disk. Consequently, it inherits the mean time to failure of that disk. For example, the mean lifetimes of the logical disks d_9^l and d_{12}^l of Figure 6.8 are 1,000,000 hours each.

If, on the other hand, a logical disk depends on multiple physical disks, then it will fail whenever any one of the physical disks fails. Hence, we can apply the harmonic sum for failure rates of independent components [30]

$$MTTF_{d^l} = \frac{1}{\frac{1}{MTTF_{d_{0'}^p}} + \frac{1}{MTTF_{d_{1'}^p}} + \cdots + \frac{1}{MTTF_{d_k^p}}} . \qquad (6.18)$$

As an example, consider applying the above two observations to the four logical disks in parity group $\mathcal{G}_1$ of Figure 6.8. This will result in the following mean lifetime for each logical disk:

$$MTTF_{d_2^l} = \quad 352,941 \text{ hours} \tag{6.19}$$

$$= \frac{1}{\frac{1}{1,000,000} + \frac{1}{1,000,000} + \frac{1}{1,200,000}} \text{ hours,}$$

$$MTTF_{d_9^l} = 1,000,000 \text{ hours,}$$

$$MTTF_{d_{12}^l} = 1,000,000 \text{ hours,}$$

$$MTTF_{d_{15}^l} = 1,200,000 \text{ hours.}$$

Recall that, if multiple logical disks map to the same physical disk, then a failure of that physical drive will concurrently render all its logical drives unavailable. For the aforementioned reason all logical disks that map to the same physical disk must be assigned to different parity groups (see Figure 6.8).

Multiple Parity Groups (Disk Merging)

Large storage systems are typically composed of several parity groups. All of the groups need to be operational for the system to function properly. If the groups are assumed to be independent, then the system can be thought of as a series of components. The overall reliability of such a configuration is

$$R_{Series}(t) = \prod_{i=1}^{\frac{D}{G}} R_i(t) \tag{6.20}$$

where D/G is the number of parity groups and $R_i(t)$ is the individual reliability function of each group.

The overall mean lifetime $MTTDL$ of a series of groups can then be derived from the harmonic sum of the individual, independent failure rates of all the components as shown in Equation 6.21 [30].

$$MTTSL = MTTDL_{System} = \frac{1}{\frac{1}{MTTDL_1} + \frac{1}{MTTDL_2} + \ldots + \frac{1}{MTTDL_{\frac{D}{G}}}} \tag{6.21}$$

Consider the simple example shown in Figure 6.8. Four parity groups are formed from 16 logical disks, which in turn have been constructed from 6 physical devices. The physical $MTTF_{disk}$ are assumed to be 1,000,000 hours (d_0^p, d_1^p, d_3^p, and d_4^p), respectively, 1,200,000 hours (d_2^p and d_5^p), corresponding to Cheetah 36 and Cheetah 73LP devices (see Table 6.1). Each logical disk inherits its $MTTF_{disk}$ from the physical drive it is mapped to. For example, $MTTF_{d_0^l} = 1,000,000$ hours.

The resulting mean lifetimes for both group $\mathcal{G}_1$ and $\mathcal{G}_2$ are $MTTSL_{\mathcal{G}_1} = MTTSL_{\mathcal{G}_2} = 694,700$ years, while $MTTSL_{\mathcal{G}_3}$ and $MTTSL_{\mathcal{G}_4}$ are 1,831,439 years each (Equation 6.15). The mean time to failure for the whole storage system is therefore

$$MTTSL_{System} = \frac{1}{\frac{2}{694,700} + \frac{2}{1,831,439}} = 251,827 \text{ years} \qquad (6.22)$$

Such an extraordinarily high reliability will minimize the chance of failure during the physical lifetime, say ten years, of the storage system.

6.5   Simulation Results

	Logical Disk Fractions	MTTF	Description
(a)	N/A	All 1,200,000 h	RAID level 5 array for comparison.
(b)	N/A	All 1,000,000 h	RAID level 5 array for comparison.
(c)	No	All 1,200,000 h	Each physical disk maps to two complete logical disks (1.0+1.0).
(d)	No	1,000,000 h and 1,200,000 h	Same as (c) but half of the physical disks have a MTTF of only 1,000,000 hours.
(e)	Yes	All 1,200,000 h	Four of the physical disks map to three logical disks each (1.0+0.5+0.5).
(f)	Yes	1,000,000 h and 1,200,000 h	Same as (e) but half of the physical disks have a MTTF of only 1,000,000 hours.

Table 6.3. Six storage configurations based on 8 physical and 16 logical disks each (no logical disks in case of (a) and (b)) and a parity group size of $G = 4$.

To verify our analytical models we compared their output with the results obtained from a simple reliability simulator that was specifically designed for this purpose. Initially the simulator was provided with a physical and logical storage configuration as well as a mapping between the two. The simulator then generated random, exponentially distributed failures based on the

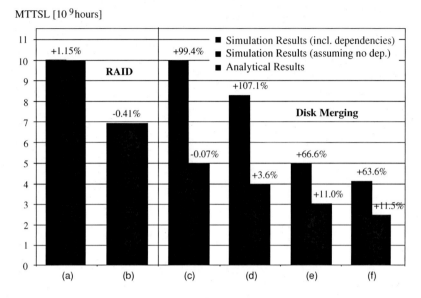

Figure 6.12. Mean time to data loss for six storage configurations based on 8 physical and 16 logical disks ((c) through (f)) and a parity group size of $G = 4$. See Table 6.3 for configuration parameters.

$MTTF_{disk}$ of each of the physical disks. A failed disk was assumed to be repaired within a fixed time of $MTTR_{disk} = 6$ hours. The simulator monitored the parity groups consisting of logical disks for two or more concurrent failures. The time between two double-failures (termed *events*) was recorded as the lifetime of the storage system. A total of 10,000 system failure and recovery cycles were simulated to obtain an average value for the mean time to service loss, $MTTSL$. Note that the simulator used the real failure rate of the disks involved at all times and hence did not make the simplifying assumption 2 of the analytical Markov model (see Section 6.4.1).

Table 6.3 lists the six different setups for the reliability tests. The physical storage system consisted of a set of eight disk drives. Six different configurations (a) though (f) were realized and tested, all based on the same eight physical disks. All physical disks were assumed to be equal in storage space and bandwidth and accommodate two logical disk drives. Such a simple storage system was chosen to be able to configure it with four different Disk Merging setups and furthermore to allow a direct comparison with a homogeneous RAID 5 setup. The six storage configurations are listed in the order of increasing complexity in Table 6.3.

Configurations (a) and (b) represented two standard, homogeneous RAID level 5 arrays with a parity group size of 4. They were introduced as a reference and baseline for comparisons with the Disk Merging variants. The difference between (a) and (b) was the assumed reliability of the disks, with $MTTF_{disk} = 1,200,000$ hours and $MTTF_{disk} = 1,000,000$ hours respectively. For configuration (c) each physical disk with a $MTTF_{disk} = 1,200,000$ hours was split into two independent logical drives for a total of 16 logical disks. Four parity groups of size 4 were formed from these 16 disks. To maximize independence, no two disks of a single parity group mapped to the same physical disk. Configuration (d) was an extension of (c) in that it assumed a reduced mean lifetime of $MTTF_{disk} = 1,000,000$ hours for half of the physical disks. Configurations (e) and (f) were similar to (c) and (d) but with the additional complexity of mapping four of the logical disks to two physical disks each, i.e., four physical disks were divided into 1.0+0.5+0.5 logical disks. Again, care was taken to maximize independence among the parity groups.

Figure 6.12 shows the resulting mean lifetimes for each of the configurations (a) through (f). The RAID 5 configuration of Figure 6.12(a) provides a $MTTSL$ of 10.1×10^9 hours. Because of the quadratic nature of Equation 6.17, the same RAID 5 system will have a reduced mean lifetime of approximately $69\% \equiv \frac{1}{1.2^2}$ if the reliability of the disks is divided by a factor of 1.2, see Figure 6.12(b). The simulation results for Figures 6.12(c) through 6.12(f) show two values each. The lower value assumes that all logical disks are independent. Therefore, a physical disk failure that causes two (or three) parity groups to fail simultaneously is recorded as two (respectively three) separate events. This value will not be achieved in practice, but was included because it is reflecting the analytical model. As illustrated, these simulation results are very close to the analytical prediction. The higher value counts multiple related occurrences as just one event, and hence is more indicative of the actual expected lifetime of a system.

The simplest and most reliable Disk Merging configuration, shown in Figure 6.12(c), achieves approximately an upper/lower reliability of 100%/50% of the $MTTSL$ of the comparable RAID 5 setup of (a). The reason for the analytical reliability of 50% is the increased number of parity groups (4) as compared with the RAID setup (2). Configuration (d) provides a lower reliability because half of the physical disks have a reduced $MTTF_{disk}$ of 1,000,000 hours. The analytical models for mixed MTTFs of Section 6.4.1 introduced some simplifying assumptions which lead to a conservative estimate of the mean lifetime. Consequently, in Figure 6.12(d) both simulation results are higher than the analytical predictions. In configurations (e) and (f), a physical disk may be part of up to three different parity groups, which further reduces the overall reliability of the system.

Collectively, Figure 6.12 shows that even though the flexibility of Disk Merging can result in a lower expected lifetime of a system as compared with RAID 5, it is still of the same order of magnitude (i.e., very high). Unsurprisingly, the increasing complexity of the configurations from (c) through (f) results in a trend of reduced mean time to data loss. Furthermore, the simplifying assumptions that were made for the analytical models with mixed MTTFs for physical disks in Section 6.4.1 are reflected in the conservative simulation results shown in Figure 6.12.

6.6   Conclusions

This chapter reviewed several classes of high availability techniques that are applicable to SM servers. All of them provide different tradeoffs between the level of reliability that may be achieved and the necessary overhead to implement it (e.g., bandwidth, memory, processing power).

Both mirroring and parity-based techniques can be applied to heterogeneous storage systems, but at the expense of introducing additional issues. These aspects are too numerous to be addressed in a single chapter. Hence, we have tried to provide a framework of fundamental observations that can build the foundation of any study that strives to investigate reliability techniques in more detail.

We have provided one example where we investigated how the class of non-overlapping parity techniques can be applied to a non-partitioning storage approach (i.e., Disk Merging). We have specified both design rules and algorithms that provide the necessary independence of logical disks among each other and across parity groups for the successful application of parity-based data redundancy.

Chapter 7

HIERARCHICAL STORAGE DESIGN FOR SM SERVERS

In this chapter, we study the physical design of Hierarchical Storage Structures (HSS) to support SM server requirements cost effectively, where a typical three-level hierarchy might consist of some memory, a number of magnetic disks, and tape jukebox with several read/write devices [61]. A copy of the data resides on the tape jukebox on permanent basis (Note: we consider tape as the tertiary storage of choice, due to its low cost per megabyte). The magnetic disk space is used to maintain the working set [45], [46] (locality) of the target application, i.e., those objects that are repeatedly accessed during a pre-specified window of time. The memory maintains pages of the files that might pertain to either traditional data or multimedia data. Given SM server requirements (i.e., throughput requirement, initial latency requirement, database size, etc.), the possible HSS configurations are numerous and results in various tradeoffs. These tradeoffs are due to the following: 1) cost-gap (i.e., the ratio between cost of the different storage devices), 2) access-gap (i.e., the ratio between access time of the different storage devices), and 3) transfer-gap (i.e., the ratio between transfer rate of the different storage devices) among the levels of the storage hierarchy. Naive designs might waste system resources and result in high system cost, or might not satisfy the user requirements. To gain a competitive edge, and to satisfy these requirements cost effectively (i.e., at the lowest possible cost), it is necessary to: 1) use a balanced storage configuration for a given set of application requirements, and 2) reduce the access cost to the storage devices by efficient physical storage implementations (e.g., using data placement techniques). In [39], a configuration planner was introduced to optimize HSS configuration, given the application throughput requirements (e.g., number of simultaneous displays of the target application), database size, and future access pattern to the objects, as inputs. The resulting configuration supports the application throughput at a minimum cost[1].

The configuration planner assumes a set of techniques to bridge the access-gap and transfer-gap between the levels of the hierarchy. The gap between the secondary storage (i.e., magnetic disk) and the main memory have been introduced in the previous chapters. The gap between the secondary storage and the tertiary storage has been neglected for the most part and it requires a new set of techniques [162], [167], [22], [163], [77], because tertiary storage has traditionally been used for off-line archiving, batch processing, and backup purposes. Moreover, magnetic tapes (the most popular tertiary storage) are sequential access devices, while magnetic disks (the most popular secondary storage) are random access devices, and hence new techniques and policies are required to bridge the gap between them.

Similar to magnetic disks, tape jukeboxes are mechanical devices. However, the characteristics of a tape jukebox is different from that of a disk drive [22]. The typical access time to data with a magnetic disk drive is in the order of milliseconds, whereas with a tape jukebox, due to the overhead of mounting and dismounting tapes from the drives and its sequential nature, access times in the order of seconds (minutes) are common. To allow for online or near-online access to tape resident data, it is necessary to bridge the 3-4 orders of magnitude access-gap between disks and tapes. For example, users cannot tolerate long startup latency, $T_{Latency}$, (i.e., the time elapsed from when the request arrives until the onset of its display) for many multimedia applications. To bridge the access-gap between tape storage and disk storage, the following solutions have been proposed: 1) to overlap disk and tape operations, 2) to use data placement on tape, 3) to apply tape scheduling, and/or 4) to reduce the miss ratio at the magnetic disk level. To support continuous display from a tape jukebox, it is necessary to retrieve and display data at a pre-specified rate, otherwise, the display will suffer from frequent disruptions and delays, termed hiccups.

The remainder of this chapter is organized as follows. In Section 7.1, we present the general trends in tape technology; the general tape storage systems characteristics; and, the characteristics of IBM 3590 tape drive. In Section 7.2, we present a framework for SM display from HSS. In Section 7.3, we present data placement techniques to reduce the access time to tape resident objects. Our conclusions are presented in Section 7.4.

---

[1]In [39], a set of techniques to bridge the access gap were assumed, and hence, with a different set of techniques the optimal storage mixture will be different.

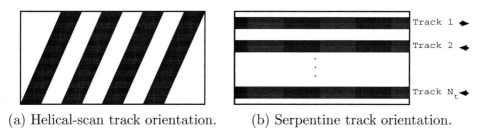

(a) Helical-scan track orientation.   (b) Serpentine track orientation.

Figure 7.1. Tape track orientations.

7.1 Overview of Tape Storage

Tape jukeboxes (or silos) consist of a number of tape cartridges, tape drives, and robot arms. The data is stored on reels of tapes housed in cartridges (or cassettes). There are a large number of different tape formats that are distinguished according to the following criteria: 1) tape track orientation, 2) tape width, and 3) cartridge size [88]. The current tape technologies primarily use one of the two tape track orientations: *helical-scan*, or *serpentine* [88], [111], [95]. With helical-scan recording, the data is recorded in diagonal stripes using a rotating drum (that contains the read/write heads) and a slow moving motor (that runs the tape over the drum). The axis of the drum is slightly slanted and, hence, the tracks formed by the passage of the heads across the tape are a series of diagonal stripes (see Figure 7.1a [88], [111], [95], [165], [49]). This technology is similar to VCR technology. Examples of helical-scan tape drives are Exabyte Mammoth 8mm drives and Sony AIT 4mm drives.

With serpentine recording, the data is placed in longitudinal tracks down the length of the tape (see Figure 7.1b [88], [111], [95], [106], [47]). The data is read by running the tape past a stationary recording assembly that contains the read/write head(s). If the number of read/write heads is smaller than the number of tracks (as is the case with most serpentine tapes) then at each pass only a portion of the tracks can be read, and multiple passes are required to read the tape in its entirety. (Note: large objects may occupy a number of tracks requiring multiple passes.) To optimize the read/write signal strength, the tape drive recording-assembly is moved slightly to establish exact track position for each tape pass. Examples of serpentine tape drives are IBM 3590, Quantum DLT8000 and Quantum SDLT320.

The common tape widths are 19mm, 8mm, 4mm, 1/2" and 1/4", where the wider the tape the wider the storage area [122], [88], [111]. The 19mm, 8mm and 4mm tapes usually use helical track orientation, whereas 1/2" and

1/4" tapes usually use the serpentine track orientation. Most of these tape formats are incompatible; however, manufacturers are increasingly trying to keep their products backward compatible, by reducing the number of cartridge sizes and having the newer drives compatible with the older formats. For example, Quantum Corporation uses the same size DLT cartridges and same basic cartridge design for all of its DLT tape drives. However, not all tape cartridge media are compatible, due to the differences in tape material and formatting used in the different tape generations.

7.1.1 Trends in Tape Storage

To increase tape capacity, manufacturers have used the following techniques: 1) increasing tape length, 2) increasing the bit density (i.e., bits per inch per track), and 3) increasing track density (i.e., number of tracks per inch). Increasing tape length is the simplest way of increasing tape capacity; however, this will result in longer search time, if the tape transport speed (i.e., the search speed of the tape) is not increased. Increasing tape length may also require tape cartridge redesign, or else the tape thickness must be reduced (which requires stronger and more stable material for tapes). It is expected that the tape thickness can be reduced by half in the next decade; however, the trend cannot be continued beyond that, due to the frailness of the tape medium [122].

Increasing the tape bit density and track density requires improvements in the recording medium, and the tape recording (i.e., the read/write head) and positioning (i.e., the servo tracking) mechanisms. It is expected that the bit density will improve from 100 kbpi in 1997 to 300 kbpi in 2007 [122]. The improvement in bit density will result in shorter fields (i.e., bits) on the tape medium, and it will require more sensitive read/write heads. As a consequence of higher bit density, the transfer rate will increase proportionally (given that the tape transport speed is maintained).

However, the most improvement is expected to come in track density, where the improvement is going to be different for the two tape track orientation formats. For helical-scan tapes, the track density is expected to improve from 2800 tpi in 1997 to 20000 tpi in 2007 [122]. For serpentine tapes, the track density is expected to improve from 750 tpi in 1997 to 20000 tpi in 2007 [122]. This 1-2 orders of magnitude in track density is expected due to the foreseeable improvements in servo tracking mechanism and closer read/write head position relative to the tape.

Given the above trends, it is expected that the tape capacity will improve by 35X for helical-scan tapes and by 150X for serpentine tapes in the next decade [122]. However, due to the flexibility of the tape medium and its shrinkage and/or expansion over time, the aerial density (i.e., bits per square inch) for the magnetic tapes will still be 2 orders of magnitude less than magnetic disks. Even though tapes have lower aerial density, the storage cost of tapes cartridges is 1-2 orders of magnitude lower than magnetic disks, due to the fact that the tape is wrapped around a reel and most of its media is hidden, and, hence, the volumetric density of tapes is higher than disks. For tapes to survive in the future (as a storage medium of choice for hierarchical structures), it is crucial that the 1-2 orders of magnitude price advantage be maintained, or else magnetic disks and/or optical disks will replace tapes as the more economical storage medium. It is expected that magnetic tapes will maintain this storage cost gap for the next decade, because magnetic tape devices take advantage of the same research used for magnetic disks [122].

One might consider the rapid progress in tape and disk storage capacity as a threat to the importance of magnetic tapes and, to some magnitude, a threat to the position of magnetic disks in storage systems. To illustrate, assume that in 1998, a database of size 10 GByte fits on a single tape (disk). If the database size grows at 60% annual rate then the database will grow at the same rate as the storage capacity, and hence, the database can always fit on a single tape (disk). If the database grows at a lower rate than the storage capacity rate, say at 40%, then the database will increasingly occupy only a small portion of a tape (disk).

However, the actual data growth is much higher than that of the storage capacity growth, where the data growth is expected to be at a higher rate than 60%, depending on the source of the statistic. For example, the National Storage Industry Consortium (NSIC) predicts the mainframe server data growth to be more than 60% annually [122], and International Data Corporation reports that the demand for corporate data storage has been growing at a rate of 80% for the past four years and it will be doubling annually for the next several years [150]. This exponential data growth is partly due to the large size of multimedia data, and the ease of generating and capturing multimedia data, specially SM data. For a 80% data growth rate, after 5 years the database will occupy 2.5 tapes (or disks), and after 10 years the database will occupy 6.1 tapes (or disks). For a 100% data growth rate after 5 years the database will occupy 4.2 tapes (or disks), and after 10 years the database will occupy 17.75 tapes (or disks).

The frequency of access to different multimedia objects (e.g., movie clips) in such storage systems is usually quite skewed [36], [188], [39], [185], i.e., a few objects are very popular while most of the rest are accessed infrequently. Furthermore, the set of popular objects may change over the course of a day. The access distribution pattern can be modelled using Zipf's law [190], which defines the access frequency of objects i to be $F(i) = \frac{c}{i^{1-d}}$, where c is a normalization constant and d controls how quickly the access frequency drops off. For example, we expect the disk cache to support the display of the popular objects, while the tape subsystem supports the display of the infrequently accessed objects. Therefore, the tape storage system services the tail-end of the Zipf distribution, which can be approximated as a uniform access distribution.

Given these trends:

1. 60-100% annual growth of data sets,

2. 1-2 orders of magnitude tape storage price advantage,

3. 3-4 orders of magnitude disk storage performance advantage, and

4. skewed data access distribution,

it is expected that tape storage devices will be used for purposes other than data backup in medium to large data repositories (such as SM servers), due to the cost benefits of tape storage.

7.1.2   General Tape Model

In this chapter, we model tape cartridges with the following parameters:

- Tape cartridge capacity (size), S_c.

- Number of tracks per tape cartridge, N_t.

- Track size, $S_t = \frac{S_c}{N_t}$ (assuming that all tracks have the same capacity)[2].

- Tape data block (segment) size, S_s[3].

- Number of blocks per track, $N_s = \frac{S_t}{S_s}$ (assuming that the track size, S_t, is a multiple of tape block size, S_s).

We assume that the tape block number and the track number are required to fully qualify a tape location. It is straightforward to map from this convention to actual physical locations on a tape.

---

[2]This assumption is used only to simplify the discussion, and it should not be viewed as a limiting factor.

[3]This parameter does not restrict our tape model to only tape devices that can support fixed block sizes. Rather, this is the read/write block size used by the application and used only for illustration purposes.

The tape drive mechanism (including the load/unload mechanism and the recording assembly) is modelled with the following parameters:

- Number of read/write heads in the recording assembly, $N_{r/w}$.

- Tape drive transfer rate, R_T: this is the aggregate transfer rate of the $N_{r/w}$ heads (note: helical-scan tapes may be assumed as a special case of serpentine track orientation with $N_t = N_{r/w}$).

- Tape load/unload position, *Pos*: the tape position when loading and unloading a tape cartridge from the tape drive, e.g.,

    - beginning of the tape, *Pos* =*beginning* (e.g., Quantum DLT and IBM 3590 drives), or

    - middle of the tape, *Pos* =*midpoint* (e.g., IBM Magstar MP drives).

- Maximum rewind time, T_{Rewind}: it is the time required to rewind from one end of the tape to the other end (e.g., T_{Rewind} for IBM 3590 tape drives is 64 seconds).

- Track switching time, T_{Track}: it is the time required to align the recording assembly from one track (or set of tracks) to another track (or set of tracks), with some minor adjustments to locate the appropriate tape block location. Note: helical-scan tapes do not incur a track switch time because there is only one (set of) track(s).

- Tape search time, $T_{Search}(i,j)$: it is the time required to search from tape location i to location j (i.e., long search). In our analysis, we assume that this time is a linear function of T_{Rewind} and the number of blocks on each track, $T_{Search}(i,j) = \frac{T_{Rewind}}{N_S} \times \mid i-j \mid$. This would generate a saw-tooth like search time from different locations. However, as shown in [86], [87], [89], [143], the saw-tooth function contains a number of discontinuities. These discontinuities are due to the specific characteristics of tape devices which are due to the different transport mechanisms used in locating tape blocks at different locations and distances.

A track group (TG) is the set of tracks that are read/written in parallel with the $N_{r/w}$ heads. These parameters characterize the general behavior of tape drives. We use them to illustrate the data placement techniques on serpentine tape devices. We verify the actual characteristics of the data placement techniques by measuring the total search times, track switching times, and rewind time (if necessary). Moreover, these parameters can be modified if necessary to better fit a specific serpentine tape device without affecting our proposed data placement techniques.

If the tape cartridge that contains the referenced data is not loaded then it is necessary to find an idle tape drive onto which the tape cartridge is loaded. If the tape cartridge is in use by some other request and/or no tape drive is available, then it is necessary to wait. In this study, we ignore the possibility that the tape cartridge and/or the tape drive are the bottlenecks. To load a tape cartridge, it requires unloading the tape from the idle drive and loading the tape with the referenced data. This time is termed the tape cartridge switch time, T_{Switch}, and is constant in most cases. We assume that the tape jukebox contains enough robot arms (and grippers), so that the tape drives do not wait idle due to contention for tape robot arms.

To retrieve an object from a loaded cartridge, two operations are performed: 1) tape and recording assembly are repositioned to the appropriate location(s), 2) data is transferred from the tape. The first operation may consist of a tape search time, a number of track switches, and/or a rewind at the end of the retrieval (if the tape is to be ejected). The total time spent in tape and recording assembly repositioning (i.e., T_{Search} and T_{Track}) is referred to as repositioning time, $T_{Reposition}$. We account for the fact that some retrieval behaviors require a rewind to Pos after reading an object (to prepare the tape for ejection), by adding this time to $T_{Reposition}$. The total access time of an object, T_{Access}, is the sum of T_{Switch} and $T_{Reposition}$ ($T_{Access} = T_{Switch} + T_{Reposition}$). Assuming that the tape is operating in a streaming mode, the time required to complete the transfer operation, $T_{Transfer_T}$, depends on the object size and the tape transfer rate, R_T ($T_{Transfer_T} = \frac{O}{R_T}$).

Parameter	Value
S_c	10 GByte
N_t	128 tracks
S_t	83.89 MByte
S_s	1 MByte
N_s	approx. 1300 blocks (segments)
$N_{r/w}$	16 for each direction (32 total)
R_T	9 MByte/s
Pos	*beginning*
T_{Rewind}	64 seconds

Table 7.1. IBM 3590 tape drive and tape cartridge specifications.

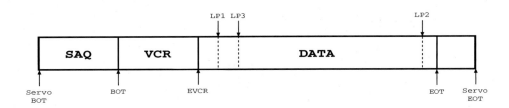

Figure 7.2. IBM3590 tape layout.

7.1.3   IBM 3590 Tape Model

To evaluate parameters that affect $T_{Reposition}$, we profiled an IBM 3590 half inch serpentine tape drive[4], where the tape length is 320 meters (1050 feet), see Table 7.1 for details. The tape capacity is 10 GByte (uncompressed), with a total of 128 tracks, resulting in approximately 84 MByte per track. The drive uses 32 read/write heads (16 in each direction) to form 8 sets of track groups (TGs). The sustained data transfer rate of the drive is 9 MB/s, and the search speed is 5 meters/second resulting in a maximum theoretical rewind time of 64 seconds ($\frac{320}{5}$).

The tape is housed in a single reel cartridge and is organized into the following regions, Figure 7.2:

1. Servo acquisition region (SAQ), which is used by the servo circuitry (i.e., the repositioning mechanism).

2. Volume control region (VCR), which is used to record format information, volume statistics information, and logical block location information.

3. Data region, which is used to store the host computer data, see [118] for details.

The servo beginning of tape (Servo BOT) and servo end of tape (Servo EOT) indicate the beginning and end of the servo information, where servo BOT is where the tape lead is attached to, and the servo EOT is attached to the cartridge reel. Beginning of tape (BOT) and end of tape (EOT) indicate the start and end of the recorded information on the tape, where this information is either volume control data or host computer data. The

---

[4]Thanks to IBM's generous gift to the Integrated Media Systems Center (IMSC) at the University of Southern California, we had free and unrestricted access to the IBM 3494 tape library that contained the IBM 3590 tape drive.

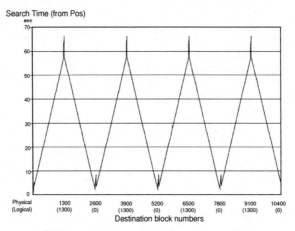

(a) Reposition time for all tracks.

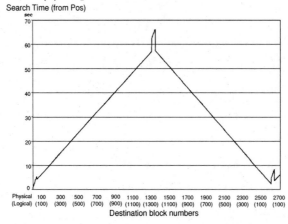

(b) Reposition time for the first two tracks.

Figure 7.3. Reposition time, $T_{Reposition}$, from *Pos* to different tape locations, for an IBM 3590 tape drive.

VCR region is identified by BOT and end of VCR (EVCR) markers, and the data region is identified by EVCR and EOT markers. There are three other markers on the data region of the tape: LP1, LP2, and LP3. LP1 is where the writing in the data region may start in the forward direction (i.e., in the direction from EVCR to EOT), and LP2 (LP3) is a warning marker that EOT is approaching in the forward (backward) direction.

To evaluate the reposition time characteristics of the IBM 3590 tape drive, we ran a *profiler* program to: 1) measure the search time, T_{Search},

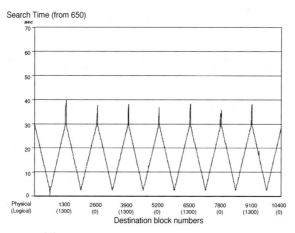

(a) Reposition time for all tracks.

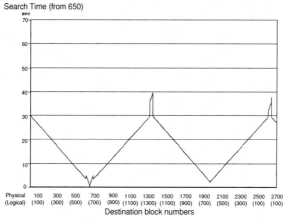

(b) Reposition time for the first two tracks.

Figure 7.4. Reposition time, $T_{Reposition}$, from 650 to different tape locations, for an IBM 3590 tape drive.

from Pos to all possible tape locations, 2) identify tape block numbers at the beginning of tracks and end of tracks, and 3) measure the track switch time, T_{Track}, between sibling tape blocks (i.e., those blocks that are physically close but far in sequential distance).

As a result of running the *profiler*, we have observed the search time from Pos to all tape locations follows a saw-tooth like function, as it was predicted, see Figure 7.3a. We also show Figure 7.3b for more details, which contains the results for the tape blocks on the first two TGs for the same tape

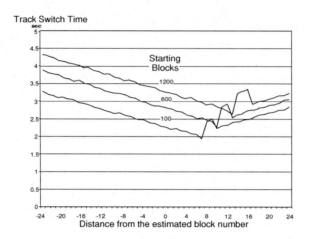

Figure 7.5. Track switch times, T_{Track}, for an IBM 3590 tape drive.

cartridge. The results of running the *profiler* on other tape cartridges are very similar, with some shift in the boundary locations due to the characteristics of specific tape and bad tape spots.

In Figure 7.3b, there are four discontinuities in the graph. The first discontinuity occurs at the beginning of the tape, i.e., close to EVCR and LP1 (small tape block numbers in Figure 7.3b), where the reposition distance is short. The second and third discontinuities occur close to the end of the tape, i.e., close to LP2 and EOT (midrange tape block numbers in Figure 7.3b). The fourth discontinuity is at the beginning of the tape, i.e., close to LP1 and LP3 (large tape block numbers in Figure 7.3b). The first discontinuity is due to the characteristics of the tape transport mechanism. When the reposition distance between block 0 and the target block is short, the full transport speed is not used to reposition the tape read/write head. The second discontinuity is due to the fact that the tape is forwarded all the way to the end before rewinding back to the target tape block location (probably past the EOT). The third discontinuity is due to the fact that the tape is not forwarded to the end before locating the appropriate tape position. We think in this case the tape reposition mechanism is able to identify that the target tape block is on a different tack, and utilizes the LP2 or some other marker to expedite the reposition process. The last discontinuity is due to the fact that the tape blocks are on two different tracks in opposite directions.

We also profiled the reposition time from the following tape locations: 1) block location 650 (physical location 650), 2) block location 1300 (physical location 1300), 3) block location 650 (physical location 1950), and 4) block location 0 (physical location 2600). Figure 7.4a shows the reposition time from middle of the first track to all possible tape locations. Figure 7.4b shows the same results in more details for the first two tracks. In this case, the reposition time is reduced almost in half, due to the shorter distance travelled. We also observe discontinuities at the track borders, which is due to the characteristics discussed above.

In Figure 7.5, we show the measured track switch time between 3 different blocks (segments) on the first track (i.e., blocks 100, 600, and 1200) and their possible siblings on track 2. The *profiler* estimates the target sibling to be $2600 - X$, where X is the block number on the first track and 2600 is the estimated number of blocks per couple of tracks. The profiler measures the track switch time between X and a number of locations close to the estimated sibling location (i.e., +/-24 block locations). Figure 7.5 shows that T_{Track} is less than 4 seconds even when the track switch is not estimated correctly (i.e., the best case is not identified correctly). Therefore, incurring an extra 4 seconds, while eliminating the more costly T_{Search}, is an attractive solution.

Our characterization of the IBM 3590 tape drive is very similar to the results reported in [89], and they can be summarized as:

- Search time from *Pos* to all blocks of the tape follows a saw-tooth like function.

- End of tracks are identified by abrupt change in search time, from both *Pos* and neighbor blocks (segments).

- Siblings have short reposition time between them (where the reposition time ranges from less than two seconds up to five seconds).

7.2 Pipelining Technique

7.2.1 Overview

To support SM display, it is necessary to retrieve and display data at a pre-specified rate; otherwise, the display will suffer *hiccups*. When considering a three level HSS, the tape storage system stores the SM repository in its entirety and it is used to display the infrequently accessed objects. Due to the long latency time associated with tape accesses and due to the mismatch between the tape production rate (i.e. transfer rate), R_T, and the display consumption rate, C (i.e., Production Consumption Ratio [61], $PCR = \frac{R_T}{R_C}$), it is necessary to: 1) use techniques to bridge the access gap between tape devices and disk devices, and 2) use techniques to guarantee a hiccup free display from tapes. In this section, we summarize techniques presented in [61] to guarantee a hiccup-free display from tapes.

With hierarchical storage organization, when a request references an object that is not disk resident, one approach might materialize the object on the disk drives in its entirety before initiating its display. In this case, assuming a zero system load, the latency time of the system is determined by the time for the tape to load the tape cartridge and reposition its read head to the starting address of the referenced object (T_{Access}), the bandwidth of the tape storage device (R_T), and the size of the referenced object (O), i.e., $T_{Latency} \approx T_{Access} + \frac{O}{R_T}$. Assuming that the referenced object is SM (e.g., audio, video) and requires a sequential retrieval to support its display, a superior alternative is to use pipelining in order to minimize the latency time. Briefly, the pipelining mechanism splits an object into s logical slices (S_1, S_2, S_3, ..., S_s) such that the display time of S_1 overlaps the time required to materialize S_2, the display time of S_2 overlaps the time to materialize S_3, so on and so forth. This ensures a continuous display while reducing the latency time because the system initiates the display of an object once a fraction of it (i.e., S_1) becomes disk resident, i.e., $T_{Latency} \approx T_{Access} + \frac{sizeof(S_1)}{R_T}$.

Another advantage of pipelining is that it enhances the *useful* utilization of resources when a user decides to abort the display of a referenced object. To illustrate, consider a user that requests an obscure (i.e., tape-resident) 30-minute video object and decides that it is not of interest after a few minutes of display. With pipelining, the display of object is overlapped with its materialization and, once the user aborts the display, the system can abort the pipeline avoiding the tape from materializing the remaining slices of the referenced object (instead, tape can be used to service some other request). Without pipelining, in addition to forcing the user to wait for a longer interval of time, the system would have had to use the tape for a longer interval time in order to stage the entire object on the disk drives.

Pipelining is a general-purpose mechanism for SM display that minimizes the latency time of the system while ensuring a continuous retrieval of the referenced object. It can be used with a tape storage device whose bandwidth is either equal to, higher or lower than the bandwidth required to display an object. The pipelining mechanism is described assuming an architecture that consists of some memory, several disk drives, and a tape storage device, Figure 7.6. We consider two alternative organization of these components: 1) memory serves as an intermediate staging area between the tape storage device, the disk drives and the display stations, and 2) the tape storage device is visible only to the disk drives via a fixed size memory. With the first organization, the system may elect to display an object from the tape

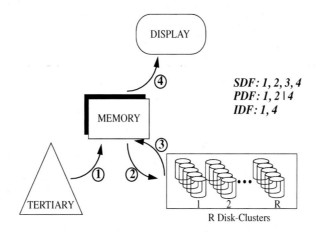

Figure 7.6. Three alternative dataflow paradigms.

storage device by using the memory as an intermediate staging area. With the second organization, the data must first be staged on the disk drives before it can be displayed. We capture these two organizations using three alternative paradigms for the flow of data among the different components:

- Sequential Data Flow (SDF): The data flows from tape to memory (STREAM 1 of Figure 7.6), from memory to the disk drives (STREAM 2), from the disk drives back to memory (STREAM 3), and finally from memory to the display station referencing the object (STREAM 4).

- Parallel Data Flow (PDF): The data flows from the tape to memory (STREAM 1), and from memory to both the disk drives and the display station in order to materialize (STREAM 2) and display (STREAM 4) the object simultaneously. (PDF eliminates STREAM 3.)

- Incomplete Data Flow (IDF): The data flows from tape to memory (STREAM 1) and from memory to the display station (STREAM 4) to support a continuous retrieval of the referenced object. (IDF eliminates both STREAM 2 and 3.)

Figure 7.6 models the second architecture (tape storage is accessible only to the disk drives) by partitioning the available memory into two regions: one region serves as an intermediate staging area between tape and disk drives (used by STREAM 1 and 2) while the second serves as a staging area between the disk drives and the display stations (used by STREAM 3 and 4).

SDF can be used with both architectures. However, neither PDF nor IDF is appropriate for the second architecture because the tape is accessible only to the disk drives. When the bandwidth of the tertiary storage device is lower than the bandwidth required by an object, SDF is more appropriate than both PDF and IDF because it minimizes the amount of memory required to support a continuous display of an object. IDF is ideal for cases where the expected future access to the referenced object is so low that it should not become disk resident (i.e., IDF avoids this object from replacing other disk resident objects). In the remainder of this section, we describe the technique from the perspective of a tertiary storage device whose bandwidth is either higher or lower than the bandwidth requirement of a SM object.

7.2.2   Streaming Media Display Using Tape Devices

The tape transfer rate, R_T, is either 1) lower or 2) higher than the bandwidth required to display an object, R_C. The ratio between the production rate of tape and the consumption rate at a display station is termed Production Consumption Ratio ($PCR = \frac{R_T}{R_C}$). When $PCR < 1$ ($PCR > 1$), the production rate of tape is lower (higher) than the consumption rate at both a display station (STREAM 4 of Figure 7.6) and the bandwidth simulated by using a single time interval per time period (STREAM 2 of Figure 7.6). In this section, we describe a pipelining mechanism that ensures a hiccup free display, while minimizing the initial latency.

$PCR < 1$

In this case, the time required to materialize an object is greater than its display time. Neither PDF nor IDF is appropriate because the bandwidth of tertiary cannot support a continuous display of the referenced object (assuming that the size of the first slice exceeds the size of memory).

Let an object X consists of n blocks[5]. The time required to materialize X is $\lceil \frac{n}{PCR} \rceil$ time periods while its display requires n time periods. If X is tape resident, without pipelining the latency time incurred to start the display of X is $\lceil \frac{n}{PCR} \rceil + 1$ time periods (see Figure 7.7). (Plus one because an additional time period is needed to both flush the last block to the disk cluster and allow the first block to be staged in the memory buffer for display, e.g., time period 11 in Figure 7.7.) To reduce this latency time, a portion of the time required to materialize X can be overlapped with its display time. This is achieved as follows. An object X is split into s logical slices ($S_{X,1}$, $S_{X,2}$, ..., $S_{X,s}$), such that the display time of $S_{X,1}$ ($T_{Display}(S_{X,1})$) overlaps the

---

[5]A block does not have to be equal in size to tape block size.

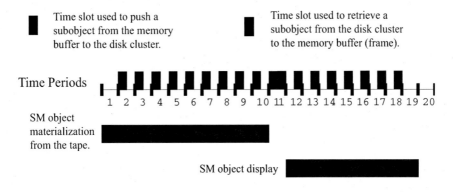

Figure 7.7. Non-pipelining approach: single disk cluster.

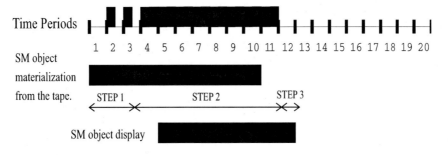

Figure 7.8. Pipelining mechanism: single disk cluster.

time required to materialize $S_{X,2}$ ($T_{Materialize}(S_{X,2})$), $T_{Display}(S_{X,2})$ overlaps $T_{Materialize}(S_{X,3})$, etc. Thus:

$$T_{Display}(S_{X,i}) \geq T_{Materialize}(S_{X,i+1}) \ for \ 1 \leq i < s \qquad (7.1)$$

Upon the retrieval of a tape resident object X, the pipelining mechanism with SDF is as follows:

STEP 1: Materialize the block(s) that constitute $S_{X,1}$ on the disk drives.

STEP 2: For $i = 2$ to s do

(a) Initiate the materialization of $S_{X,i}$ from tape device onto disks[6].

---

[6] During this step, at least a portion (not necessarily all) of $S_{X,i}$ is materialized on the disk cluster.

(b) Initiate the display of $S_{X,i-1}$.

STEP 3: Display the last block[7].

The duration of STEP 1 determines the latency time of the system. During STEP 2, while the subsequent slices are materialized from the tape device, the disk resident slices are being displayed. STEP 3 displays the last block materialized on the disk clusters.

To illustrate, assume a system with a single disk cluster ($R = 1$) as in Figure 7.8. Each time period is partitioned into two time slots ($N=2$)[8]. Assume that object X consists of 8 blocks. If $PCR = 0.8$, then the time required to materialize the object is 10 time periods ($\lceil \frac{n}{PCR} \rceil$). Figure 7.7 shows the materialization of the object from the tape device followed by its display without the use of the pipelining mechanism, yielding a latency time of 11 time periods. Figure 7.8 shows how the pipelining mechanism overlaps the display of X with its materialization. In this case, the incurred latency time is 4 time periods while the system ensures a continuous display of X.

The number of time periods required for each step is computed as follows. From Figure 7.8 it is obvious that STEP 3 requires no time slots. Furthermore, the duration of STEP 2 corresponds to the display time of $n-1$ blocks that requires n time periods. To compute the number of time periods for STEP 1 ($TC_{STEP1}(X)$), we subtract the total time required for STEP 2 (n time periods) from the time required to flush X to the disk clusters in its entirety ($\lceil \frac{n}{PCR} \rceil + 1$ time periods):

$$TC_{STEP1}(X) = \lceil \frac{n}{PCR} \rceil - n + 1 \qquad (7.2)$$

During STEP 2, the portion of the object materialized by the pipelining mechanism is $\lfloor PCR * (n-1) \rfloor$. The remainder of the object must constitute $S_{X,1}$:

$$S_{X,1} = n - \lfloor PCR * (n-1) \rfloor \qquad (7.3)$$

The granularity of $S_{X,1}$ is in terms of blocks. The size of $S_{X,1}$ is important because it determines the latency incurred when X is referenced. The size of the slices is different and can be computed as a function of $S_{X,1}$. This is not presented because their size has no impact on the latency time of the system.

---

[7]The last block has already been staged in the memory frame.

[8]Recall, a time period is partitioned into fixed time slots, with each slot corresponding to the retrieval time of a block from the disk drive.

$PCR > 1$

In this case, the bandwidth of tape exceeds the bandwidth required to display an object. Therefore, Equation 7.1 is satisfied when each slice of an object consists of a single block. Moreover, the layout of each object on the tape storage device is sequential and device independent because a complete block can be materialized during each time period.

Two alternative approaches can be employed to compensate for the fast production rate: either 1) multiplex the bandwidth of the tape drive among several requests referencing different objects (on the same tape cartridge), or 2) increase the consumption rate of an object by reserving more time intervals per time period to render that object disk resident. The first approach wastes the tape drive bandwidth because the device is required to reposition its read head multiple times. The second approach utilizes more memory resources in order to avoid the tape device from repositioning its read/write head, and to guarantee the round robin assignment of the blocks to the disk drives.

Multiplexing: One approach to compensate for the high bandwidth of tape drive is to multiplex its bandwidth among several requests, providing each request referencing an object X with a bandwidth equal to R_C. We start by describing this approach for PDF paradigm. Subsequently, we demonstrate how the utilization of the tape storage device can be maximized using a combination of both PDF and IDF.

Assuming that the tape storage device is multiplexed among j distinct requests, and a new request arrives increasing the total number of requests to k (i.e. $j + 1$). The following shows the PDF paradigm for the newly arrived request:

STEP 1: Materialize the first portion of the object referenced by the newly arrived request into memory buffers (a portion consists of one or more blocks).

STEP 2:

(a) Each memory buffer is displayed and flushed to a disk cluster at the same time.

(b) Next portions of each of the k referenced objects are transferred to the memory buffers.

STEP 3: If the materialization of one of the k objects (say X) completes then the tertiary either sits idle and waits for the arrival of a new request or services another request based on the availability of its resources and the bandwidth requirement of the object referenced by the new requests.

The number of objects that can be multiplexed simultaneously (k) depends on R_T and the time required to reposition the read/write head of tape among these requests[9]. Assume that a fixed size memory is allocated to each of the k requests, where the size of memory is a multiple of the block size, say $z \times size(B)$ where z is an integer. The time required to display the memory buffer for each object X (i.e., $\frac{z \times size(B)}{R_C}$) should be greater than or equal to the total time required to: 1) multiplex the tertiary among $k - 1$ other requests (i.e., $(k - 1) \times \frac{z \times size(B)}{R_T}$), 2) materialize the next portion of X (i.e., $\frac{z \times size(B)}{R_T}$), and 3) the time required for the tertiary to reposition its read/write head among the k requests (i.e., $\sum_{i=1}^{k} T_{Reposition}(o_i)$). Thus, multiplexing must satisfy the following constraint on behalf of each of the k requests:

$$\frac{z \times size(B)}{R_C} \geq k \times \frac{z \times size(B)}{R_T} + \sum_{i=1}^{k} T_{Reposition}(o_i) \qquad (7.4)$$

If this constraint is violated then PDF cannot guarantee a continuous display of an object. In this case, the system may employ the SDF paradigm as described earlier because each of the k requests observes a PCR less than 1. For the rest of this section, we describe PDF assuming that the constraint posited in Equation 7.4 is satisfied.

Equation 7.4 assumes that k time slots are reserved per time period. Moreover, the k reserved time slots should be positioned so that the system flushes the blocks to the clusters in the same manner (round-robin) as it would have been read if the object was disk resident (each of the z blocks read on behalf of a request is flushed and displayed simultaneously, one per time period). The upper bound on the number of required time slots is $\lfloor PCR \rfloor$; no additional time slots should be allocated as they cannot be used.

An interesting property of Equation 7.4 is the effect of memory buffers. By increasing z, more data is transferred every time the read head of the tape is repositioned. This enables the system to multiplex requests that require the read head of tape to travel a longer distance and incur a higher repositioning time.

A static scheduler can determine if a new request can be added to the k currently multiplexed requests based on the available resources (time slots

---

[9]We restrict multiplexing to objects resident on the same tape cartridge.

and memory) and the idle time of the tape device. To multiplex k objects using PDF, k time slots and $k \times z$ memory frames are required. The idle time of the tape when multiplexed among k requests (derived using Equation 7.4) is:

$$IdleTime = \frac{z \times size(B)}{R_C} - \sum_{i=1}^{k} T_{Reposition}(o_i) - \frac{z \times k \times size(B)}{R_T} \qquad (7.5)$$

If the constraint posited in Equation 7.4 is satisfied on behalf of each multiplexed request, then $IdleTime$ will be a number either greater than or equal to zero. The minimum value for k is one. In this case, the tape is used for $\frac{z \times size(B)}{R_T}$ and sits idle for the duration of time computed by Equation 7.5. This process is repeated $\frac{n}{z}$ times. This shows that in the worst case the system can employ the PDF paradigm with one time slot and one memory frame (the required resources to display an object). Upon the arrival of a new request, the scheduler may encounter alternative scenarios: 1) the additional reposition time required to service the new request renders the idle time of the tape to be a negative number (this can also happen when the new request results in a k that is greater than $\lfloor PCR \rfloor$), 2) there is insufficient memory to service the new requests, 3) there are insufficient time slots to materialize the referenced object, and 4) a combination of the first three scenarios. In the first two cases (and any combination that includes these two cases), the system has no choice and must schedule the new requests to be served at some point in the future when one of the k active requests completes.

For the third scenario, the scheduler may employ the IDF paradigm (in combination with PDF) to harness the full bandwidth of the tape storage device without materializing the object on the disk drives. The IDF paradigm is specially useful for requests that reference objects whose frequency of access is so low that the system elects not to materialize them on the disk drives. Assuming that the system services m requests using the IDF paradigm, the multiplexing approach must guarantee the following constraint on behalf of each $(k + m)$ request:

$$\frac{z \times size(B)}{R_C} \geq (k + m) \times \frac{z \times size(B)}{R_T} + \sum_{i=1}^{k+m} T_{Reposition}(o_i) \qquad (7.6)$$

where $k + m \leq \lfloor PCR \rfloor$. The system is pure PDF (IDF) when $m = 0 (k = 0)$.

With the current tape storage technologies, the multiplexing technique is impractical due to the following reasons: 1) it wastes tape bandwidth (a valuable resource), and 2) it negatively impacts the storage reliability. We leave the explanation of these disadvantages as exercises.

Non-Multiplexing: The second approach accommodates the high band-width of tape drive by increasing the consumption rate of the system. This is achieved by dedicating additional time slots per time period to materialize the referenced object (increasing the rate of data flow in STREAM 2), with one time slot dedicated to display the object (STREAM 3). This technique is appropriate for neither PDF nor IDF because both techniques require the production rate of the tertiary (STREAM 1) to be approximately the same as C (STREAM 4)[10]. However, it is appropriate for SDF because it uses the disk space as a staging area to materialize the object (STREAM 2) at a rate of R_T while retrieving (STREAM 3) and displaying (STREAM 4) it at a rate of C, eliminating the constraint enforced by both PDF and IDF. With SDF, the display of an object is initiated once the first block becomes disk resident.

Assuming k is the number of time slots allocated per time period to materialize the object, the ideal case is when k is equal to $\lceil PCR \rceil$ because it renders the consumption rate of the disk cluster to be either higher or equivalent to R_T. If $k < PCR$, then some extra memory is required to temporarily buffer the portion that cannot be flushed to the clusters. In this case, the memory resident portion continues to accumulate as a function of time until the object becomes disk resident in its entirety. This section describes analytical models to determine the number of required time slots and memory frames used to materialize the blocks on the disk clusters. We start by assuming that k is a constant for the total number of time periods required to materialize the reference object. Subsequently, we relax this assumption and consider the case where the value of k fluctuates from one time period to another.

Assuming k time slots are available per time period, consider a vertical assignment of time slots across the clusters: the time slots are assigned among the clusters in a round-robin manner starting with an available cluster. Thus, h_c time slots are dedicated to cluster c where $\lfloor \frac{k}{R} \rfloor \leq h_c \leq \lceil \frac{k}{R} \rceil$ (i.e., the k time slots are divided equally among the R clusters). When $h_c > 0$, the clusters are adjacent due to round-robin assignment. Obviously, k should be $\sum_{c=1}^{R} h_c$ and the physical upper bound enforced on k is number of time slots (N) times number of clusters (C), $N \times C$. Assuming the vertical assignment, we now calculate the maximum amount of memory required to prevent overflow. During each time period, $Min(k, PCR)$ blocks are transferred from memory to disk clusters. Assuming the object consists of n blocks, then the number

---

[10]Multiplexing (as described in Section 7.2.2) decreases rate of data flow in STREAM 1 in order to employ both PDF and IDF.

of time periods required to render the object disk resident in its entirety is:

$$q = \left\lceil \frac{n}{Min(k, PCR)} \right\rceil \tag{7.7}$$

Moreover, the number of time periods required to read the object from tape is $\frac{n}{PCR}$. If $k < PCR$ then the production rate will be higher than the consumption rate, requiring more time periods to flush the object onto the disk drives as compared to the number of time periods required to read it from tape. The difference between these two components ($\frac{n}{Min(k,PCR)} - \frac{n}{PCR}$) determines the portion of an object that becomes memory resident once the tape drive completes servicing this request. To compute the exact number of blocks that are memory resident, it is sufficient to multiply this difference by $Min(k, PCR)$:

$$MaxMem = n - \frac{n \times Min(k, PCR)}{PCR} \tag{7.8}$$

$MaxMem$ represents the maximum amount of memory required in one or more time periods[11]. The amount of memory required for each time period is computed as follows: 1) compute the fraction of an object that is disk resident per time period, 2) compute the fraction that remains on the tape storage device per time period, 3) subtract the size of an object from the sum of item 1 and 2. The number of blocks that are tape resident at time period i, for each time period i where $i = 0, 1, ..., q$ is:

$$Ter_i = Max(n - (i \times PCR), 0) \tag{7.9}$$

where PCR is the number of blocks retrieved from the tape per time period, i is the number of time periods elapsed since the initiation of the materialization, and n is the number of blocks that constitute the referenced object. Therefore, $i \times PCR$ is the number of blocks that are retrieved from the tape thus far. By deducting this from n, we compute the number of blocks that are still tape resident. The maximum function is used to avoid Equation 7.9 from producing a negative value once the tape drive has completed the retrieval of the object. When the materialization first starts (time period 0), $Ter_0 = n$.

The number of blocks that reside on the disk clusters at time period i is:

$$Disk_i = Min(Disk_{i-1} + Min(k, PCR), n) \tag{7.10}$$

---

[11]In computing $MaxMem$, we assumed that different fragments of different blocks are transferred from memory to disk clusters simultaneously. Therefore, the available memory should be managed in frames, where the size of a frame corresponds to the size of a fragment; otherwise, the available memory may become fragmented, reducing its utilization.

Figure 7.9. Intra-object layout (placement) of the block on the tape.

where $Disk_{i-1}$ represents the accumulated blocks that reside on the disk clusters at time period $i - 1$, and $Min(k, PCR)$ is the number of blocks added to disk cluster at time period i. The minimum function is used to avoid Equation 7.10 from producing a value greater than n once the object has transferred to the disk clusters in its entirety. When the materialization first starts, $Disk_0=0$.

Parameter	Value
R	3
PCR	1.5
$n = \lceil \frac{size(o_x)}{size(B)} \rceil$	12
$U_{Cluster}$	2

Table 7.2. System parameters for the example.

Hence, the number of blocks that are memory resident at time period i is:

$$Mem_i = n - (Ter_i + Disk_i) \qquad (7.11)$$

To illustrate the application of the above analytical models, and to discuss how this method supports the round-robin assignment of the blocks to the disk clusters, consider the following examples. Table 7.2 contains the parameters for all the examples. Figure 7.9 shows the layout of object on tape, and based on the PCR, the portion read per time period[12]. Thus, it requires 8 time periods to read X from tape. Round-robin assignment of X on the three cluster system is shown in Table 7.3.

---

[12]This is the intra-object layout (placement) of the blocks on the tape, however, the intra-cartridge placement can follow *CP w/ sharing*, *CP w/o sharing*, or *WARP* as shown in Section 7.3.

Cluster 0	Cluster 1	Cluster 2
X_0	X_1	X_2
X_3	X_4	X_5
X_6	X_7	X_8
X_9	X_{10}	X_{11}

Table 7.3. Desired assignment of blocks to a three cluster system.

Example 7.2.1: Assume $k = 1$ time slot is dedicated per time period to materialize X. Equation 7.7 determines that 12 ($q = 12$) time periods are required before X becomes completely disk resident. Equation 7.8 determines that the maximum amount of required memory corresponds to the size of four blocks ($MaxMem = 4$). Equations 7.9 to 7.11 are used to compute the amount of memory required during each time period (Table 7.4). The numbers in the first three rows of the table correspond to the number of blocks that reside on either the tape, disk or memory. The last two rows of Table 7.4 show the block(s) that become either memory or disk resident at each time period respectively. The schedule for allocating the clusters (to flush the blocks onto disk drives) is identical to the one for displaying them.

Table 7.4 shows that the maximum memory required is 4 blocks during time period eight (time period $\lfloor \frac{n}{PCR} \rfloor$), and it is the same as that computed by Equation[13] 7.8. □

Example 7.2.2: By dedicating two time slots per time period to materialize X ($k = 2$), the time required to materialize X is reduced to 8 time periods ($q = 8$) and no extra memory is required ($MaxMem = 0$). Note that because $k > PCR, k - PCR$ (i.e., 0.5) of the disk bandwidth allocated during a time interval is wasted (see Table 7.5 and Figure 7.10). □

We now extend the discussion to consider the case where k is no longer a constant. The scheduler may vary k for different time periods to achieve the best memory/time-slot combination based on its perspective on the future status of resources. Obviously, Equations 7.7 and 7.8 are no longer valid. Furthermore, Equation 7.10 should be revised to:

$$Disk_i = Min(Disk_{i-1} + Min(k_i, PCR + Mem_{i-1}), n) \qquad (7.12)$$

---

[13]If n is not divisible by PCR, then the last portion of the object is less than $Min(k, PCR)$. This is not captured in Equation 7.8; however, Equation 7.11 compensates to compute the exact amount of required memory per time period.

Time period (i)	Ter_i	$Disk_i$	Mem_i	Flushed to Disk	Memory Resident
0	12	0	0	-	-
1	10.5	1	0.5	X_0	$\frac{1}{2}X_1$
2	9	2	1	X_1	X_2
3	7.5	3	1.5	X_2	$X_3,\frac{1}{2}X_4$
4	6	4	2	X_3	X_4,X_5
5	4.5	5	2.5	X_4	$X_5,X_6,\frac{1}{2}X_7$
6	3	6	3	X_5	X_6,X_7,X_8
7	1.5	7	3.5	X_6	$X_7,X_8,X_9,\frac{1}{2}X_{10}$
8	0	8	4	X_7	X_8,X_9,X_{10},X_{11}
9	0	9	3	X_8	X_9,X_{10},X_{11}
10	0	10	2	X_9	X_{10},X_{11}
11	0	11	1	X_{10}	X_{11}
12	0	12	0	X_{11}	-

Table 7.4. Example 7.2.1.

Time period (i)	Ter_i	$Disk_i$	Mem_i	Flushed to Disk
0	12	0	0	-
1	10.5	1.5	0	$X_0,\frac{1}{2}X_1$
2	9	3	0	$\frac{1}{2}X_1, X_2$
3	7.5	4.5	0	$X_3,\frac{1}{2}X_4$
4	6	6	0	$\frac{1}{2}X_4,X_5$
5	4.5	7.5	0	$X_6, \frac{1}{2}X_7$
6	3	9	0	$\frac{1}{2}X_7,X_8$
7	1.5	10.5	0	$X_9,\frac{1}{2}X_{10}$
8	0	12	0	$\frac{1}{2}X_{10},X_{11}$

Table 7.5. Example 7.2.2.

where k_i is the number of time slots available during time period i. To describe $Min(k_i, PCR+Mem_{i-1})$, note that at the ith time period, Mem_{i-1} blocks are memory resident and ready for consumption. In addition, PCR blocks are produced per time period by the tape. Thus, if enough time slots are available per time period, all of these ($PCR+Mem_{i-1}$) can be consumed. Otherwise, the provided number of time slots in that period (k_i) determines the number of consumed blocks. In Equation 7.10, since k is fixed in all q time periods, if $k > PCR$ then $Mem_i = 0$ for all $i = 1, 2, ..., q$. This results in $Min(k, PCR + Mem_{i-1}) = PCR$. However, if $k \leq PCR$ then independent of the value of Mem_{i-1}, $Min(k, PCR + Mem_{i-1}) = k$. This explains why Mem_{i-1} does not appear in Equation 7.10.

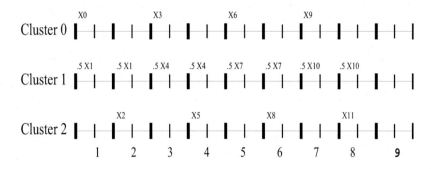

Figure 7.10. Schedule for Example 7.2.2.

In general, the number of available time slots per time period determines the amount of memory required per time period. This technique does not waste the bandwidth of the tape drive and it keeps the tape in streaming mode, however, it requires more resources as compared to multiplexing[14]. Moreover, if k_i is greater than $PCR + Mem_{i-1}$ then $k_i - (PCR + Mem_{i-1})$ defines the fraction of a time interval wasted during time period i.

7.3 Data Placement Techniques

7.3.1 Overview

Data placement techniques can be classified into the following two schemes: 1) prediction-based [146], [28], [31], and 2) retrieval-based [48]. The former estimates the expected access pattern by analyzing the past references, the characteristics of the stored objects, and/or the association between the different objects. The latter analyzes the characteristics of the object retrieval to match it with the characteristics of the tape devices. Prediction-based placement requires information about the type of queries expected and their frequency, whereas retrieval-based placement requires no prior knowledge about the type of queries and their frequency. These techniques can be used to optimize data placement on a single tape cartridge (i.e., intra-media optimization) and/or optimize data placement on all tape cartridges in a jukebox (i.e., inter-media optimization).

[146] considers the placement of multidimensional arrays, by describing techniques to cluster data into multidimensional tiles (i.e., chunks) that are accessed together. To further improve the access time, techniques to re-

---

[14]Scheduler may be forced to do multiplexing due to lack of resources.

order and replicate these chunks are considered. In [28], data placement of multidimensional arrays is also considered. The authors develop algorithms to partition the datasets into clusters, based on anticipated future access. These clusters can be a reorganization of the original dataset. The goal is to minimize the number of clusters read from the tertiary storage system. In [31], a framework is developed to optimize data placement on tapes using object *heat* (access frequency), for VOD application. It is shown that optimal data placement will have the hottest objects placed on the same tape cartridge. Moreover, these objects should be placed such that the hotter objects are closer to the *Pos*.

[48] uses a data striping technique that requires no knowledge of access patterns; rather, it is intended to take advantage of multiple tapes that are synchronized to be read in parallel. It has been shown that tape striping techniques (i.e., arrays of tapes) can be used to reduce the response time. Due to the long access times, the required tape synchronization, and the low number of drives per tape jukebox, the striping works well for moderate system loads and very large object sizes.

All the above studies, except for [48], use prediction to optimize data placement for future access patterns; however, this type of optimization has the following disadvantages: 1) as access patterns change over time, the data placement may become sub-optimal, 2) disk level caching affects the observed access pattern at the tape level, and 3) unanticipated access patterns may observe long access-times.

In this section, we present three data placement techniques: 1) contiguous placement with track sharing (*CP w/ track-sharing*), 2) contiguous placement without track sharing (*CP w/o track-sharing*), and 3) Warp ARound data Placement *WARP* [37], [40], [41], [38]. The first technique is a traditional data placement technique used for tapes. It places one object after the other on a tape, and hence, multiple objects may share a track (see Figure 7.11a). The number of tracks occupied (or partially occupied) by an object, t_{no}, is a function of the position of the start of the object, its size, and track size. To simplify the discussion, we assume t_{no} to be only a function of the latter two terms, $t_{no} = \lceil \frac{O}{S_t} \rceil$. The second technique is a modification of the first. It does not allow two objects to share the same track (see Figure 7.11b). The objective of this technique is to improve the reposition time for certain types of object retrievals, by eliminating the need to locate the first block of all objects on a tape after loading a tape cartridge. *WARP* is different than the proposed data placement techniques presented in [146], [28], [31], because it exploits the special characteristics of the serpentine tape technology to optimize the data placement. This optimization results in a

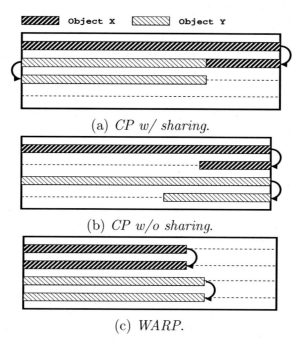

(a) *CP w/ sharing.*

(b) *CP w/o sharing.*

(c) *WARP.*

Figure 7.11. Data Placement Techniques.

data placement technique that is independent of the retrieval orders or access patterns. Moreover, the variance between the best, the worst, and the average reposition-time is very small, allowing for a priori prediction of the access-time behavior by time sensitive applications, such as real-time applications and multimedia applications. The objective of *WARP* is to eliminate T_{Search} entirely by: 1) placing the object on even number of tracks, and 2) adjusting the amount of data on each track group (TG), such that every two tracks hold *almost* the same amount of data, i.e., same amount of data is allocated to each track wrap group (TWG).

These constraints will result in minimizing the distance between the head of the object (i.e., the first block) and its tail (i.e., the last block), see Figure 7.11c. If the first block is placed on the load/unload position, *Pos*, then the tape will be at *Pos* at the end of each object retrieval. This is desirable since most drives require the tape to be positioned at *Pos* before unloading it. Moreover, by having the first and last blocks of all objects in *Pos* then very little reposition time is incurred when retrieving multiple objects from the same tape cartridge. In essence, the observed tape length is optimal for each object when *WARP* is utilized, i.e., the object is wrapped around prior to reaching the physical end of the tape. The *CP w/o sharing* and *WARP* data placement techniques might require a larger number of tracks than t_{no}, due to the added restrictions on the start position of the object head. The number of tracks required by these techniques must be an even number, and hence, the number of tracks required, t'_{no}, is equal to t_{no}, when t_{no} is even, and it is equal to $t_{no} + 1$ otherwise.

$$t'_{no} = \begin{cases} t_{no}; \text{ when } t_{no} \text{ is even} \\ t_{no} + 1; \text{ when } t_{no} \text{ is odd} \end{cases} \qquad (7.13)$$

Without loss of generality and to simplify the discussion, we assume a single read/write head per recording assembly, $N_{r/w} = 1$, and we assume an object of size O to be placed on a tape consists of n equi-sized blocks (b_0, b_1, ..., b_{n-1}), where each block size is equal to a block size, S_s. To analyze the different data placement techniques, we assume two modes of retrieval: Single Object Retrieval (SOR), and Multiple Object Retrieval (MOR). With SOR, one object is read from a single tape, hence, we expect the reposition time of each object to consist of a reposition time from Pos to the beginning of the object, a number of track switches, and a rewind time from the end of the object to Pos. With MOR, multiple objects are read from a single tape, hence, we expect the reposition time to consist of a reposition time from the end of the previously read object to the beginning of the requested object, and a number of track switches. In the remainder of this section, we describe logical algorithms[15] for the three data placement techniques and analyze their associated reposition times (under SOR and MOR) and wasted space, for Pos = beginning.

7.3.2 Contiguous Placement with Track Sharing (*CP w/ sharing*)

Contiguous placement with track sharing (*CP w/ sharing*) places one object after another contiguously on the tape. This is a simple data placement that facilitates long transfers and simplifies transferring data from older tape technologies to newer tape technologies (i.e., no re-organization is required)[16]. The data placement technique follows these constraints:

1. The first data block of each object, b_0, is placed in the tape block immediately after the previous object. (Note: b_0 of the first object is placed at the load/unload position, Pos.)

---

[15]The actual physical data placement implementation is tape device dependant.

[16]No re-organization is required if we assume that no device dependent intra-object placement optimization is used to improve system resource utilization, as described in Section 7.2.2

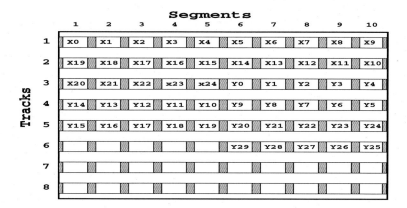

Figure 7.12. *CP w/ sharing* objects X and Y, *Pos = beginning.*

2. The difference between the tape block position of any two data blocks
   b_i and b_{i+1} is either 0 or 1, depending on the tape block location. At
   the end of a track (i.e., tape block N_s on odd tracks and tape block 1 on
   even tracks), the next block is placed on the next track with a distance
   of 0 tape blocks (i.e., the same tape block number is used while the
   track number is incremented by one). At any other tape block position,
   the next block is placed on the next higher tape block number when
   placing data on an odd numbered track, and on the next lower tape
   block number when placing data on an even numbered track.

To illustrate, consider a tape cartridge C that consists of 8 tracks, $N_t = 8$,
where each track consists of 10 tape blocks, $N_s = 10$. Given that there is only
one read/write head on the recording assembly, consider the placement of an
object X that consists of 25 data blocks (i.e., $n = 25$) on C, see Figure 7.12.
The distance (in tape blocks) between any two adjacent data blocks is either
0 or 1. For example, the distance between block X_8 and X_9 is 1 and between
X_9 and X_{10} is 0. Next, consider the placement of a second object Y that
consists of 30 data blocks (i.e., $n = 30$) on the same tape cartridge, see
Figure 7.12. The first block of object Y (Y_0) comes immediately after the
last block of object X (X_{24}).

Now consider the placement of the two objects on a serpentine tape with
$Pos = middle$, Figure 7.13. The distance between any two adjacent data
blocks is still either 0 or 1 tape blocks. However, there is a shift in the start
position for both objects, X and Y. The first block of X (X_0) is placed
at tape block 6 on track 1. The distance between blocks X_9 and X_{10} is
1, because they are not at the track boundaries anymore. The distance

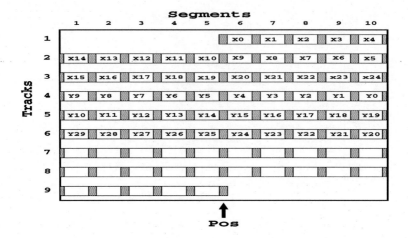

Figure 7.13. *CP w/ sharing* of objects X and Y, *Pos = middle.*

	Wasted Space	
	average	worst
Pos = beginning		
CP w/ sharing	0	0
CP w/o sharing	N_s	$2N_s - 1$
WARP	N_s	$2N_s - 1$
Pos = midpoint		
CP w/ sharing	0	0
CP w/o sharing	$\frac{N_s}{2}$	$N_s - 1$
WARP	$\frac{N_s}{2}$	$N_s - 1$

Table 7.6. Approximate wasted space for the different data placement techniques.

between X_4 and X_5 is 0, because they are now at the track boundaries. The placement of the second object Y starts at tape block location 10 on track 4.

By definition, *CP w/ sharing* optimizes for space, and hence, no space is wasted per object (see Table 7.6). However, some space is wasted per tape due to the intra-tape fragmentation, and this wasted space is: $= S_C - \lfloor \frac{O}{S_C} \rfloor \times O$. With SOR, on average, it is necessary to: 1) reposition the read/write head 1/2 tape length to locate the beginning of the object (from *Pos*), 2) perform $(t_{no} - 1)$ track switches, and 3) reposition the read/write head 1/2 tape length to *Pos*, resulting in approximately $T_{Rewind} + (t_{no} - 1)T_{Track}$ total

	$T_{Repostion}$	
	Average	Worst
$Pos = beginning$		
CP w/ sharing	$T_{Rewind} + (t_{no} - 1)T_{Track}$	$2T_{Rewind} + t_{no} \times T_{Track}$
CP w/o sharing	$\frac{1}{2}T_{Rewind} + (t_{no} - 1)T_{Track}$	$T_{Rewind} + (t_{no} - 1)T_{Track}$
WARP	$(t'_{no} - 1)T_{Track}$	$(t'_{no} - 1)T_{Track}$
$Pos = middle$		
CP w/ sharing	$\frac{1}{2}T_{Rewind} + (t_{no} - 1)T_{Track}$	$T_{Rewind} + t_{no} \times T_{Track}$
CP w/o sharing	$\frac{1}{4}T_{Rewind} + (t_{no} - 1)T_{Track}$	$\frac{1}{2}T_{Rewind} + (t_{no} - 1)T_{Track}$
WARP	$(t'_{no} - 1)T_{Track}$	$(t'_{no} - 1)T_{Track}$

Table 7.7. Approximate reposition time for the different data placement techniques in *SOR* mode.

	$T_{Repostion}$	
	Average	Worst
$Pos = beginning$		
CP w/ sharing	$\frac{1}{3}T_{Rewind} + (t_{no} - 1)T_{Track}$	$T_{Rewind} + t_{no} \times T_{Track}$
CP w/o sharing	$\frac{1}{2}T_{Rewind} + (t'_{no} - 1)T_{Track}$	$T_{Rewind} + (t'_{no} - 1)T_{Track}$
WARP	$(t'_{no} - 1)T_{Track}$	$(t'_{no} - 1)T_{Track}$
$Pos = middle$		
CP w/ sharing	$\frac{1}{3}T_{Rewind} + (t_{no} - 1)T_{Track}$	$T_{Rewind} + t_{no} \times T_{Track}$
CP w/o sharing	$\frac{1}{4}T_{Rewind} + (t_{no} - 1)T_{Track}$	$\frac{1}{2}T_{Rewind} + (t_{no} - 1)T_{Track}$
WARP	$(t'_{no} - 1)T_{Track}$	$(t'_{no} - 1)T_{Track}$

Table 7.8. Approximate reposition time for the different data placement techniques in MOR mode.

reposition time (see Table 7.7). With MOR, on average, it is necessary to: 1) reposition the read/write head 1/3 tape length to locate the beginning of the object (from the end of the previously read object), and 2) perform $(t_{no} - 1)$ track switches, resulting in approximately $\frac{1}{3}T_{Rewind} + (t_{no} - 1)T_{Track}$ total reposition time (see Table 7.8). This variation in reposition time between SOR and MOR is due to the fact that an object may occupy any location on the tape, leading to a uniform distribution of the object heads and tails on the tape. Therefore, the average distance between *Pos* and a head or tail of an object is half a tape length (SOR case), and the average distance between any head and tail is one third of tape length (MOR case). In the worst case, the distance between *Pos* and the head of an object is one tape length and

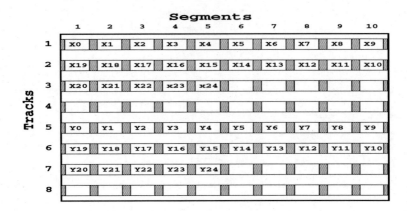

Figure 7.14. *CP w/o sharing* of objects X and Y, *Pos = beginning.*

between the tail of an object and *Pos* is one tape length (SOR case), and the distance between the head and tail of two objects is one tape length (MOR case). This results in an approximately $2T_{Rewind} + (t_{no} - 1)T_{Track}$ worst case reposition time for SOR, and $T_{Rewind} + (t_{no} - 1)T_{Track}$ worst case reposition time for MOR.

When *Pos* is in the middle (*Pos = middle*), the average distance between any head or tail to *Pos* is reduced to $\frac{1}{4}$ of a tape length (SOR case). However, the distance between object heads and tails is not affected (MOR case). This results in an approximately $\frac{1}{2}T_{Rewind} + (t_{no} - 1)T_{Track}$ average reposition time for SOR, and $\frac{1}{3}T_{Rewind} + (t_{no} - 1)T_{Track}$ average reposition time for MOR. In the worst case, the distance between *Pos* and the head of an object is $\frac{1}{2}$ tape length and between the tail of an object and *Pos* is $\frac{1}{2}$ tape length (SOR case), and the distance between the head and tail of two objects is one tape length (MOR case). This results in an approximately $T_{Rewind} + (t_{no} - 1)T_{Track}$ worst case reposition time for SOR, and $T_{Rewind} + (t_{no} - 1)T_{Track}$ worst case reposition time for MOR.

7.3.3 Contiguous Placement without Track Sharing (*CP w/o sharing*)

Contiguous placement without track sharing (*CP w/o sharing*) is very similar to *CP w/ track-sharing*. The only difference between the two is that *CP w/o track-sharing* restricts the position of the first block of each object to be at *Pos*. The objective of this technique is to improve the reposition time for SOR, by eliminating the need to search for the first block of the objects. This restriction results in objects not sharing tracks. The data placement

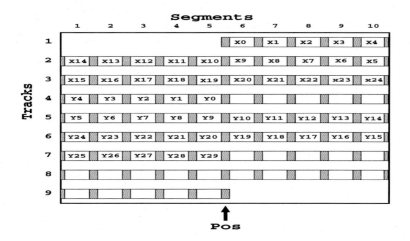

Figure 7.15. *CP w/o sharing* of objects X and Y, *Pos = middle.*

technique follows these constraints:

1. The first data block of each object, b_0, is placed in the tape block positioned at the load/unload position, *Pos*.

2. The difference between the tape block position of any two data blocks b_i and b_{i+1} is either 0 or 1, depending on the tape block location. At the end of a track (i.e., tape block N_s on odd tracks and tape block 1 on even tracks), the next block is placed on next track with a distance of 0 tape blocks (i.e., the same tape block number is used while the track number is incremented by one). At any other tape block position, the next block is placed on the next higher tape block number when placing data on an odd numbered track, and on the next lower tape block number when placing data on an even numbered track.

To illustrate, consider a tape cartridge C that consists of 8 tracks, $N_t = 8$, where each track consists of 10 tape blocks, $N_s = 10$. Given that there is only one read/write head on the recording assembly, consider the placement of an object X that consists of 25 data blocks (i.e., $n = 25$) and object Y that consists of 30 data blocks (i.e., $n = 30$) on C, see Figure 7.14. The first blocks of objects X and Y are placed on the load/unload position, *Pos*, and the distance between any two blocks of an object is either 0 or 1. For example, the distance between block X_8 and X_9 is 1 and between X_9 and X_{10} is 0.

Now consider the placement of the two objects on a serpentine tape with $Pos = middle$, Figure 7.15. The distance between any two adjacent data blocks is still either 0 or 1 tape blocks. However, there is shift in the start position for both objects, X and Y. The first block of X (X_0) is placed at tape block 6 on track 1. The distance between blocks X_9 and X_{10} is 1, because they are not at the track boundaries anymore. The distance between X_4 and X_5 is 0, because they are now at the track boundaries. The placement of the second object Y starts at tape block location 6 on track 4.

The major drawback to CP w/o $sharing$ is the wasted space. When $Pos = beginning$, one track is wasted per object on average and two tracks are wasted per object in the worst case. When $Pos = middle$, $\frac{1}{2}$ track is wasted per object on average and one track is wasted per object in the worst case (see Table 7.6).

When $Pos = beginning$, on average, regardless of the retrieval mode it is necessary to: 1) perform $(t_{no} - 1)$ track switches, and 2) reposition the read/write head 1/2 tape length back to Pos, resulting in $\frac{1}{2}T_{Rewind} + (t_{no} - 1)T_{Track}$ average reposition time (see Table 7.7). There is no difference between the two retrieval modes because the position of the head of the object is restricted to be at Pos. In the worst case, the read/write head is repositioned one tape length back to Pos after completing an object transfer, resulting in an approximately $T_{Rewind} + (t_{no} - 1)T_{Track}$ worst case reposition time. We use t_{no} instead of t'_{no}, because the former is the number of tracks occupied by the object and the later is the number of tracks reserved for the object. When $Pos = middle$, the distance between the tail and Pos is $\frac{1}{4}$ tape length on average, and $\frac{1}{2}$ in the worst case, resulting in an approximately $\frac{1}{4}T_{Rewind} + (t_{no} - 1)T_{Track}$ average reposition time and $\frac{1}{2}T_{Rewind} + (t_{no} - 1)T_{Track}$ worst case reposition time (see Table 7.8).

7.3.4   Wrap ARound Placement (WARP)

The objective of $WARP$ is to minimize the distance between the first block, b_0, and the last block, b_n, of an object. To minimize this distance, the number of tracks used to hold the object must be even (so that the block placement wraps around), and the number of blocks per track has to be almost equal on all tracks. The smallest number of tracks required to hold an object of size O is t'_{no}. This is the optimal number of tracks for $WARP$. A smaller even number of tracks cannot hold the object in its entirety, and a larger even number will result in unnecessary wasted space and track switches. In the best case, no extra track switching is required and the rewind distance is 0. In the worst case, there is an extra track switching and a maximum of

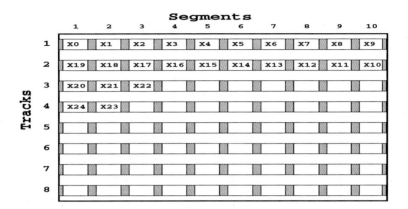

Figure 7.16. Wrap ARound data Placement (*WARP*) with an alternative optimization, *Pos* = *beginning*.

$t_{no} - 1$ tape block adjustment is required during an object retrieval.

To further optimize *WARP*, it is possible to reduce the number of blocks on the last $\lceil \frac{n}{t'_{no}} \rceil \times t'_{no} - n$ tracks (excluding track t_{no}) by one block, and add these blocks to the t'_{no} track. This will result in having $\lceil \frac{n}{t'_{no}} \rceil - 1$ blocks on the last $\lceil \frac{n}{t'_{no}} \rceil \times t'_{no} - n$ tracks, and this in turn will result in a single tape block adjustment during the object retrieval. This optimization will result in the following constraints:

1. The first data block, b_0, is placed at the load/unload position, *Pos*.

2. On the first $(t'_{no}(1 - \lceil \frac{n}{t'_{no}} \rceil) + n)$ tracks:

   - $\lceil \frac{n}{t'_{no}} \rceil$ blocks are placed starting from tape block 1 on odd numbered tracks and starting from tape block $\lceil \frac{n}{t'_{no}} \rceil$ on even numbered tracks.

   - The difference between the tape block number of any two data blocks b_i and b_{i+1} is either 0 or 1, depending on the location of the blocks. At tape block $\lceil \frac{n}{t'_{no}} \rceil$ on odd tracks and tape block 1 on even tracks, the next block is placed on the next track with a distance of 0 tape blocks (i.e., same tape block number). At any other tape block position, between 1 and $\lceil \frac{n}{t'_{no}} \rceil$, the next block is placed on the next higher tape block number when placing data on an odd numbered track, and on the next lower tape block number when placing data on an even numbered track.

3. On the last ($\lceil \frac{n}{t'_{no}} \rceil \times t'_{no} - n$) tracks:

- $\lceil \frac{n}{t'_{no}} \rceil - 1$ blocks are placed starting from tape block 1 on odd numbered tracks and starting from tape block $\lceil \frac{n}{t'_{no}} \rceil - 1$ on even numbered tracks.

- The difference between the tape block number of any two data blocks b_i and b_{i+1} is either 0 or 1, depending on the location of the blocks. At tape block $\lceil \frac{n}{t'_{no}} \rceil - 1$ on odd tracks and tape block 1 on even tracks, the next block is placed on the next track with a distance of 0 tape blocks (i.e., same tape block number). At any other tape block position, between 1 and $\lceil \frac{n}{t'_{no}} \rceil - 1$, the next block is placed on the next higher tape block number when placing data on an odd numbered track, and on the next lower tape block number when placing data on an even numbered track.

With this optimization, in the worst case, there will be an extra track switching[17] (see Table 7.7 for details). To illustrate, consider the placement of object X on tape C, using *WARP*.

An alternative way to employ *WARP* is to apply contiguous placement on the first $t'_{no} - 2$ tracks and apply *WARP* to the remainder of the object on the last two tracks. This will result in having $(t'_{no} - 2)N_S$ blocks on the first $t'_{no} - 2$ tracks. The remainder of the object, $n - (t'_{no} - 2)N_S$ blocks are placed on the last two tracks, using the above *WARP* algorithm, i.e., the last $n - (t'_{no} - 2)N_S$ blocks are treated as a single object. To illustrate this method, consider the placement in Figure 7.16. The advantage of this adjustment in data distribution on the tracks are: 1) all of the wasted space is collected into the last two tracks, and 2) the distance from *Pos* to the start of the wasted space is minimized. Therefore, it might be possible to utilize this wasted space for other objects and observe a shorter reposition time. However, it is important to note that tape devices are mostly append-only devices. For example, if an application starts writing on track 3 at tape block 5, then all previously data written from that location to the end of the tape is over-written (i.e., not accessible). To utilize this space for other objects, it is necessary to schedule the data placement of the object in advance.

To illustrate *WARP* data placement with $Pos = middle$, see Figure 7.17. In either load/unload positions, the average and worst reposition times are approximately $(t'_{no} - 1)T_{Track}$, regardless of the retrieval mode. This dra-

---

[17]Earlier, we demonstrated that a small number of tape block adjustments, when combined with track switches, might result in an extra 1-2 seconds of latency.

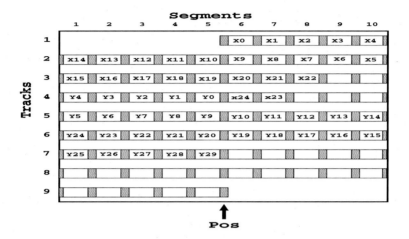

Figure 7.17. Wrap ARound data Placement (*WARP*), *Pos = middle*.

matic reduction in reposition time[18] comes at an added storage cost. The amount of wasted space with *WARP* is, however, similar to *CP w/o sharing* (see Table 7.6).

7.3.5 Cost Model

Obviously, *WARP* trades space for time, and due to this wasted space the improvement in $T_{Reposition}$ might not always be justified. This waste in storage space might result in: 1) higher number of tape switches (which in turn might lead to a higher T_{Access}), and 2) higher $\$MB_T$. In this section, we describe an analytical model to calculate the performance ($T_{Latency}$) and the cost effectiveness ($\$MB_T$) of the data placement techniques.

We use two metrics to compare *WARP* and *CP w/ sharing*: 1) average initial latency when accessing tape resident objects ($T_{Latency}$), and 2) cost per MByte of storage ($\$MB_T$). The $T_{Latency}$ metric is the average wait time before servicing a user request, when the referenced object is tape resident. By using pipelining for SM object, the initial latency becomes the tape access time, T_{Access}, and the time necessary to retrieve the first portion of the SM object. When $PCR \geq 1$, we can assume that $T_{Latency} \approx T_{Access}$, otherwise, $T_{Latency} \approx T_{Access} + \frac{(1-PCR) \times O}{R_T}$ (see Section 7.2 for details).

The $\$MB_T$ metric describes how economical the tape subsystem is in supporting the application database size, DB, and bandwidth requirements,

---

[18]The reduction is dramatic because we replace T_{Search} with T_{Track} and there are 1-2 orders of magnitude difference between them.

$Band_{Requirement}$. The $\$MB_T$ depends on the total tape subsystem cost, $Cost_{Tape}$, and the database size DB ($\$MB_T = \frac{Cost_{Tape}}{DB}$). The $Cost_{Tape}$ consists of: 1) tape storage cost, $Cost_{Storage}$ (i.e., tape media cost), 2) tape bandwidth cost, $Cost_{Bandwidth}$, and 3) tape library cost. The tape library cost is directly proportional to the number of tape cartridges, tape drives, and robot arms. We assume that the tape library cost is embedded into the tape cartridge and tape drive costs, and hence, $Cost_{Tape}$ is:

$$Cost_{Tape} = Cost_{Storage} + Cost_{Bandwidth}. \qquad (7.14)$$

The $Cost_{Storage}$ is the cost of the cartridges (media) necessary to hold the database in its entirety, and it depends on the database size DB, the effectiveness of the data placement technique to pack the objects onto a tape cartridge, and the tape cartridge cost $Cost_{Cartridge}$. Obviously, $WARP$ wastes storage space to improve $T_{Reposition}$. The amount of space wasted per object with $WARP$ is:

$$waste = t'_{no} \times S_t - O, \qquad (7.15)$$

where[19] O is the object size and S_t is the track size. $CP\ w/\ sharing$ does not waste space per object, however, some space might be wasted per tape due to the intra-tape fragmentation. The $Cost_{Storage}$ depends on the number of cartridges required to hold the database, DB. With $WARP$, the $Cost_{Storage}$ is:

$$Cost_{Storage}(WARP) = \left\lceil \frac{\frac{DB}{O}}{\lfloor \frac{N_t}{t'_{no}} \rfloor} \right\rceil \times Cost_{Cartridge}. \qquad (7.16)$$

In Equation 7.16, the first term is the number of tape cartridges necessary to hold the database of size DB and object sizes of O, given tape cartridge specification. The number of objects that can be stored on a tape cartridge depends on the number of tracks on a tape cartridge, and the number of tracks required by each object (which depends on the track size and object size), and as the number of track density increases then the cost of implementing $WARP$ decreases. With $CP\ w/\ sharing$, the total tape storage cost is:

$$Cost_{Storage}(CPw/sharing) = \left\lceil \frac{\frac{DB}{O}}{\lfloor \frac{S_C}{O} \rfloor} \right\rceil \times Cost_{Cartridge}. \qquad (7.17)$$

---

[19]We assume equi-sized objects to simplify the discussion.

In this case, the number of objects that can be stored on tape is directly proportional to the tape cartridge size and the object size.

The $Cost_{Bandwidth}$ depends on: 1) the bandwidth requirement of the application, $Band_{Requirement}$, 2) the effective bandwidth of a tape drive, $Band_{Effective}$, and 3) the tape drive cost, $Cost_{Drive}$:

$$Cost_{Bandwidth} = \left\lceil \frac{Band_{Requirement}}{Band_{Effective}} \right\rceil \times Cost_{Drive}. \tag{7.18}$$

In Equation 7.18, the first term is the number of drives necessary to support the $Band_{Requirement}$, where the $Band_{Requirement}$ is dictated by the configuration planner or the system administrator. The $Band_{Effective}$ depends on the effectiveness of the data placement technique in reducing T_{Access}:

$$Band_{Effective} = \frac{O}{T_{Access} + \frac{O}{D_T}}. \tag{7.19}$$

Depending on the application requirement, it is possible to use the maximum (worst case), minimum (best case), or average T_{Access} when calculating the $Band_{Effective}$. Here, we will consider the average case.

The average T_{Access} depends on the portion of requests serviced as MOR and SOR:

$$T_{Access} = p \times T_{Reposition}^{MOR} + (1 - p) \times (T_{Reposition}^{SOR} + T_{Switch}), \tag{7.20}$$

where p is the probability of servicing the request as MOR, and $T_{Reposition}^{MOR}$ and $T_{Reposition}^{SOR}$ are the reposition times observed when servicing requests as MOR and SOR respectively.

The probability p of servicing a request as MOR depends on: 1) the expected access distribution of references to tapes, 2) number of tapes, and 3) number of objects per tape. In [36], it was shown that the frequency of access to different clips (objects) is usually quite skewed for a SM system, i.e., a few number of objects are very popular while most of the rest are accessed infrequently. Furthermore, the set of popular objects may change over the course of a day. The access distribution pattern can be modelled using Zipf's law [190], which defines the access frequency of objects i to be $F(i) = \frac{c}{i^{1-d}}$, where c is a normalization constant and d controls how quickly the access frequency drops off. In [39], it was shown that in a well-configured HSS, the disk cache is expected to support the display of the popular objects, while the tape subsystem supports the display of the infrequently accessed objects. Therefore, the tape storage system services the tail-end of the Zipf distribution, which can be approximated as a uniform access distribution.

In case of *CP w/ sharing*, the number of tapes to hold the database and number of objects per tape are $\lceil \frac{DB}{O}/\lfloor \frac{S_C}{O} \rfloor \rceil$ and $\lfloor \frac{S_C}{O} \rfloor$ respectively. In case of *WARP*, the number of tapes to hold the database and number of objects per tape are $\lceil \frac{DB}{O}/\lfloor \frac{N_t}{t'_{no}} \rfloor \rceil$ and $\lfloor \frac{N_t}{t'_{no}} \rfloor$ respectively. Assuming a uniform access distribution, then the probability of requesting an object from a loaded tape is

$$p(CPw/sharing) = \frac{\lfloor \frac{S_C}{O} \rfloor}{\lceil \frac{DB}{O}/\lfloor \frac{S_C}{O} \rfloor \rceil}, \qquad (7.21)$$

for *CP w/ sharing*, and

$$p(WARP) = \frac{\lfloor \frac{N_t}{t'_{no}} \rfloor}{\lceil \frac{DB}{O}/\lfloor \frac{N_t}{t'_{no}} \rfloor \rceil} \qquad (7.22)$$

for *WARP*.

To illustrate, consider a tape library that uses 10GByte IBM 3590 tape cartridges, and a 100GByte data base size. Assuming average object size of 500MByte, then *CP w/ sharing* requires approximately 10 tape cartridges, while *WARP* requires 50 tapes to hold the same database. This has two negative effects on the tape storage system: 1) higher media cost, and 2) more tape switches. In this case, *WARP*'s media cost is 5 times higher than the media cost of *CP w/ sharing*. *WARP* also increases the number of tape switches from 0.9% to 0.98% of all requests to the tape storage system. As average object size increases (i.e., up to the size of two tracks $\approx$ 2.6GB), the number of tapes required for *CP w/ sharing* stays constant, while the number of tapes required for *WARP* decreases. This is due to the lower wasted space. Even though *WARP* increases the number of tape switches, the overall access time is expected to still be lower than the access time of *CP w/ sharing*. This is due to the following facts. First, *WARP*'s access time for SOR retrieval is either lower than or similar to *CP w/ sharing*'s access time for MOR retrieval. Second, *WARP*'s access time for MOR retrieval is lower than *CP w/ sharing*'s access time for MOR retrieval. Finally, for reasonable sized databases (i.e., hundreds to thousands of tapes), the difference between $p(CPw/sharing)$ and $p(WARP)$ is very small.

7.3.6   Performance Results

Reposition Time Measurements for the Data Placement Techniques Using IBM3590 Tape Drive

To evaluate the performance of the data placement techniques presented earlier in this section, we have implemented these techniques on an IBM

3590 tape drive (as described in Section 7.1), housed in an IBM 3494 tape library and connected to an IBM R/S 6000 workstation front-end. We opted to implement the techniques, rather than simply simulating the system, for the following two reasons. First, it is our experience that tape drive specifications do not report all necessary details and do not always match actual system performance. For example, we have not been able to identify the time required to perform a track switch, i.e., T_{Track}, in any of the tape specifications we have come across. Therefore, we have spent some time evaluating and characterizing the IBM 3590 tape drive (see Section 7.1). Second, we wanted to verify that *WARP* can be implemented and the performance gains are consistent with our predictions.

The implementation of *CP w/ sharing* is straightforward. It requires us to write one object after another on the tape without paying attention to the characteristics of the tape being used. However, the implementation of the other two data placement techniques requires some information about the characteristics of the tape, such as: 1) the blocks (segments) at *Pos* (per track), and 2) siblings on different tracks. For the implementation of *CP w/o sharing*, we require item 1, while *WARP* requires the identification of both items. There are two methods of identifying these important characteristics: 1) use a *profiler*, or 2) use an adaptive data placement algorithm. The former technique is an off-line method, where it first identifies the characteristics of the main boundaries of a tape cartridge and then uses this information for the placement of the data. We have found that it is not necessary to completely profile the tape prior to data placement, because the improvements are marginal and do not justify the high cost of running the complete *profiler*. The latter technique adapts to the specific characteristics of the tape and it optimizes the placement (e.g., finds the appropriate siblings for *WARP*) while it is in the process of the data placement. Here, we show the results using the *profiling* method.

To evaluate the different placement techniques, we use single object retrieval (SOR), and multiple object retrieval (MOR), as described earlier in this section. With SOR, we assume that a single object is requested from a loaded tape cartridge with the read/write heads at *Pos*, and it is necessary to rewind back to *Pos* after reading the object (to prepare the tape for ejection). We believe this is the most common mode of operation, when the disk cache absorbs most of hot objects (i.e., the locality of references). With MOR, we assume that there are 20 random requests to objects on a loaded tape, where a new request is generated immediately after serving the current request. In this mode of operation, it is not possible to use scheduling, because the queue contains at most one object. We have also assumed

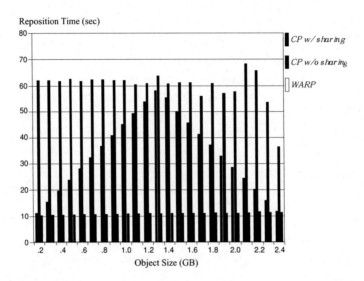

Figure 7.18. Average reposition times of SOR for *CP w/ sharing*, *CP w/o sharing*, and *WARP*.

a situation where there are a small number of requests for a single tape, i.e., there are 3-4 requests pending for a number of objects on a tape cartridge. In this case, it is possible to use scheduling to reduce the average reposition time for the requests. For all the cases, we varied the object sizes from 200 MByte up to 2.4 GB.

Figure 7.18 compares the average reposition time for the three data placement techniques, with SOR. We observed that *WARP* constantly outperforms the other two data placement techniques. The average $T_{Reposition}$ is reduced by 71%-84% (2.5-5 times improvement) as compared to *CP w/ sharing*, and is reduced by 5%-81% as compared to *CP w/o sharing*. Even though the improvements over *CP w/o sharing* are not always significant, it is important to note that both *WARP* and *CP w/o sharing* waste the same amount of space. Therefore, *WARP* is always superior to *CP w/o sharing*, and hence we will not consider *CP w/o sharing* further. Figure 7.19 compares maximum, minimum, and average $T_{Reposition}$ for *CP w/ sharing* and *WARP*'s maximum $T_{Reposition}$ observed across all object sizes, with SOR. We observe that *WARP*'s worst case across all objects is higher than the minimum $T_{Reposition}$ of *CP w/ sharing* in only two cases (i.e., 200 MByte and 2.1 GByte object sizes). Another interesting observation is that the difference between maximum and minimum $T_{Reposition}$ decreases as object size increases from 200 MByte - 1.3 GB, and then it increases for objects larger

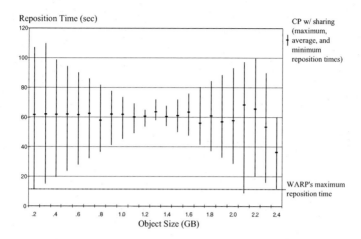

Figure 7.19. Maximum, minimum, and average reposition times of SOR for *CP w/ sharing* and *WARP*'s maximum reposition time across all objects.

than 1.3 GByte. This phenomena is due to the interaction between the object size and the IBM 3590 tape track size. For object size of 1.3 GByte, each object occupies a single track, and hence, either the head of the object is close to *Pos* and its tail far from it, or the head is far from *Pos* and its tail close to it.

Figure 7.20 compares the average $T_{Reposition}$ for *CP w/ sharing* and *WARP*, with MOR. Again, *WARP* consistently has a lower $T_{Reposition}$ than that of *CP w/ sharing*. The average $T_{Reposition}$ is reduced by 57%-81%. Again, in Figure 7.21, we compare the maximum, minimum, and average $T_{Reposition}$ for *CP w/ sharing* and *WARP*'s maximum $T_{Reposition}$ observed across all objects, with MOR. In this case, the minimum $T_{Reposition}$ of *CP w/ sharing* is lower than *WARP*'s worst case for almost all object sizes. This is due to the fact that in the best case, for *CP w/ sharing*, the head of the next object is adjacent (or very close) to the tail of the previously read object, and hence, only a short repositioning is necessary.

We compare *WARP*'s worst case with the maximum, minimum, and average cases of *CP w/ sharing* (for both SOR and MOR) to show the effectiveness of *WARP* under all conditions. We did not show the maximum, the minimum, and the average $T_{Reposition}$ of *WARP*, due to the small differences between them. We have observed that the difference between maximum and minimum $T_{Reposition}$ for *WARP* (for all object sizes) to be between 2.93 and 3.36 seconds (with MOR) and 4.15 and 4.23 seconds (with SOR). This low variance between the best, the worst, and the average $T_{Reposition}$ allows for

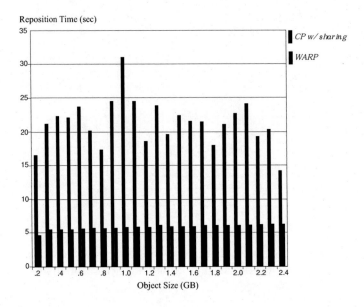

Figure 7.20. Average reposition times of MOR for *CP w/ sharing*, and *WARP*.

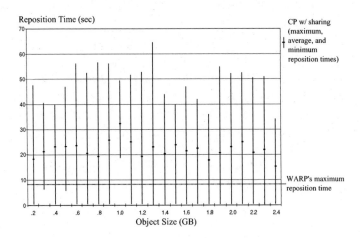

Figure 7.21. Maximum, minimum, and average reposition times of MOR for *CP w/ sharing* and *WARP*'s maximum reposition time across all objects.

a priori prediction of the access behavior, which is ideal for SM servers due to the real-time characteristic of SM objects.

Figure 7.22 compares the average $T_{Reposition}$ of *WARP* with the maximum, the minimum, and the average $T_{Reposition}$ of *CP w/ sharing* with

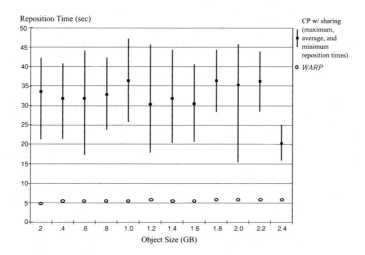

Figure 7.22. Comparison of *WARP* average reposition time and *CP w/ sharing*'s best, worst, and average reposition times (with scheduling).

MOR, when there are 3-4 requests to be scheduled. For objects of size 200 MByte up to 1 GByte, we assumed a queue of 4 requests, and a queue of 3 requests for larger objects. The reason we restricted the size of the queue to 4 requests for small objects is due to the complexity of finding the optimal schedule for larger queues (i.e., the complexity is exponential to the size of the queue). We restricted the size of the queue to 3 requests for large objects because in this case the tape contains only a handful of objects. Again, *WARP* consistently observes a lower $T_{Reposition}$, even when compared to the optimal schedule. For all the experiments, we executed each scenario a number of times, and the reported numbers are averaged out among them. All of these observations are consistent with the predictions we made in Section 7.3.

Performance of the Data Placement Techniques using IBM3590 Tape Drive

To measure the performance of *WARP*, we assume a tape library system that uses IBM 3590 tape drives (as in the previous section), where the tape switch time is 15 seconds ($T_{Switch} = 15$ s) and the sustained transfer rate is 9 MB/s ($D_T = 9$ MB/s). We assume $PCR \geq 1$, and hence, $T_{Latency} \approx T_{Access}$.

In Figure 7.23, we show the percentage of the space wasted per object, for *WARP*. As expected, the amount of wasted space and the percentage of waste decrease as the object size increases, until the size of the object equals

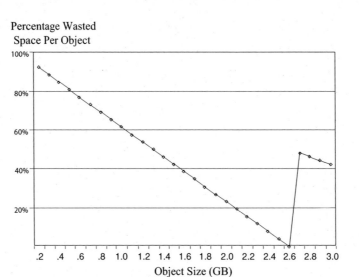

Figure 7.23. Percentage space wasted per object with *WARP*, for IBM 3590.

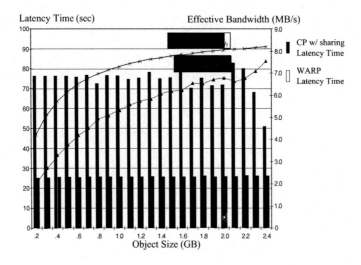

Figure 7.24. Expected average $T_{Latency}$ and $Band_{Effective}$ for IBM 3590 tape drive for *CP w/ sharing* and *WARP* data placement techniques, assuming a 1 TB *DB*.

two tracks (in this case). After that, *WARP* requires two more tracks (i.e., total of 4 tracks) to hold the object, and hence, the amount of wasted space and the percentage of wasted space swing back up.

To measure the expected $T_{Latency}$ and $Band_{Effective}$, we assume a $DB = 1$ TB. In Figure 7.24, using the bars and the left y-axis, we show the expected average $T_{Latency}$, and using the lines and the right y-axis, we show the expected average $Band_{Effective}$ for a single IBM 3590 tape drive. Again as expected, *WARP* consistently outperforms *CP w/ sharing* by having a lower $T_{Latency}$ and higher $Band_{Effective}$, even though, with *WARP*, more tape switches are performed. The reductions in $T_{Latency}$ range from 49%-69%, and the improvements in $Band_{Effective}$ range from 9%-108%. Results are similar for larger database sizes; however, as the database size shrinks, the performance advantage of *WARP* is reduced.

7.4   Conclusion

A growing number of applications store, maintain, and retrieve large volumes of multimedia data, where the data is required to be available online or near-online [162], [163], [77], [108]. Example applications include video-on-demand servers, web-servers, multimedia databases, scientific applications, and e-learning applications. The rapid progress in mass storage technology is enabling designers to implement these large data repositories cost effectively using hierarchical storage structures (HSS), where an HSS consists of a number of fast (expensive) and slow (cheap) storage devices.

In this chapter, we focused on the design of HSS that consists of a three-level hierarchy, with some memory, a number of magnetic disks, and tape juke box with several read/write devices. We focused on tape juke boxes as the tertiary device of choice, due to its popularity as a backup device and due to its low storage cost. In a typical HSS, the tape juke box holds the SM database in its entirety, while the magnetic disks are used to maintain the active set of SM objects, i.e., those objects that are repeatedly accessed during a pre-specified window of time. The memory maintains pages of the files that might pertain to either traditional data or multimedia data.

In Section 7.1, we presented the general trends in tape storage technology, the general tape storage systems characteristics, and the characteristics of IBM 3590 tape drive. Even though magnetic tape devices are mechanical devices similar to magnetic disk drives, their characteristics are different than that of magnetic disk drives. The typical access time to data with a magnetic disk drive is in the order of milliseconds, whereas with a tape jukebox, due to the overhead of mounting and dismounting tapes from the drives and its sequential nature, access times in the order of seconds (minutes) are common. To allow for online or near-online access to tape resident data, it is necessary to bridge the 3-4 orders of magnitude access-gap between disks and tapes.

In Section 7.2, we present a general-purpose pipelining technique for SM display that minimizes the latency time of the system while ensuring a continuous retrieval of the referenced object. This technique can be used with a tape storage device whose bandwidth is either equal to, higher or lower than the bandwidth required to display an object. When the tape storage bandwidth is higher than the display bandwidth, we show that multiplexing of the tape device bandwidth is not practical, due to the wasted (valuable) tape bandwidth and due to its impact on the reliability of the tape media.

A trend in serpentine tape technology indicates an increase in the number of tracks per cartridge. Consequently, the read/write heads require sweeping the tape a number of times in forward and backward directions in order to read a large object. In Section 7.3, we have introduced a new data placement technique for serpentine tapes, termed *WARP*, that takes advantage of this zig-zag sweeping characteristic to reduce the access time to tape resident data. As a result, the access gap between disk storage and serpentine tapes is reduced significantly to allow for online or near-online access to tape resident data, within HSS. Moreover, given the expected sub-second track switch time, T_{Track}, in the near future, this novel data placement technique is expected to observe an even better performance.

We investigated the performance and cost effectiveness of *WARP* and two traditional data placement techniques, in the context of SM servers using tape systems with IBM 3590 tape drives. We observed 49%-69% (1.96-3.23 times) reduction (improvement) in $T_{Latency}$, when using *WARP*. Moreover, the variance between the best, the worst, and the average $T_{Reposition}$ (and in turn $T_{Latency}$) is very small, allowing for a priori prediction of $T_{Latency}$ behavior for multimedia applications. However, $\$MB_T$[20] of *WARP* is sometimes higher and sometimes lower than that of *CP w/ sharing*, depending on the application requirements (i.e., database size and tape bandwidth requirements). *WARP* for the most part is more expensive for small objects (i.e., objects 500 MByte or smaller), while it is either cheaper or similar in price for large objects (i.e., objects 1.5 GByte or larger). For medium sized objects (i.e., objects between 500 MByte and 1.5 GByte), the difference between $\$MB_T$ for the different data placement techniques vary depending on the application requirements. Using the two metrics, a hierarchical storage manager or a file system may judicially decide the appropriate placement technique, given the object size and application requirements.

In general, HSS should not be operated in thrashing mode, where a large

---

[20]Note: $\$MB_T$ depends on the total tape subsystem cost, $Cost_{Tape}$, and the database size, DB, where $Cost_{Tape} = Cost_{Storage} + Cost_{Bandwidth}$.

number of requests wait to be served by the tape jukebox. System administrators should fix this problem when it arises, by introducing additional disk storage to accommodate the working set of the SM application. This is due to the following observations:

- The overall system performance would be very poor because the tape jukebox becomes a bottleneck.

- Tape jukeboxes are designed to store large volumes of data and not to service frequently accesses to SM objects.

- The reliability of tape jukeboxes is inversely proportional to their usage (i.e., operating a system in a thrashing mode will decrease the MTTF of the tape jukebox).

Due to the above observations and due to the fact that multiplexing is not practical for tape jukeboxes, we believe that tape scheduling should have limited impact on the HSS design for SM servers.

Chapter 8

DISTRIBUTED SM SERVERS

Distributed SM applications (e.g., VOD) are expected to provide service to a large number of clients often geographically *dispersed* over a metropolitan, country-wide, or even global area. With a naive design, employing only one large Centralized Streaming Media Server (CSMS) to support the distributed clients results in inefficient resource allocation that renders the design virtually impractical. For instance, the overall bandwidth requirement to implement an interactive VOD system with such a design is estimated to be as high as 1.54 Pb/s for the continental United States [123]. On the other hand, using a set of distributed but independent local CSMSs, i.e., a distributed design without resource sharing, is also proven to be inefficient. In fact, the idea of connecting the resources into networks, such as the Internet, was originally motivated by this problem. The full economic potential of the distributed media applications will not be realized unless *cost-effective* solutions are achieved for serving the SM to dispersed users.

In this chapter, we investigate issues involved in designing a Distributed Streaming Media Server (DSMS) to support VOD applications efficiently. We focus on VOD applications, however, the techniques are general and can be extended to support other types of distributed SM applications. In Section 8.1, we present the components of a DSMS. In Section 8.2, we describe a novel DSMS architecture, termed *Redundant Hierarchy (RedHi)*. In Section 8.3, we present our concluding remarks and future research directions.

8.1 Solutions for Distributed Streaming Applications

To support distributed streaming applications, one group of researchers has focused on various techniques to either reduce bandwidth requirements of individual media streams, with methods such as *smoothing* [141], *staging* [4], and *bandwidth renegotiation* [98], or to reduce overall bandwidth requirements of multiple streams via aggregation (or *statistical multiplexing*), for example with *batching* and *multicasting* [23], [7]. As an orthogonal solution, other researchers have proposed distribution of the service to manage dispersedness of clients; i.e., employing a number of CSMSs each to serve clients located in a certain locality and interconnecting them via a high-speed network infrastructure to be able to share/exchange the SM objects. It is important to note that although serving clients locally will result in a dramatic reduction in total *communication bandwidth* requirement of the system, unless servers are able to share the objects across the network, the huge aggregated *storage capacity* requirement of the servers will render the no-sharing approach impractical. Systems designed based on this approach [57], [79] (DSMS approach), are shown to be able to provide the optimal solution, i.e., the minimum *communication-storage* cost for distributed SM applications [147], [11]. In this chapter, we focus on the latter solution.

8.1.1   Components of Distributed Streaming Media Servers

A distributed VOD system can usually be considered as a rich repository of SM objects with a large set of dispersed clients. Objects should be served to the clients on their demand. Generally, a DSMS that realizes such a VOD system is designed as a network with hierarchical (or tree) topology, with individual CSMSs as the nodes and network links as the edges of the hierarchy. Nodes are assumed to be able to store a limited number of SM objects, and stream a finite number of SM objects. Meanwhile, network links are expected to guarantee the specific QOS requirements of SM communications. Currently, there are two alternative underlying telecommunication infrastructures that can provide such guaranteed services: 1) circuit-switching networks, such as SONET, which provide *dedicated* physical and/or logical links, and 2) packet-switching networks that support specific guaranteed services (such as "Integrated Services" [32] or "Differentiated Services" [18] for IP-based networks, e.g., the Internet, or CBR and VBR services for ATM networks [112]) to emulate characteristics of the dedicated links by enforcing statistical and/or deterministic guarantees.

The nodes at the leaves of the hierarchy, termed *head-ends*, are points of access to the system. In practice, clients are connected to the head-ends (usually located in local central offices or COs) via broadband local-access networks, e.g., xDSL or cable network. When a request for an object arrives at a head-end, if the object is available in its local storage the head-end serves the client, or else the request will be forwarded to the higher levels of the hierarchy and eventually some other node that has the object stored locally will serve the client by streaming the object through the hierarchy and

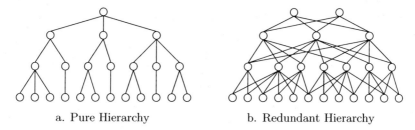

a. Pure Hierarchy   b. Redundant Hierarchy

Figure 8.1. Pure Hierarchy vs. Redundant Hierarchy

finally via the head-end to the client. As this brief description of the system structure and functionality conveys, a DSMS network should also consist of a middleware component for resource management. The middleware is supposed to address two different orthogonal issues:

1. *Object placement*: Static and/or dynamic mapping of objects onto the DSMS network nodes (the storage space) so that the overall communication storage cost of the system is optimal. Many researchers have addressed this problem, also known as Media Asset Mapping Problem (MAMP), by introducing analytical models that consider user access patterns [127] in addition to communication and storage constraints to obtain optimized object distribution and replication (or caching) policies [134], [140], [107], [96].

2. *Object delivery*: On client demand and in real-time, locating the replicas of the objects within the DSMS network, and selecting the appropriate replica and allocating system *streaming* resources (i.e., node and link bandwidth) for object delivery so that high resource utilization is achieved.

8.2   RedHi: A Typical Distributed Streaming Media Server

In this section, we focus on developing a novel DSMS architecture, termed *Redundant Hierarchy (RedHi)* DSMS. As far as a CSMS can support the required storage and streaming capabilities, choice of the CSMS for the DSMS network nodes does not affect the DSMS design. Similarly, if the DSMS middleware is implemented at the application layer, as far as the links (either physical or logical, dedicated genuinely or by emulation) provide the required guaranteed service, the DSMS design will be independent of the underlying telecommunication infrastructure. Therefore, the *network topology* and the *resource management middleware* are the two main components that characterize a DSMS architecture. In the remainder of this section, we describe these components of our proposed DSMS architecture.

8.2.1 Network Topology

In [155], we extended the pure hierarchy to a new topology termed *Redundant Hierarchy*, or briefly *RedHi* (see Figure 8.1b). RedHi is defined as follows:

Definition 8.2.1: Leveled Graph. "Leveled Graph" $G = (\mathcal{V}, \mathcal{E}, n)$ is a graph (with $\mathcal{V}$ as the set of nodes and $\mathcal{E}$ as the set of edges) that satisfies the following two conditions:

1) There exists a partition $\{\mathcal{V}_i \mid i \in 1..n,\ n > 1\}$ of $\mathcal{V}$ such that $\forall e \in \mathcal{E}$, if $e = (u, w)$, $\exists j \in 2..n$ such that $u \in \mathcal{V}_{j-1}$ and $w \in \mathcal{V}_j$. In such a case, u is a *parent* node of w, and w is a *child* node of u.

2) $\forall v \in \mathcal{V}_1, \exists e \in \mathcal{E}$ such that $e = (v, w)$ and v is a parent node of w. Moreover, $\forall v \in \mathcal{V}_{i \in 2..n}, \exists e \in \mathcal{E}$ such that $e = (u, v)$ and v is a child node of u.

The number n is the number of *levels* (i.e., partition members) in the leveled graph G. If $v \in \mathcal{V}_1$, v is a *root* node of G. Nodes that have no children are termed *head-ends* or *leaves* of G.

Definition 8.2.2: Pure Hierarchy. A "Pure Hierarchy" is a leveled graph in which:

1) there is one and only one root, and
2) every node, except the root, has one and only one parent node.

Definition 8.2.3: Redundant Hierarchy (RedHi). A "Redundant Hierarchy" is a leveled graph in which:

1) there are at least two roots, and
2) every node, except the roots, have at least two parent nodes.

RedHi relaxes the hard degree-1 parent connectivity restriction with pure hierarchy to be degree-2 or more. Consequently, there is a higher potential for load-balancing among nodes and links of the DSMS network (both in the absence and presence of node and/or link failures); hence, higher streaming resource utilization and cost-efficiency. The redundancy in RedHi is not of bandwidth, but of number of links. In other words, the aggregated bandwidth of the links connecting a node to its parents can be the same as the bandwidth of the link connecting the node to its single parent in a pure hierarchy. Therefore, RedHi does not impose higher bandwidth requirements, but only requires higher connectivity. In fact, even if we are restricted to use dedicated physical links for DSMS network, RedHi structure is quite compatible with redundant hierarchical organization of the current telecommunication networks such as the Internet because COs (Central Offices), POPs (Points of Presence), and ISPs (Internet Service Providers), which are best potential locations for DSMS nodes, are usually redundantly connected to several larger parent nodes in the hierarchy [21], [20].

8.2.2   Resource Management

Here, we focus on an object delivery scheme that exploits the characteristics of the RedHi topology to realize its potential advantage, i.e., higher utilization of streaming resources. Consider a set of shared SM objects distributed among nodes of a DSMS network based on an object placement scheme[1]. Several replicas of an object may exist in the network at a time, and replicas may *dynamically* be inserted and/or deleted according to some object replication (or caching) policy. As a client request arrives to *read* an object, an "object delivery scheme" is required to locate replica(s) of the object, and select and allocate the *best* path within the DSMS network (sourcing in one of the object replicas) to deliver the object to the client. This resource management scheme should select the best path so that the overall utilization of the streaming resources (i.e., node and link bandwidth) of the DSMS network is high.

Many distributed object processing applications are not latency- and/or overhead-tolerant. In such systems, full decentralization of the resource management is inefficacious and impractical. However, with SM applications, the additional start-up latency and resource management overhead (attributed to decentralization) is negligible as compared to the duration and resource requirements of SM objects; e.g., less than 2 seconds of start-up latency for a 1-hour movie, or 10b/s of communication overhead for a 1Mb/s bandwidth movie (or equivalently, less than 5KB of total communication overhead for a 0.5GB storage-size movie). On the other hand, one can achieve optimal robustness and scalability with decentralization of the management. Here, assuming the RedHi topology and a fixed object replication scheme, we introduce an object delivery scheme that *collapses* three mechanisms, namely *object location*, *path selection*, and *resource reservation*, into one fully *decentralized* delivery mechanism. Decentralization of the resource management meets our scalability and robustness objectives optimally, whereas collap-

---

[1]Here, we assume objects are not decomposable; hence, object is the grain size for distribution.

sion of the mechanisms helps satisfying the start-up latency and overhead constraints. To achieve high resource utilization, the object delivery scheme selects and allocates the best streaming path to serve each request based on our proposed cost function, which considers 1) the overall amount of the streaming resources engaged by selecting each streaming path, and 2) the relative loads of the paths. The object delivery scheme also comprises various optional object location and resource reservation policies. Finally, this scheme is designed as an application layer resource management middleware for the DSMS architecture to be independent of the underlying telecommunication infrastructure.

We have extended the NS (Network Simulator) networking event simulator [1] to incorporate our object delivery scheme. We conducted several experiments via simulation to evaluate performances of the proposed optional policies and to verify that our object delivery scheme meets our objectives. Results of our experiments show that our delivery scheme with a RedHi DSMS network can achieve up to 50% improvement in resource utilization over a pure hierarchy.

The remainder of this section is organized as follows. First, we discuss the design objectives of the object delivery scheme. Second, we explain our design approach to meet the required objectives. Third, we define the functionality of our object delivery scheme. Fourth, we describe our simulation model and present the results of our experiments. Finally, we provide our conclusions and discuss our future directions.

Design Objectives

Optimization of resource utilization is the main objective of an object delivery scheme. In addition to resource utilization optimization, the mechanisms used for object delivery should meet some other objectives, the most important of which are scalability, robustness, minimum start-up latency, and minimum resource management overhead. We believe that unlike many other parallel/distributed object processing applications [102], such as distributed object programming models (e.g., CORBA), PRAMs (Parallel Random Access Machines) [53], and mobile object tracking systems [10], which must strive to minimize the latency and/or resource management overhead of access to the objects as their highest priority objectives, with object streaming applications, due to often large duration and high resource requirements of SM objects these two objectives are generally less important. For instance, with a VOD system the user reneging behavior due to intolerable start-up latency is often modeled by distributions with expected values estimated as high as several minutes [36], [3], [91]; e.g., in [91], the user reneging behavior is modeled by a normal distribution with expected value $\mu = 5min$ and standard deviation $\sigma = \mu/3$. Thus, any start-up latency in the order of seconds will result in negligible user reneging rate. Besides, as compared to the high demands of SM objects for system resources (i.e., memory, CPU-time, and bandwidth), the amount of resources used for resource management purposes is relatively small. Therefore, since our object delivery scheme is designed for a DSMS network, we consider these two objectives, namely minimum start-up latency and minimum resource management overhead, as the least important ones. Below, we order the required objectives based on their importance and provide a more detailed definition for each objective:

1. *Resource utilization optimization*: the object delivery scheme should realize the higher streaming resource utilization advantage of the RedHi topology, which is considerable both in the absence and presence of node and/or link failures. We evaluate the utilization by the conventional *client blocking ratio* (or simply *blocking*) measure; the less the blocking, the higher the resource utilization. Of course, with low blocking and high resource utilization the media server system is more cost-effective/profitable.

2. *Scalability*: the object delivery scheme should remain practical for large-scale applications; particularly, resource management performance bottlenecks should be avoided to allow extending the DSMS network by adding nodes and links as required.

3. *Robustness*: the object delivery scheme should be tolerant towards the failures of the resource management system components. Particularly, if the resource management scheme is designed to run on top of the same DSMS network used to deliver the actual SM objects, failures of the network components, i.e., nodes and links, should have the least possible impact on the performance of the resource management system.

4. *Start-up latency alleviation*: the object delivery scheme should decrease the object streaming start-up latency to the point that client reneging rate due to latency is negligible.

5. *Resource management overhead alleviation*: Memory, CPU-time, and bandwidth requirements of the resource management imposed as overhead to the DSMS system should be negligible as compared to requirements of the actual content of the system, i.e., SM objects.

This ordering of the objectives has an important impact on the main decision with the design of the object delivery scheme, i.e., positioning strategy (centralized or decentralized) for the control data that are used for resource management purposes.

Design Approach

For resource management tasks such as object location, path cost evaluation, and resource allocation/reservation, the object delivery scheme should maintain a set of static and/or dynamic state data, termed *metadata*, about system resources (i.e., objects, nodes, and links). The metadata can be either maintained at a central location in the DSMS network, or distributed among the network nodes. At one extreme case, the metadata are fully distributed (or *decentralized*) so that each node only stores the metadata relevant to the local resources (i.e., the node itself, the adjacent links, and the objects stored at that node). At the other extreme, a central management point becomes both a performance bottleneck and a single point of failure. However, full decentralization of the metadata results in optimal scalability and robustness for the object delivery scheme. This is because management load is shared to the extreme, and impact of each failure is reduced to the minimum (only metadata relevant to the locality of a certain failed node or link may be unavailable). Meanwhile, with decentralization, management of the *metadata*, on its own, may add to the response time and resource requirements of the resource management system; hence, metadata decentralization possibly results in higher start-up latency and resource management overhead (particularly CPU-time and bandwidth overhead, but not memory overhead).

However, the increase in start-up latency and resource management overhead attributed to the decentralization of metadata is negligible as compared to the duration and resource requirements of SM objects. Therefore, we design our delivery scheme assuming fully decentralized metadata maintenance to achieve optimal scalability and robustness. In this case, each node executes a set of three *distributed* algorithms to perform object location, path selection, and resource reservation tasks. To meet the other two (less important) objectives, namely start-up latency and resource management overhead, many distributed resource management systems allow *replication* of the metadata besides decentralization; i.e., there might be several replicas of the same metadata located in various positions in the system. Replication potentially facilitates access to the metadata; hence, it decreases overall response time and resource requirements of the resource management system. However, since metadata may be dynamically updated, replication introduces the consistency issue, which either decreases the accuracy of the resource management system, or, if addressed, results in additional overhead. With our design, we use task *collapsion* instead of metadata replication. Our object delivery scheme collapses the three tasks required for object delivery into a single delivery mechanism that performs the tasks integrated. As compared to serial execution of the three distributed algorithms, integrated execution helps minimizing the start-up latency. Moreover, the collapsion mechanism uses multi-purpose communication (e.g., single packet sent for object location and path cost evaluation at the same time) to reduce the resource management overhead (particularly CPU-time and bandwidth overhead due to decentralization).

In brief, decentralization of the mechanism meets the higher priority scalability and robustness objectives optimally, whereas collapsion of the tasks helps satisfy the start-up latency and resource management overhead constraints. Next, we focus on describing the *functionality* of the object delivery mechanism developed based on this approach, and explain how it achieves its main objective, i.e., high resource utilization.

Mechanism

Our object delivery scheme runs as an application layer middleware on a DSMS network. The comprehensive set of assumptions that we make with respect to the DSMS network is as follows:

- The DSMS network has a RedHi topology.

- Nodes of the DSMS network are CSMSs (possibly heterogeneous), each with limited streaming bandwidth and storage space (to cache some SM objects locally) known to the middleware. Nodes execute both the resource management middleware and the stream server.

- Links of the DSMS network (logical, physical, or both) are provided by some underlying telecommunication infrastructure that guarantees availability of a fixed limited bandwidth known to the middleware. Resource management control traffics and SM objects *share* the links. Streaming resources are reserved for media streams but control traffics are served based on the best-effort service model[2]. The middleware

---

[2] Since our object delivery mechanism uses *soft-state* (for the control state of the mechanism and not the metadata state of the resources) it does not necessarily assume guaranteed service without lost/delayed packets on the communication links. Therefore, optionally the resource management mechanism can use best-effort service for communication between nodes to eliminate resource management communication overhead on the DSMS network.

considers the links as point-to-point connections for duplex communication between adjacent pairs of nodes.

- We assume a simple object distribution and caching policy: the entire set of objects are located at the root(s) of the network when DSMS is initiated; as an object is served through a path from an object server node to a head-end, the object is cached at all intermediate nodes on the path based on the LRU (Least Recently Used) caching policy [164].

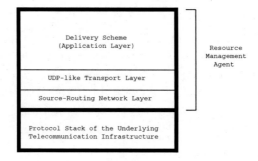

Figure 8.2. Protocol stack at a DSMS node.

Our middleware consists of a set of peer resource management agents mapped one-to-one onto the DSMS network nodes. Each agent comprises a protocol stack on top of the network links (see Figure 8.2), which makes the middleware independent of the nature of the links: the network layer supports source routing, the transport layer executes a UDP-like protocol with minimum required functionality, and the application layer implements the actual delivery mechanism. Here, we focus on the functionality of the application layer, which is the main contribution of this paper. Implementation of other layers is trivial and based on the standard approaches[3].

The agent associated with each node exclusively maintains the local metadata relevant to the node resources; i.e., objects stored in local storage, available bandwidth of the node, and available bandwidth of the links that connect the node to its child nodes. As the agent running on a head-end receives a request from a client for streaming an object, the agents running on the network nodes collectively execute a *distributed* mechanism to 1) locate all nodes of the DSMS network that have the object stored locally, 2) select one of those nodes as the object server and select the "best" path through

---

[3]Our source code that implements these protocols is available at ftp://zahedaan.usc.edu/GMeN/simulation.

Data Field	Definition	Packet Type
type	packet type	all
headend_ID	ID of the head-end which receives the request	all
sequence_no	unique sequence number assigned to the request by the head-end	all
path	path	all
object_ID	ID of the requested object	all
response	result of the object location query (REJECT, FOUND, or NOT-FOUND)	response
local_seqno	local sequence number assigned to the query packets replicated at a node	query, response
path_length	path length (number of nodes traversed in the path)	query, response
path_freeBW	amount of free streaming resources of the path	query, response

Table 8.1. Data fields of the packets.

the DSMS network for streaming the object from the object server to the head-end so that overall utilization of the streaming resources of the network (i.e., nodes and links bandwidth) is optimized by load balancing, and 3) reserve required streaming resources along the selected path so that the server can start streaming the object to the client. Below, first we provide an overview on the life cycle of a request as served by this object delivery mechanism. Subsequently, we focus on the particular characteristics of each of the three tasks collapsed into the mechanism.

Request Life Cycle   As a head-end receives a *request* from a client to read an object (e.g., as an RTSP request [149]), its agent generates a *query* packet and propagates/floods it to the network. Propagation of the query packet is performed by *selective* flooding based on a propagation policy that determines the coverage of the propagation. Instances of the flooded query packet traverse different paths and incrementally evaluate the cost of the paths as they are forwarded from agent to agent. Some query packets reach the nodes that store a replica of the requested object locally. Those nodes send *response* packets back along the same path traversed by the corresponding query packets to report availability of the object replica as well as cost of the path to reach the replica. The agents along the path compare the responses to select and reserve the best path and filter out the rest of the responses so that finally the head-end receives the response that indicates the best path. The path is already reserved and the selected object server starts streaming the requested object as soon as it receives a *start* packet from the head-end.

Data Field	Definition
headend_ID	ID of the head-end which initiated the query
sequence_no	unique sequence number assigned to the request by the head-end
querier_ID	ID of the neighbor agent that has forwarded the privileged query
querier_localseqno	local sequence number assigned to the query packet by the querier (the neighbor)
path_length	path length (number of nodes traversed by the query packet)
path_freeBW	amount of free streaming resources of the path traversed by the query packet)
reserve_count	number of paths for which the node resources are reserved (reserved once only)
reserve_status	if local resources are reserved for this request (RESERVED or NOT-RESERVED)
status	state of the request (RESPONDED or NOT-RESPONDED)
lifetime	time-out to purge the query state

Table 8.2. Data fields of the query state.

Agents use four types of packets to communicate and to execute the object delivery algorithm: query packet, response packet, release packet, and start packet. Data fields included in these packets are illustrated in Table 8.1. As indicated in the table, the combination of the two fields sequence_no and headend_ID, existing at all types of the packets, uniquely identifies the packets generated in response to the same object request[4]. Below, we summarize functionality of the delivery algorithm by discussing how packets are processed at an agent.

Figure 8.3 shows how an agent processes a *query packet* received from a neighbor. The agent creates and maintains a *query state* for each request as it receives the first query packet (see Table 8.2 for the contents of the query state). The agent may receive several query packets about the same request from different paths at different times. It only updates the query state if the received query has traversed the path with the minimum cost up to the current time; such a query is termed the current *privileged* query. Other queries are immediately rejected. A query is interpreted as: "What is the best path to reach a replica of the requested object through you (the agent)?". As a new privileged query is received, if a replica of the requested object is available locally and local resources are also enough to

---

[4]We use a *lollipop* sequence number space [130]. The space is large enough so that the sequence number does not wrap around during the lifetime of a request.

```
while (lifetime > 0) {
    ReceiveQuery(new_query);

        if (request is already RESPONDED)
            SendResponse(new_query, REJECT);
        else if (path_length > 2× network diameter)
            SendResponse(new_query, REJECT);
        else if (the requested object is available in local storage)
                if (do not have enough local streaming resources to stream the
                    object)
                    SendResponse(new_query, REJECT);
            else {
                SendResponse(new_query, FOUND);
                if ((reserve_count == 0) {
                    ReserveLocalResources(object_ID);
                    reserve_status = RESERVED;
                }
                reserve_count++;
            }
        else // the requested object is not available in local storage
            if ((cannot forward the query to any other agent) OR
                (have already received a query packet from a path that costs
                less))
                    SendResponse(new_query, REJECT);
            else {
                if (not the first privileged query)
                    SendResponse(old_query, REJECT);
                if ((reserve_count == 0) AND
                    (RESERVATION_MODE == ERP)) {
                    ReserveLocalResources(object_ID);
                    reserve_status = RESERVED;
                    reserve_count++;
                }
                UpdateQueryState(new_query);
                ForwardQuery(new_query);
                ResetResponseTimeout();
            }
}

PurgeQueryState();
```

Figure 8.3. Processing of a query packet at an agent.

stream the object, the agent sends a response packet in reply to the query
to report 1) the path between the head-end and the local node, and 2) the
total cost of the path. Otherwise, it replicates the privileged query, updates

the path parameters in the replicated packet (i.e., `path`, `path_length`, and `path_freeBW`), and selectively floods it to other neighbors, querying them for the same object. The forwarded query packets are marked with a local sequence number, `local_seqno`, which increases each time the query state is updated with a new privileged packet. Therefore, in case multiple queries have already been forwarded, the response packets received are identifiable. Eventually, when a privileged query is replied by sending a response packet to the querier, `status` of the request is set to RESPONDED (see Figure 8.4). Those query packets received while query state is in RESPONDED mode are rejected. Also note that to damp possible lost packets, if the length of the path traversed by the received query packet is more than two times the network diameter, the query is rejected. The query state will be maintained for a duration indicated with `lifetime`, which is long enough for the delivery algorithm to end (with our experiments, we set `lifetime` to the streaming duration of the requested object).

Figure 8.4 illustrates how an agent processes the *response packets* received from neighbors. As an agent receives the first query packet and forwards it to its neighbors, it generates a *response state* to keep track of the responses from neighbors (see Table 8.3 for the contents of the response state). The agent waits to receive the response packets in reply to its query at most for an amount of time equal to `time_out`. Hence, in a way the response state is a soft state and failure to receive responses from some neighbors does not confuse the algorithm. The `time_out` value is set to 2 times the maximum RTT between the agent and a root node. If the received response packets report finding replicas of the requested object, the agent generates a response packet containing the path to reach a replica with the minimum cost, sets the `local_seqno` filed of the packet to `querier_localseqno`, and sends it back to the querier of the current privileged query. If a new privileged query is received and forwarded before response is sent to the last privileged query, the response state will be reset and the agent will wait to receive responses for the new current privileged query. In such a case, responses to the old forwarded queries are ignored because those queries are rejected as the new privileged query arrives (see Figure 8.3). As mentioned before, when the current privileged query is responded, the query state of the request is set to RESPONDED mode and the response state of the request is purged. Agents ignore the responses relevant to a request for which they do not have a response state.

With recursive execution of the processes that we have just described, agents can cumulatively locate the replicas of the requested object and select the best path to reach a replica. Using *release packets*, reservation of the

while (`time_out` > 0) AND
    (response to the current privileged query is not received from all
      addressees)) {
    $ReceiveResponse(new\_respnse)$;

    if ((local sequence number of response packet is equal to `local_seqno`
        in response state) AND
        // ignore the responses to the old privileged queries; they have
        // already been rejected
      (response $! =$ `REJECT`) AND
        // ignore `REJECT` responses
      (response $! =$ `NOT − FOUND`) AND
        // ignore `NOT − FOUND` responses
      (cost of the path advertised in the response packet is less than
       minimum cost)
        // ignore non-optimal paths
)
        $UpdateResponseState(new\_response)$;

    if ((local sequence number of response packet is not equal to
        `local_seqno` in response state) OR
      (cost of the path advertised in the response packet is more than
       minimum cost)
        $SendRelease(new\_response)$;
}
// sending the aggregated response to the current privileged query packet
if (have not received any `FOUND` response for the current privileged query) {
    $SendResponse($`NOT − FOUND`$)$;
    if (`RESERVATION_MODE` $==$ `ERP`) {
        $ReleaseLocalResources($`object_ID`$)$;
        `reserve_status` $=$ `NOT − RESERVED`;
        `reserver_count` $= 0$;
    }
} **else** {
    if (`RESERVATION_MODE` $==$ `ERO`)
        if (do not have enough local streaming resources to stream the
          object) {
            $SendRelease(new\_response)$;
            $SendResponse($`REJECT`$)$;
        }**else** {
            $ReserveLocalResources($`object_ID`$)$;
            `reserve_status` $=$ `RESERVED`;
            `reserve_count`$++$;
        }
    else // `RESERVATION_MODE` $==$ `ERP`
        $SendResponse($`FOUND`$)$;
}
`status` $=$ `RESPONDED`;
$PurgeResponseState()$;

Figure 8.4. Processing of a response packet at an agent.

Data Field	Definition
headend_ID	ID of the head-end which initiated the query
sequence_no	unique sequence number assigned to the request by the head-end
addressee_IDs	IDs of the neighbor nodes that receive the forwarded query
local_seqno	local sequence number assigned to the current privileged query packet
path	path to reach the replica advertised in the response packet
path_length	length of the path to reach the replica advertised in the response packet
path_freeBW	amount of free streaming resources of the path to reach the replica advertised in response
time_out	time-out for receiving response packets (soft-state to consider possible failures)

Table 8.3. Data fields of the response state.

best path is also performed integrated with the other two tasks. Since there are two options for the implementation of reservation, we delay discussion of this part of the mechanism until the *Resource Reservation* section. As the head-end agent receives responses from its neighbors, similar to regular agents it processes the response packets to select the best path (see Figure 8.4). Successively, it sends a source-routed *start packet* to the selected object server and requests for streaming the object through the selected path. Finally, when streaming of the object is terminated, head-end sends a release packet to the object sever through the same path. All agents along the path release the streaming resources reserved for the object as they receive and forward the release packet.

Object Location   As a routing mechanism, *flooding* is well-known for 1) tremendous robustness, by taking every possible path in parallel, 2) ability for concurrent communication with all nodes of a network, which is particularly useful in distributed applications, and 3) minimum delay, by taking the path with minimum delay among other paths. All these characteristics are highly desirable for object location. Of course, the overhead imposed by flooding is a prohibitive disadvantage that restricts its use with most distributed object processing applications. However, as we explained before, SM applications are overhead-tolerant.

Since our application object location is integrated with other tasks, we need to modify the typical flooding algorithm to adapt it for our delivery mechanism. Typically, flooding is used to inform/update nodes of a network with some piece of information. Therefore, as soon as one replica of the

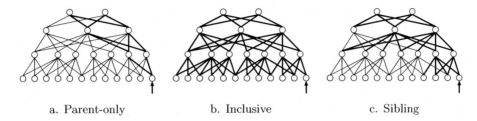

a. Parent-only   b. Inclusive   c. Sibling

Figure 8.5. Coverage of the propagation policies.

flooded packet is received by a node, the node can ignore other replicas. Since with our delivery mechanism replicas of the flooded query packet evaluate the cost of the paths that they traverse in addition to querying the nodes for the requested object, the nodes must differentiate the replicas based on their paths. When a node receives new privileged queries, it resets its query state and repeats the query process by forwarding the new query to its neighbors, even though it might have already received and forwarded query packets for the same request but from more costly paths (see Figure 8.3). Recursively, this approach allows other nodes to re-evaluate their queries based on the cost of the path introduced by the new query packet, so that progressively cost of the selected path is optimized.

Moreover, we introduce several propagation policies for *selective* flooding, to be able to control the coverage of the object location and study its effect on the resulting utilization. The wider the coverage, the nearer the resource utilization to the optimal case, because more paths and object servers are investigated. On the other hand, wide-spread object location results in higher start-up latency and resource management overhead. The three choices of propagation policy are as follows:

1. *Parent-only*: With this propagation policy, nodes only forward the queries to their parents. Therefore, the coverage of the object location is limited to the the ancestors of the head-end that receives the request for the object (see Figure 8.5a).

2. *Inclusive*: This propagation policy is equivalent to non-selective flooding, i.e., nodes forward the query packets to all their neighbors except the one from which the query is received. Therefore, all replicas of the requested object are located by covering the entire network (see Figure 8.5b).

3. *Sibling*: This propagation policy is an extension of the parent-only

policy. With this policy, a node that receives a query from one of its children forwards it to 1) the siblings of the child from which the query is received, and 2) its own parents. However, if a node receives the query from one of its parents, it does not forward the query at all (see Figure 8.5c). Sibling policy is simply one logical intermediate case, with wider coverage for object location as compared to the parent-only policy but not as wide as the full coverage of the inclusive policy.

Our experiments show that the sibling policy (as compared to the inclusive policy) results in near-optimal utilization while it imposes much less overhead and latency with its limited coverage.

Path Selection   Nodes actively cooperate to select the best path from the head-end to a replica in a distributed fashion. The query packets incrementally evaluate the cost of the path that they traverse as they are forwarded from agent to agent. Evaluation of the path cost during propagation of the query packets allows the nodes to select the best path from the head-end to the node and damp the query packets that have not traversed the best path (see Figure 8.3 for query processing). On the other hand, the response packets sent in reply to the queries also contain path cost values that enable the node to select the best path from the node to a replica (see Figure 8.4 for response processing).

Our cost function evaluates a path based on two parameters: 1) length of the path (`path_length`), and 2) amount of the free streaming resources available along the path (`path_freeBW`). To contrast costs of two paths, first they are compared on the total amount of streaming resources required to stream an object through the paths. This value is proportional to the length of the path[5]; hence, we use `path_length` as the measure for this comparison. Second, if the paths have the same length, the cost function breaks the tie condition based on the amount of free streaming resources of the paths; the more free resources are available along a path, the less cost is associated with the path, so that we can avoid blocking fairly loaded paths and balance the load among the parallel paths. We propose two measures to assess the amount of free streaming resources in a path: 1) *bottleneck*, i.e., minimum amount of free bandwidth available among the components (nodes and links) of a path, and 2) *average*, i.e., average amount of free bandwidth available at those components. The two parameters representing the cost of a path, P, are formally defined in Definition 8.2.4.

---

[5]Particularly, it is equal to $B \times (2n - 1)$, where B is the bandwidth of the object and $n =$`path_length`.

Definition 8.2.4: Cost Function Parameters Assuming $P = N \cup L$, where[6]

$$N = \{n | n \text{ is a node of the DSMS network, } n \in P\}$$
$$L = \{l | l \text{ is a link of the DSMS network, } l \in P\}$$

then

`path_length` $= \|N\|$

`path_freeBW` $= \begin{cases} \text{path\_freeBW}_{bottleneck} \\ or \\ \text{path\_freeBW}_{average} \end{cases}$

where

`path_freeBW`$_{bottleneck} = min(\{FreeBandwidth(i) | i \in P\})$

`path_freeBW`$_{average} = \dfrac{\sum_{i \in P} FreeBandwidth(i)}{\|P\|}$

With this cost function, we never select longer paths over available shorter paths. Since in DSMS networks with leveled topologies (such as RedHi or pure hierarchy) head-ends are the leaves at the lowest level, any path sourcing in higher levels should go through the lower levels to reach a head-end; therefore, sharing loads of the short paths with long paths does not necessarily result in better overall utilization. We try load balancing by considering the path traffic only when the possible path options have the same length. Our cost function can be used to evaluate the cost of the paths incrementally. As the query packet is forwarded from agent to agent, each agent updates the cost parameters of the path by considering the effect of appending the local link and node to the path. Finally, it is important to note that our resource management system, as an application layer middleware, applies the cost function to optimize utilization of the streaming resources. Therefore, the cost function does not need to consider the network level costs, such as link delay, or the actual financial cost of the resources along the path.

Since 1) with flooding we evaluate *all* possible paths restricted to the DSMS nodes explored during object location (i.e., all paths within the subnet of the network that is covered by the particular propagation policy applied), and 2) with our cost function we are able to minimize the resources allocated to respond each request (by selecting the shorter paths as well as balancing

---

[6]Similar to the bandwidth of a link, bandwidth of a node is defined as the total number of bits of information that a node can serve per time unit.

the load among the paths with the same length), we expect our path selection approach can realize the higher utilization of the RedHi topology. Our experiments verify our expectations. We have also performed experiments to compare performance of `path_freeBW`$_{bottleneck}$ with `path_freeBW`$_{average}$.

Resource Reservation   Reservation of the streaming resources by the application layer middle-ware is a bookkeeping task required to ensure load balancing as well as uninterrupted streaming. Reservation and release of the resources are performed by updating the metadata maintained locally. Generally, there are two alternative approaches for resource reservation: 1) *pessimistic*: to pre-reserve all resources that might possibly be required for the selected path during the path selection process, and to release the extra resources only after the path is actually selected, and 2) *optimistic*: to start reserving the required resources after path selection is finalized, hoping the resources are still available. With the pessimistic approach, pre-reservation guarantees availability of the required resources when the path is selected. However, with pre-reservation since the selected path is not known yet, resources are over-reserved. Therefore, concurrent requests underestimate the available streaming resources, which might result in selecting inappropriate paths. On the other hand, with the optimistic approach resources are not over-reserved, but the resources required for the selected path might have been pre-allocated to other concurrent requests.

The extreme pessimistic and optimistic approaches are not appropriate for our mechanism. The extreme pessimistic approach results in a high blocking ratio. Meanwhile, with the extreme optimistic approach reservation cannot collapse into the mechanism, but it should be performed in serial with other delivery tasks. We propose two moderated policies for resource reservation: Early Release Pessimistic (ERP), and Early Reserve Optimistic (ERO). With ERP, resources are still pre-reserved during query propagation, but unlike the extreme pessimistic approach, the extra resources are released early *during* the path selection process (and not after the path is selected), as soon as they are disqualified from being in the best path. With ERO, resources are reserved early during response processing, before path selection is finalized. Both of these policies suggest pre-reservation, but ERP is still more pessimistic as compared to ERO because with ERP reservation is performed during query propagation while ERO reserves resources during response processing.

Agents decide on pre-reserving the local resources as they receive query packets (with ERP policy) or response packets (with ERO policy). Release packets are used to release the extra resources afterwards. Figure 8.6 shows

$ReceiveRelease(new\_release)$;

if (reserve_status == RESERVED) {
    reserve_count−−;

    if (reserve_count == 0) {
        $ReleaseLocalResources$(object_ID);
        reserve_status = NOT − RESERVED;
    }
}
if (not the last hop in the **path** included in the release packet)
    $ForwardRelease(new\_release)$;

Figure 8.6. Processing of a release packet at an agent.

how an agent processes a *release packet*. A release packet is source-routed from a node towards an agent that has found a replica of the requested object stored locally and has sent a FOUND response in reply to an earlier query. The release packet releases all resources reserved for the corresponding request along the path. The agents that are included in several reserved paths reserve the resources only once and release the resources only if they receive release packets for all the paths.

Level(l)	CSMSs			Links to higher level	
	Nodes ($\mathcal{V}_{level}$)	Bandwidth (in Mb/s)	Storage (in GB)	Edges	Bandwidth (in Mb/s)
1	$\{0\}$	400	250	$\emptyset$	-
2	$\{1,2,3\}$	300	160	$\{(i,j) \mid i \in \mathcal{V}_2, j \in \mathcal{V}_1\}$	300
3	$\{4,..,12\}$	200	40	$\{(i,j) \mid i \in \mathcal{V}_3, j \in \mathcal{V}_2, j = (i-1)/3\}$	200
4	$\{13,..,39\}$	100	10	$\{(i,j) \mid i \in \mathcal{V}_4, j \in \mathcal{V}_3, j = (i-1)/3\}$	100

Table 8.4. Pure Hierarchy Configuration.

The reservation process is performed based on either ERP or ERO policy as follows (see Figures 8.3 and 8.4):

1. *Early Release Pessimistic (ERP)*: With ERP, an agent reserves the required local sources as it receives the first query packet about the request. As response packets in reply to the forwarded queries are received, the node sends release packets back along all paths from which FOUND responses are received except the best path selected at the node. If the agent does not receive any FOUND response, it releases

Level(l)	CSMSs			Links to higher level	
	Nodes ($\mathcal{V}_{level}$)	Bandwidth (in Mb/s)	Storage (in GB)	Edges	Bandwidth (in Mb/s)
1	$\{0,40\}$	200	250	$\emptyset$	-
2	$\{1,2,3\}$	300	160	$\{(i,j) \mid i \in \mathcal{V}_2, j \in \mathcal{V}_1\}$	150
3	$\{4,..,12\}$	200	40	$\{(i,j) \mid i \in \mathcal{V}_3, j \in \mathcal{V}_2,$ $j = (i-1)/3$ or $j = (i-1)/3+1,$ if $j > 3$ then $j = 1\}$	100
4	$\{13,..,39\}$	100	10	$\{(i,j) \mid i \in \mathcal{V}_4, j \in \mathcal{V}_3,$ $j = (i-1)/3$ or $j = (i-1)/3+1,$ if $j > 12$ then $j = 4\}$	50

Table 8.5. RedHi Configuration.

the local resources reserved for the request and rejects the query.

2. *Early Reserve Optimistic (ERO)*: With ERO, an agent does not reserve local resources until it receives FOUND responses in reply to its queries. Thus, the resources are only reserved when the agent actually finds a path to reach a replica of the requested object, although the path might not be selected as the best path at last. Releasing the extra resources is performed similar to ERP, by sending release packets in reply to all FOUND responses but the one that advertises the path with minimum cost. If the agent does not find enough local resources to reserve for the requested object, it releases the resources on the selected path and rejects the query.

Our experiments show that ERO results in lower blocking ratio; hence, higher utilization.

Performance Evaluation

We have extended the NS (Network Simulator) networking event simulator to incorporate our object delivery scheme[7]. We conducted several experiments via simulation to 1) evaluate performances of our proposed optional object location policies, cost functions, and resource reservation policies, and 2) verify that our object delivery scheme meets our objectives.

With our experiments, we simulated a 4-level 3-ary pure hierarchy, and a 4-level redundant hierarchy (RedHi) with $d = 2$ as its degree of redundancy (i.e., the number of parents for each node). The detailed setup and configuration parameters of these hierarchies are shown in Table 8.4 and Table

---

[7]The source code for our extension of NS is available at ftp://zahedaan.usc.edu/GMeN/simulation.

8.5, respectively. The configuration parameters are selected to estimate the capabilities of the products available in the market. Particularly, head-ends are assumed to have storage capacity and streaming bandwidth of 10 GB and 100 Mb/s, respectively, both of which are compatible with off-the-shelf PCs used as CSMS [187]. Since head-ends are numerous, to reduce cost of the DSMS, in practice low-cost nodes are used as head-ends. At higher levels (i.e., levels with less l), which have less number of nodes, available resources per node are scaled up. The links of the hierarchies have bandwidth x-multiple of 50 Mb/s to estimate SONET OC-x carriers. Fan-in of each node (total bandwidth of the links to parents) is equal to bandwidth of the node itself. Total link/node bandwidth available at each level decreases in higher levels of the hierarchies because we presume more resources are required to stream the objects cached at lower levels of the hierarchies. The total amounts of streaming resources assigned to the two hierarchies are identical; RedHi redundancy is in connectivity rather than streaming resources. Our simulator assumes delay with normal distribution[8] for links of the hierarchies. Considering the propagation delay of typical terrestrial long-haul links, the normal distribution is defined with expected value $\mu = 70ms$ and standard deviation $\sigma = \mu/5$.

Parameter	Definition	Value
T_M	Duration of an object	1 hours
S_M	Storage size of an object	0.5 GB (MPEG-4 compressed)
B_M	Bandwidth requirement of an object	1 Mb/s
N_H	Number of head-ends	27
B_H	Bandwidth of a head-end	100 Mb/s
T_S	Duration of each simulation execution (virtual)	100 hours

Table 8.6. Simulation parameters.

A static population of 500 objects (e.g., movies) is assumed. All objects are initially stored at the root node(s) of the hierarchies, from which they are distributed through the DSMS network and replicated at other nodes. We do not exclude statistics collected during warm-up period of object distribution from experiment results; we presume possible impact of warm-up is the same for all simulation executions. To capture the popularity profile of the objects, we assume Zipf distribution[9] with skew factor $\theta = 0.271$. It is shown that

---

[8] $f(x) = \frac{1}{\sigma\sqrt{2\pi}}e^{-(x-\mu)^2/2\sigma^2}$

[9] With a Zipf distribution, if the objects are sorted according to the access frequency, then the access frequency for the ith movie is given by $f_i = \frac{C}{i^{1-\theta}}$, where θ is the skew factor of the distribution and C is the normalization constant.

this distribution accurately models the popularity of rental movies [36]. All the objects are assumed to have identical specifications (see Table 8.6).

In an ideal case, optimistically a hierarchy can serve up to L_{max} requests during a simulation execution:

$$L_{max} = \frac{B_H}{B_M} \times \frac{T_S}{T_M} \times N_H$$

To emulate α percent load (i.e., $\frac{\alpha}{100} \times L_{max}$ requests) for a hierarchy, the ith head-end generates requests by discrete simulation of Poisson distribution[10] with λ_i as the expected number of requests generated during the simulation execution. λ_i parameters are randomly selected so that:

$$\Sigma_{i=1}^{N_H} \lambda_i = \frac{\alpha}{100} \times L_{max}$$

For each execution of the simulation, we evaluate resource utilization in the hierarchies based on the blocking ratio of the requests, B:

$$B = \frac{\text{Number of rejected requests during } T_S}{\text{Total number of requests generated during } T_S} \times 100$$

A request is rejected if DSMS cannot find any path with enough streaming resources to serve the request. Also, we define I_B as the improvement of the blocking ratio in RedHi as compared to the pure hierarchy:

$$I_B = B_{Pure} - B_{RedHi}$$

With our experiments, we used *inclusive* propagation with RedHi and *parent-only* propagation with the pure hierarchy. We used `path_freeBW`$_{average}$ as the cost function, and *ERO* as the reservation policy, unless stated otherwise. Besides, since the results of our experiments are often quite consistent (standard deviation is always less than 5% of the average value), we only report on the average values of the resulting quantities and exclude higher moments.

Performance of Optional Policies and Cost Functions   We proposed three propagation policies for object location, two variations of a cost function to evaluate load of the streaming paths, and two policies for resource reservation. We conducted some comparative experiments to assess relative performance of these techniques under various loads. We delay discussing the propagation policies until next section, where we study their impact on resource utilization.

---

[10]$\text{Pr}[k$ requests arrive in time interval $T_S]=\frac{\lambda_i^k}{k!}e^{-\lambda_i}$, where $\lambda_i = \mu T_S$ and μ is the mean arrival rate in request per time unit.

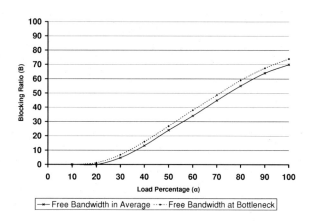

a. Cost Functions

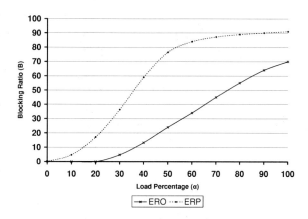

b. Reservation Policies

Figure 8.7. Comparative evaluation of the optional techniques.

path_freeBW$_{bottleneck}$ and path_freeBW$_{average}$, the two variations of our cost function, are used for load balancing between paths with identical lengths. Based on the definition, path_freeBW$_{bottleneck}$ only considers the load on the bottleneck of a path whereas path_freeBW$_{average}$ evaluates the load of the entire path. Figure 8.7a shows that `path_freeBW`$_{average}$ marginally outperforms `path_freeBW`$_{bottleneck}$ by capturing more data about the available resources along the path.

We discussed two pessimistic (ERP) and optimistic (ERO) resource reservation policies. Figure 8.7b depicts the relative performance of these policies. Since with ERP resources are vastly over-reserved at all nodes that receive at least one privileged query packet, total available resources of the DSMS network are underestimated by the object delivery mechanism. As a result of this underestimation of resources, more requests are rejected when ERP is used for reservation; hence, there is a higher blocking ratio. Although ERO outperforms ERP under the entire range of the load, since with higher loads DSMS resources exhaust, as load increases relative advantage of ERO over ERP decreases.

Resource Utilization: RedHi vs. Pure Hierarchy   In the two subsequent sections, we report on the relative resource utilization of the hierarchies assuming 1) no link failures, or 2) possible link failures, respectively.

Without Link Failures   Figure 8.8a depicts the blocking ratio of several hierarchy-setups under various loads: the pure hierarchy with parent-only propagation policy as compared to RedHi with parent-only, sibling, and inclusive propagation policies. As expected, blocking ratio is an increasing function of the load in all cases. The number of rejected requests depends on both the load (proportionally) and the available resources of the system (reciprocally). However, as the load increases available system resources decrease. Therefore, the number of rejected requests grows faster than the load; hence, increasing blocking ratio.

Blocking ratio of the pure hierarchy has the highest growth rate. As the capability of resource sharing increases by 1) employing a highly-connected topology (RedHi), and 2) applying propagation policies with wider coverage, due to statistical multiplexing of the resources the growth rate of the blocking ratio decreases; hence, a large improvement on B with the average loads. On the other hand, with higher loads since system resources are close to exhaustion, resource sharing is less beneficial. Therefore, while the pure hierarchy is already saturated, the blocking ratio with other setups grows rapidly. As a result, the offset between setups decreases such that at L_{max} the offset is marginal. If we increase the load further (of course any additional load is certainly rejected) all systems will asymptotically converge to $B = 100\%$.

Figure 8.8b depicts the relative improvement of the blocking ratio at RedHi with various propagation policies as compared to the pure hierar-

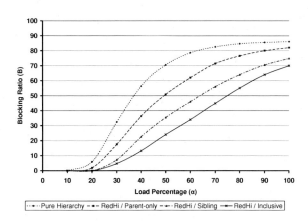

a. Various hierarchy-setups

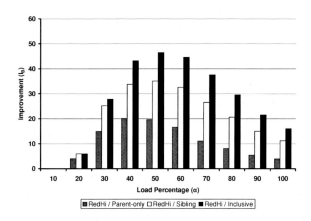

b. RedHi utilization improvement over the pure hierarchy

Figure 8.8. Resource utilization (without link failures).

chy (with the parent-only propagation policy). On the one hand, this figure shows the benefit of using the RedHi topology to achieve higher resource utilization (RedHi/parent-only setup vs. the pure hierarchy/parent-only setup). On the other hand, the advantage of applying the propagation policies with

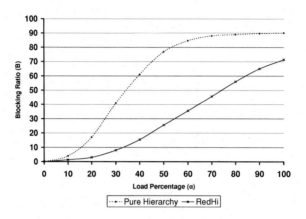

a. The pure hierarchy vs. RedHi

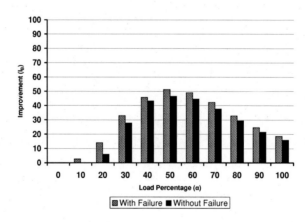

b. RedHi utilization improvement: with failure vs. without failure

Figure 8.9. Resource utilization (with link failures).

wider coverage is illustrated by observing the relative increase in improve-ment of B with RedHi/parent-only, RedHi/sibling, and RedHi/inclusive se-tups. Particularly, it is important to note that although the sibling policy has much less coverage as compared to the inclusive policy, applying this propagation policy results in fairly high improvement in B.

With Link Failures   Since the RedHi topology has a higher connectivity as compared to the pure hierarchy, we expect even higher improvement in B when some of the network links fail. Assuming links of the DSMS network are physical links (i.e., logical and physical topologies of the DSMS network are the same[11]), we simulate the link failures by disabling the links selected randomly with a Poisson arrival time distribution with $\lambda = 10$ as the expected number of link failures during T_S (i.e., $MTBF = \frac{T_S}{\lambda} = 10$ hours). Therefore, on average nearly 25% of the links in the pure hierarchy and 12.5% of the links in RedHi fail during the simulation. The $MTTR$ for each link is set to 2 hours.

Figure 8.9a depicts the blocking ratio of the RedHi/ inclusive setup as compared to the pure hierarchy/parent-only setup, both under the failure condition. The behavior of the two setups is similar to the no-failure condition (see Figure 8.8a) with an expected relative increase in B for both setups. However, as illustrated in Figure 8.9b, the improvement in B due to using RedHi/inclusive (instead of the pure hierarchy/parent-only) is more under the failure condition as compared to the no-failure condition. This growth in improvement of B can be attributed to the higher connectivity of the RedHi that results in less probability for node isolation.

Start-up Latency and Communication Overhead   We conducted some experiments to verify that the start-up latency and the communication overhead with our proposed delivery mechanism is acceptable.

Figure 8.10a depicts the average start-up latency for each request in various hierarchy setups. The start-up latency for a request is the duration of the time period between arrival time of the request and the starting time of the streaming. The average latency is calculated over all requests arrived during T_S and served successfully. As illustrated in the figure, the start-up latency reasonably grows as connectivity of the topology (from the pure hierarchy to RedHi) and the coverage of propagation policy increase. However, even for the RedHi/inclusive setup the latency is still less than 2 seconds, which as compared to the 5-minute expected user reneging time is negligible.

Figure 8.10b shows the total amount of the control data exchanged between the resource management agents in various hierarchy setups during the entire simulation (T_S). Similarly, the more the network connectivity and the propagation coverage, the more the communication overhead. Obviously, the RedHi/inclusive setup involves more communication between the

---

[11]In the case of the DSMS networks with logical links, a single physical link failure may result in multiple logical link failures.

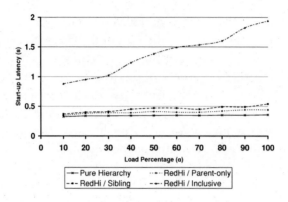

a. Average start-up latency

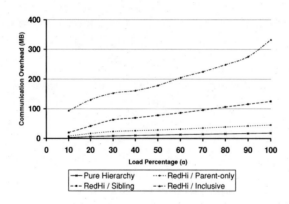

b. Average communication overhead

Figure 8.10. Start-up latency and communication overhead with the proposed delivery mechanism.

network agents and results in the largest amount of overhead. With the RedHi/inclusive setup, the communication overhead per each successfully served request amounts to:

$$\frac{350 \times 10^6 \times 8}{L_{max} \times (1 - \frac{B_{L_{max}}}{100})} = \frac{350 \times 10^6 \times 8}{27 \times 10^4 \times (1 - 0.7)} \approx 35Kb$$

Considering 1-hour duration of the objects of our simulation model, this overhead is equivalent to $\frac{35Kb}{3600s} \approx 10b/s$ per request, which is certainly negligible for a 1Mb/s bandwidth object. Even if we need to consider the bursty nature of the overhead, the worst-case 35Kb communication overhead per request is negligible as compared to the size of the typical SM objects. If the size of the objects were of the same order of the size of this overhead, we would prefer to use File Transfer (e.g., ftp) to serve the objects rather than serving them continuously.

8.3   Conclusion and Future Work

In this chapter, we described an object delivery scheme for a DSMS network assuming the RedHi topology. The unique characteristics of this scheme are as follows:

- Given the unique features of SM objects, our object delivery scheme is designed as a fully *decentralized* resource management system that guarantees scalability and robustness as well as negligible start-up latency and resource management overhead.

- Our object delivery scheme *collapses* three different resource management tasks into one *distributed* mechanism that executes all tasks integrated.

- Our object delivery scheme is designed as an *application layer* resource management middle-ware that is independent of the underlying telecommunication infrastructure of the DSMS network. It does not assume any specific capabilities (e.g., multicasting and anycasting) for the telecommunication infrastructure except for providing point-to-point communication between adjacent nodes of the DSMS network.

With our object delivery scheme, we introduced a cost function as well as various object location and path reservation policies. We performed extensive experiments via simulation to evaluate performances of these policies as well as overall performance of our object delivery scheme. Results of our experiments show that:

- *Inclusive*, `path_freeBW`$_{average}$, and *ERO* outperform other propagation, cost evaluation, and reservation techniques, respectively.

- Our delivery scheme with a RedHi DSMS network can achieve up to 50% improvement in resource utilization over a pure hierarchy.

- The start-up latency and the communication overhead per each request can be as low as 2 seconds and 10b/s, respectively.

This study can be extended in two ways. First, our object delivery scheme is designed independent of the object placement scheme applied with DSMS. Assuming object placement can be enforced by a single organization that has full control and authority over the servers and their corresponding contents, one can potentially improve the efficiency of the object delivery by intelligent distribution of the metadata. Assuming a particular object placement policy, the flood-based search for the objects can be replaced by a routing-based technique to decrease both the management overheard and start-up latency of access to the objects. Recent advent of the structured peer-to-peer networks, also called DHTs (Distributed Hash Tables), is an evidence of this research trend. Second, with our object delivery scheme we assume a fixed streaming network topology that is used to communicate both the actual content and the control data. This scenario is appropriate when we assume a single large-scale organization, such as Akamai, manages the DSMS and designs the DSMS network by locating its servers at appropriate locations at the backbone level in the Internet. However, with some other, probably more interesting, scenarios one should not assume a fixed topology; for example, with small businesses that intend to provide streaming service to their customers but cannot afford leasing high-bandwidth backbone links, or with unstructured peer-to-peer data sharing applications such as Gnutella where the management network topology cannot be effectively designed and is dynamically changing. Recent advent of QoS support techniques at the application layer, e.g., Differentiated Services and Integrated Services, paves the way to provide streaming services according to the later scenarios. Extending the techniques we introduced here, and/or investigating new management techniques to support DSMS over various network topologies other than RedHi is a challenging research problem to be addressed.

Chapter 9

SUPER-STREAMING

Up to now, we have made the following two major design decisions:

1. We strove to fix the delivery rate of SM objects to their display bandwidth requirement[1], R_c. This has been done in order to minimize the client side buffer requirement. In this chapter, we denote this paradigm as *Streaming*.

2. We assumed a client-server model, which we term a *two-level architecture* in this chapter.

Although these decisions were necessary before to focus on the design of centralized SM servers, we now believe that the field is mature enough to challenge both of these design decisions. Our argument to challenge the former is that with current technological advancements, it is not far-fetched to assume clients with large and inexpensive disk/memory storage whether their end system is a PC, a PC/TV or a set-top box. Hence, we propose a *Super-streaming* paradigm that supports the delivery of SM objects at a rate higher than their display bandwidth requirement by utilizing the client site buffer (i.e., disk and/or memory). Note that super-streaming subsumes streaming. The second decision, a two-level architecture, has been relaxed in Chapter 8.

In this chapter, we start by describing our super-streaming paradigm assuming a two-level architecture. Subsequently, we extend it to an m-level architecture such as RedHi discussed in Chapter 8. To realize the advantage of the super-streaming paradigm, consider the following simple scenario. Assume a two-level architecture where two nodes are connected to each other via an OC-3 channel (with 155 Mb/s bandwidth capacity). One node is a SM server and the other node is a client (e.g., a set-top box). Now assume

---

[1]In this chapter, we assume constant bit-rate (CBR) (i.e., the display bandwidth R_c of a single object is fixed for the entire display duration). To support variable bit-rate objects, a CBR system can employ one of the techniques described in [5].

that the client requests a two hour MPEG-2 movie with a display bandwidth requirement of 4 Mb/s. To support this client, two delivery approaches have been discussed in the literature. First, to utilize 4 Mb/s of the OC-3 channel and stream the video to the user [109]. Second, to utilize the entire 155 Mb/s of the channel and dump the video to the client's disk buffer as fast as possible [29]. The first approach has the advantage of not requiring any disk buffer at the user site (as opposed to 3.6 GB with the second approach). Instead, the second approach has the advantage of utilizing the entire link bandwidth and freeing it up after two minutes (as opposed to two hours with the first approach). Either of these approaches might be reasonable depending on the system load and client resources. In this chapter, we propose a super-streaming paradigm designed not only to cover these two extreme cases but also everything in between. In Section 9.1, we describe a super-streaming paradigm for a *two-level* architecture. In Section 9.2, we extend our discussion of super-streaming to a multi-level (*m-level*) architecture. In Section 9.3, we provide our conclusions and future research directions.

9.1   Two-Level Architecture (Client-Server Model)

In this section, we describe our super-streaming paradigm assuming a *two-level* architecture. First, we describe a technique to extend SM servers so that they can support super-streams, i.e., the objects that are delivered faster than their display rate. Next, we discuss two alternative resource management policies to assign extra server, network and client site resources to requests in order to support super-streaming. Finally, we propose two alternative admission control policies to deal with the extra resources occupied by active super-streams in order to accept or reject new requests in need of these resources.

   Figure 9.1 depicts a client server VOD system. As argued in [25], the use of client-side buffer (memory and/or disk) has a number of advantages such as:

1. Reducing the effect of delivery rate and consumption rate mismatch.

2. Making the server and network real-time requirements less stringent.

3. Allowing the server a more flexible delivery, thus improving the server load balancing.

4. Providing a more efficient support for some VCR operations.

In this section, we discuss how the flexibility in delivery (i.e., the ability to deliver SM objects at a rate higher than their consumption rate) can be exploited to improve the system throughput.

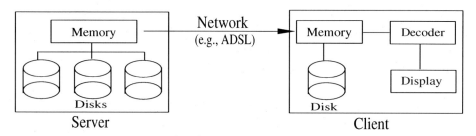

Figure 9.1. The client-server architecture.

In Figure 9.1, the client is connected to the multimedia server via a dedicated network line such as ADSL[2]. A client submits a request to the server, and subsequently, the request is either served immediately, or rejected if it requires resources that are not available. With streaming, once the request is accepted, the encoded data is delivered to the client at the rate of consumption R_c. After a temporary stage in memory, the data is decoded and then displayed.

Super-streaming, however, attempts to utilize otherwise idle resources (during off-peak periods), to expedite the delivery of SM objects hoping to make more resources available to future requests. With super-streaming, objects can be delivered at a rate higher than their display rate. Therefore, a client side buffer is required to store the data that is delivered but not yet displayed. The maximum amount of data that super-streaming can buffer at the client can be computed using Equation 9.1, where R_l is the bandwidth on the link connecting the client to the server, R_c is the object display bandwidth requirement and $Object_{size}$ is the size of the object.

$$max\_buffer\_size = \frac{(R_l - R_c) * Object_{size}}{R_l} \qquad (9.1)$$

Trivially, if the client buffer size is greater than or equal to the maximum buffer size (max_buffer_size) computed by Equation 9.1, the system can deliver the object to the client at R_l without the risk of data loss. However, if the client buffer size is less than max_buffer_size, special care is needed to avoid overflowing the client buffer. That is, the system can still initiate

---

[2]Although other types of connections are possible, ADSL appears to be the strongest candidate [16].

a super-stream with a delivery rate of R_l, but needs to downgrade it to a regular stream with a delivery rate of R_c once the buffer is full.

9.1.1   Multiple Delivery Rates

Typically, SM servers can support a single delivery rate R_c, per stored SM object. However, to support the super-streaming paradigm, SM servers are required to support higher delivery rates for those objects that are super-streamed to the client. In this section, we propose a technique to extend the SM servers to support super-streams.

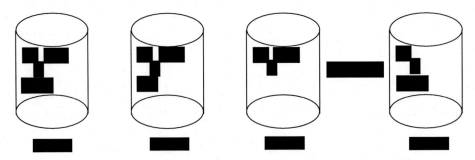

Figure 9.2. Round-robin assignment.

Centralized SM servers, such as [109], are typically multi-disk systems that employ stripping techniques to improve load balancing and throughput. To store an object X, the system partitions it into f subobjects: X_1, X_2, ..., X_f (see [69] for the computation of the size of a subobject). These subobjects are then placed on disks (each disk can itself be a cluster of disk drives) in a round-robin manner (Figure 9.2). The time to display a subobject is termed a *time interval*. To support simultaneous display of several objects, a time interval is divided into slots, with each slot corresponding to the retrieval time of a subobject from a disk. Once object X is referenced, its display employs a cycle-based approach: the server sends the first subobject of X to the user during the first time interval and its display is initiated at the beginning of the second time interval. During the second time interval, the server sends X_2 to the user which initiates its display at the beginning of the third time interval. This process is repeated until all subobjects of X have been retrieved and displayed. By employing this approach, the server is able to deliver X to the user at the rate of consumption. This system is not geared toward delivering X at any other rate. However, with super-streaming, the server may need to deliver X at three or four times R_c. To achieve this, we propose a multiple delivery rate technique.

Suppose that a server needs to deliver the object X at a rate of $2R_c$. In this case, two subobjects of X need to be delivered per time interval. An immediate solution might be to send consecutive subobjects per time interval, i.e., send X_1 and X_2 in the first time interval, X_3 and X_4 in the second time interval, and so on. However, due to round-robin assignment of the subobjects to disks, this technique will result in a complicated scheduling. An alternative technique is to partition X (logically) into two equi-size portions, (A=$X_1..X_{of-1}$) and (B=$X_{of}..X_f$) (Figure 9.3a). Subsequently, to support a rate of $2R_c$, two independent streams are initiated (S_1 starting at X_1) and (S_2 starting at X_{of}). This way, the system can follow its regular scheduling technique by treating these two as independent streams. Note that since the client will consume X at the rate of R_c, the out of sequence arrival of X_{of+i}'s subobjects will be buffered for later consumption and will not interfere with sequential consumption of X_{1+j}'s subobjects.

A scheduling problem may arise when the start of the second stream X_{of} falls in a slot that is occupied by other streams. This is because the system cannot delay the delivery of the second stream until a slot becomes available. In this case, the server can compensate by increasing the length of either the first portion or the second portion. Increasing the length of the second portion is not an option since it may result in a buffer overflow at the client side. By increasing the length of the first portion, the system shifts the start of the second stream to the first available slot. To illustrate, in Figure 9.3, the start of the second stream X_{of} falls on an occupied slot. The system compensated by shifting the start of the second portion to the subsequent subobject of X that falls on the first available slot (say X_{of+s} in Figure 9.3b).

This solution can be generalized to enable SM servers to support any multiple of R_c. That is, if a rate of nR_c is required, and the server has enough resources to support this rate, the object can be partitioned into n equi-size portions and n independent streams are initiated from the start of each of these portions.

As mentioned before, with super-streaming, if the client buffer size is not as large as max_buffer_size (see Equation 9.1) then nR_c cannot be supported for the entire delivery. In this case, the system starts nR_c delivery and then reduces the rate to R_c. This can be achieved by enforcing a larger first portion for the referenced object X and smaller $n-1$ portions. The total size of the last $n-1$ portions is hence determined by the client buffer size.

It is important to point out that the proposed solution is not only applicable to our application but can be employed by other applications such

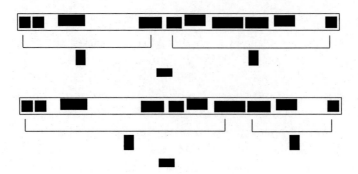

Figure 9.3. Object Partitioning.

as news-on-demand. With news-on-demand, it is essential to make news available to users as fast as possible. Therefore, the system must be able to download new news clips at $5R_c$ or $10R_c$. Since reading and writing continuous media objects can be considered as dual problems, finding a solution to one automatically solves the other. To transfer an object at a rate of nR_c, n reading streams at the sender and n writing streams at the receiver can be established.

The multiple delivery rate solution we proposed is restricted to delivering the object at nR_c where n is an integer. Supporting non-integer multiple rates is a more challenging task which we plan to tackle as our future research.

9.1.2   Resource Management

The amount of resources that can be utilized by super-streams depends on a number of factors such as: available server, link, and client bandwidths as well as the client buffer capacity. Hence, a policy is required to identify idle resources and assign them to the eligible super-streams. With our two-level architecture, each client has its own buffer and a dedicated link connecting it to the server. Hence, the only resource under competition is the server bandwidth. Therefore, the goal of the resource management policy is to maximize the utilization of this shared resource.

One policy may examine the system resources upon the arrival of a new request and assign the maximum amount of resources to this request. The request holds on to these resources (unless downgraded)[3] until it terminates. When the request terminates the resources are released. We termed this approach *Assign and Release* (*A&R*). The main drawback of this approach

---

[3]We discuss the act of getting back resources from an active request (i.e., downgrading) as part of our admission control policy in Section 9.1.3.

is that it does not utilize any resources released due to other requests downgrades or terminations (i.e., these resources will remain idle until the arrival of a new request).

An alternative approach, termed *Assign and Reassign (A&RA)*, not only assigns the maximum amount of resources to newly arriving requests but also attempts to reassign any resources made available by the downgrading or the departure of other active requests. To achieve this, the system labels requests capable of utilizing more resources than what they are actually holding as *upgradable*. The system then reacts to the release of resources by locating an upgradable request and assigning the free resources to it. The system will continue to do so until it runs out of either free resources or upgradable requests. As mentioned earlier, the goal of resource management in the two-level architecture is to maximize the utilization of the server and thus have no reason to favor one upgradable request over the others. Therefore, if more than one upgradable request exist, the system can select one at random. If the selected request is not capable of utilizing all the free resources, the system will assign the maximum amount of resources to this request and allocate the rest to another upgradable request(s). The performance of these alternative approaches is compared in Section 9.1.5 and the results demonstrate the superiority of *A&RA*.

9.1.3 Admission Control

With streaming, admission control is straightforward: when a new request arrives and the system has the required resources, then the request is accepted; otherwise it is rejected. Employing super-streaming, however, introduces new admission control alternatives that did not exist with streaming. In this section, we discuss these admission control alternatives.

With super-streaming, a newly arrived request may need resources that are utilized by some super-streams. In this case, the system may adopt one of the following approaches:

1. *Non-preemptive*: reject the new request.

2. *Preemptive*: downgrade one or more super-streams to free enough resources in order to accept the new request.

Although simple, the non-preemptive approach does not treat requests fairly. It rejects the new request, while allowing super-streams to occupy more resources than required. Therefore, employing this approach may lead to a high rejection rate.

By downgrading super-streams, the preemptive approach is attempting to achieve fair treatment of requests. However, this approach is more challenging. With this approach, the admission control policy needs to decide: 1) if downgrading one or more super-streams will free enough resources to admit the new request, and 2) which super-streams to downgrade in the presence of multiple candidates. Employing $A\&RA$ resource management within the two-level architecture ensures that no resources remain free in the existence of an upgradable request. Therefore, the admission control policy can locate a super-stream at random and downgrade it to free enough resources to admit the new request. This will continue until the request is admitted or the system runs out of super-streams, thus causing the rejection of the new request. Note that the admission control policy does not actually start downgrading super-streams until it ensures that these downgrades will lead to the admission of the new request. It is also important to point out that when a super-stream is downgraded to a regular stream, it maintains the minimum amount of resources needed to sustain the quality of display expected by the user. For the remainder of this section, preemptive admission control policy is assumed. However, the performance of the two policies will be compared within our m-level architecture in Section 9.2.

9.1.4   Analytical Model

To compare the performance of super-streaming with that of streaming, we developed an analytical model. We made the following simplifying assumptions to develop our model (these assumptions are relaxed in the simulation study):

1. All objects stored in the system have identical size and consumption rate R_c.

2. The delivery ratio

$$r = \frac{R_l}{R_c} \tag{9.2}$$

   is an integer.

3. Client side buffer size is greater than or equal to the object size and its transfer rate is greater than or equal to R_l.

4. Poisson inter-arrival and exponential display duration are assumed for requests.

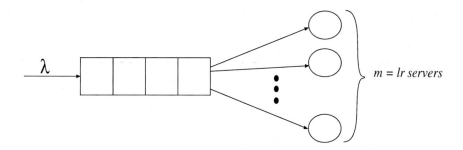

Figure 9.4. M/M/m queue.

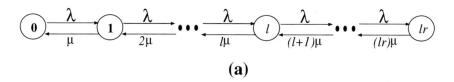

(a)

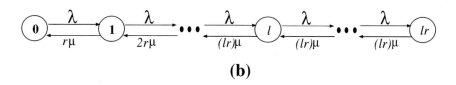

(b)

Figure 9.5. M/M/m queue.

The two-level architecture shown in Figure 9.1 can be modeled after the M/M/m queuing system shown in Figure 9.4. The multimedia server is represented by the single-queue multi-server system with m=lr server each of which is capable of supporting R_c. With Streaming, the data is delivered at the consumption rate. Therefore, each accepted request is assigned to one and only one of the m servers in the M/M/m system. This is reflected in the state diagram shown in Figure 9.5a. The system has maximum queue length $= 0$ and thus only $lr+1$ states (request are either accepted or rejected but never queued). The two main performance criteria we are going to investigate in this section are the probability of rejecting a request P_{rej} and the server utilization. Using the state diagram in Figure 9.5a, both performance measures can be calculated as follows:

$$p_1 = (\frac{\lambda}{\mu}) \times p_0$$

$$p_2 = (\frac{\lambda}{\mu})^2 \times \frac{1}{2!}p_0$$

$$p_k = (\frac{\lambda}{\mu})^k \times \frac{1}{k!}p_0 \quad (0 < k \leq lr)$$

since,

$$\sum_{k=0}^{lr} p_k = 1$$

$\Rightarrow$

$$\sum_{k=0}^{lr} (\frac{\lambda}{\mu})^k \times \frac{1}{k!}p_0 = 1$$

$\Rightarrow$

$$p_0 = [\sum_{k=0}^{lr} (\frac{\lambda}{\mu})^k \times \frac{1}{k!}]^{-1}$$

So the system utilization is:

$$\rho = \sum_{k=0}^{lr} p_k \times \frac{k}{lr} = (\frac{\lambda}{\mu})(\frac{1}{lr}) \sum_{k=0}^{lr-1} (\frac{\lambda}{\mu})^k \times \frac{1}{k!}p_0 \qquad (9.3)$$

Since the system starts rejecting requests only when the maximum number of concurrent requests m=lr is reached, the probability of rejection equals the probability of state lr:

$$P_{rej} = P_{lr} = (\frac{\lambda}{\mu})^{lr} \times (\frac{1}{(lr)!})P_0 \qquad (9.4)$$

However, with $A\&RA$ (Section 9.1.2), requests are assigned the maximum amount of resources possible. Since the client side buffer is assumed to be greater or equal to the object size, the first l request will be assigned the lr servers in the M/M/m system. When the $l+1$ request arrives, the *preemptive* admission control policy will downgrade one of the active requests and reassign some of its resources to the new request. This will continue until the maximum number of concurrent requests m is reached (state lr in Figure 9.5b). The behavior of $A\&RA$ is demonstrated by the state diagram in Figure 9.5b. Note that $A\&RA$ reassigns resources of departing requests.

Therefore, the server will continue to operate at its maximum capacity as long as there are l or more requests in the system. The probability of rejection and the utilization of super-streaming can be calculated as follows:

$$p_k = (\frac{\lambda}{r\mu})^k \times \frac{1}{k!}p_0 \quad (0 < k \leq l)$$

$$p_k = (\frac{\lambda}{lr\mu})^{k-l}p_l \quad (l < k \leq lr)$$

$$\sum_{k=0}^{lr} p_k = 1$$

$\Rightarrow$

$$\sum_{k=0}^{l}(\frac{\lambda}{r\mu})^k \times \frac{1}{k!}P_0 + \sum_{k=1}^{lr-l}(\frac{\lambda}{lr\mu})^k \times \frac{1}{l!}(\frac{\lambda}{r\mu})^l P_0 = 1$$

$\Rightarrow$

$$P_0 = [\sum_{k=0}^{l}(\frac{\lambda}{r\mu})^k \times \frac{1}{k!} + \sum_{k=1}^{lr-l}(\frac{\lambda}{lr\mu})^k \times \frac{1}{l!}(\frac{\lambda}{r\mu})^l]^{-1}$$

$\Rightarrow$ So the system utilization is:

$$\rho = \sum_{k=0}^{l} p_k \times \frac{k}{l} + \sum_{k=l+1}^{lr} p_k \tag{9.5}$$

and the probability of rejection is:

$$P_{rej} = P_{lr} = (\frac{\lambda}{lr\mu})^{lr-l} \times (\frac{\lambda}{r\mu})^l \times \frac{1}{l!}p_0 \tag{9.6}$$

With this analytical model, streaming can be considered as a special case of super-streaming when $r=1$.

Trivially, the performance of the system depends on the value of r. The higher the value of r the faster it is for the multimedia server to reach its maximum delivery rate and the higher the probability that it stays at this rate. This is depicted in Table 9.1 and 9.2. As the value of r increases, the improvement in both server utilization and probability of rejection is evidence. For example, at $Load=85\%$ as r increases from 1 to 5 the rejection percentage decreases from 8.59% to 0.82%, more than a 90% improvement.

Load(%)	r=1	r = 2	r = 5	r = 10
55	0.5475	0.5500	0.5500	0.5500
60	0.5941	0.5999	0.6000	0.6000
65	0.6382	0.6495	0.6499	0.6500
60	0.6790	0.6988	0.6997	0.6998
75	0.7158	0.7467	0.7490	0.7493
80	0.7485	0.7927	0.7973	0.7979
85	0.7770	0.8354	0.8432	0.8445
90	0.8017	0.8734	0.8849	0.8870
95	0.8229	0.9057	0.9203	0.9232
100	0.8411	0.9318	0.9480	0.9512

Table 9.1. The effect of the delivery ratio r on the server utilization.

Load(%)	r=1	r = 2	r = 5	r = 10
55	0.46	0.01	0.00	0.00
60	0.98	0.02	0.00	0.00
65	1.81	0.07	0.01	0.01
60	3.00	0.19	0.04	0.03
75	4.56	0.44	0.13	0.09
80	6.44	0.91	0.34	0.26
85	8.59	1.71	0.82	0.65
90	10.92	2.95	1.68	1.45
95	13.38	4.66	3.13	2.83
100	15.89	6.82	5.20	4.88

Table 9.2. The effect of the delivery ratio r on rejection percentage.

9.1.5   Performance Evaluation

We conducted a simulation study in order to: 1) verify our analytical mode, 2) study the impact of buffer size on the performance of the system, and 3) compare the performance of our alternative resource management policies. First, we describe our simulation setup and then report the results for each case in turn.

Simulation Setup

Poisson distribution was used to simulate the requests' inter-arrival and execution periods. Each experiment was repeated multiple times with different seeds for the random number generators. The following system parameters were used: R_c=1.5 Mb/s, R_l=3.0 Mb/s and thus r=2, average object length=100 minutes, and server bandwidth= 30 Mb/s.

Load(%)	Server Utilization			Rejection(%)		
	Analytical	Simulation	Diff (%)	Analytical	Simulation	Diff (%)
55	0.5475	0.5467	0.15	0.46	0.44	4.35
60	0.5941	0.5934	0.12	0.98	0.93	5.10
65	0.6382	0.6378	0.09	1.81	1.76	2.76
60	0.6790	0.6780	0.15	3.00	2.97	1.00
75	0.7158	0.7150	0.11	4.56	4.48	1.75
80	0.7485	0.7482	0.04	6.44	6.38	0.93
85	0.7770	0.7767	0.04	8.59	8.51	0.93
90	0.8017	0.8013	0.05	10.92	10.78	1.28
95	0.8229	0.8224	0.06	13.38	13.26	0.90
100	0.8411	0.8406	0.06	15.89	15.77	0.76

Table 9.3. Simulation vs. Analytical (streaming).

Load(%)	Server Utilization			Rejection(%)		
	Analytical	Simulation	Diff (%)	Analytical	Simulation	Diff (%)
55	0.5500	0.5491	0.16	0.01	0.01	0.00
60	0.5999	0.5989	0.17	0.02	0.02	0.00
65	0.6495	0.6486	0.14	0.07	0.07	0.00
60	0.6988	0.6978	0.14	0.19	0.18	5.28
75	0.7467	0.7458	0.12	0.44	0.42	4.55
80	0.7927	0.7918	0.11	0.91	0.88	3.30
85	0.8354	0.8344	0.12	1.71	1.68	1.75
90	0.8734	0.8722	0.14	2.95	2.93	0.68
95	0.9057	0.9047	0.11	4.66	4.63	0.64
100	0.9318	0.9306	0.13	6.82	6.82	0.00

Table 9.4. Simulation vs. Analytical (super-streaming).

To create a more realistic environment, we relaxed some of the assumptions listed in Section 9.1.4 as follows:

1. Object size varies from 80 to 120 minutes (with a uniform distribution).

2. Client buffer size from 0% to 100% of the average object size.

Simulation results

The analytical model we presented is valid only under the assumptions listed in Section 9.1.4. However, to compare the performance of the two delivery approaches in a more realistic environment, we built a simulation model that enabled us to relax some of these assumptions. To examine the stability

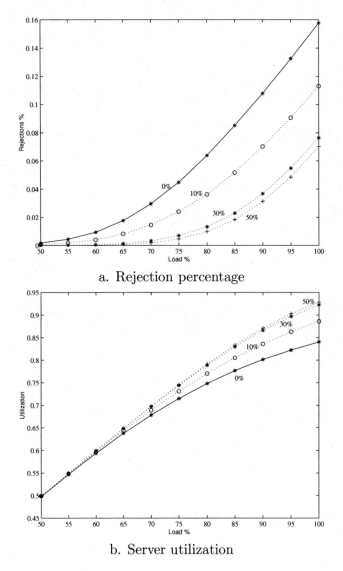

a. Rejection percentage

b. Server utilization

Figure 9.6. The impact of client buffer size.

of our simulation model we run a number of experiments with the same assumptions we used in our analytical model and compared the two results.

Tables 9.3 and 9.4 show the comparison between our analytical and simulation results. Our analytical model can be verified by the very low value of Diff(%) (the percentage difference between the two sets of results).

Assuming a client side buffer with a size greater or equal to the object size had a major impact on the performance results obtained in Section 9.1.4. To understand the effect of the client buffer size on the performance of the system, we ran a number of experiments varying the buffer size from 0% to 100% of the average object size. In Figure 9.6, the x-axis represents the system load (100% indicates that the system is operating at maximum capability). The y-axis represents the percentage of requests rejected (Figure 9.6a) or server utilization (Figure 9.6b) for the system employing streaming (with buffer size equals 0% of object size) and super-streaming (with buffer size equals 10%, 30% or 50% of object size). As expected, Figure 9.6a shows that the number of rejections is inversely proportional to the client buffer size. For example, at $Load=85\%$, as the buffer size increases from 0% (streaming) to 10% and 30%, the rejection percentage improves by 40% and 62% respectively. The client buffer size determines the amount of data that the super-stream can buffer (i.e., delivered ahead of its display). Therefore, as the client buffer size increases, the probability that idle resources are utilized by super-stream increases and thus the better performance. The corresponding increase in utilization is shown in Figure 9.6b.

We performed a number of experiments to evaluate the performance of our resource management policies. Figure 9.7 demonstrates the superiority of $A\&RA$. Utilizing the resource of departing request maximizes server utilization thus reducing the rejection percentage. For example, at $Load=85\%$, $A\&RA$ rejected 35% less request than $A\&R$. The effect of delivery ratio r was demonstrated by the analytical model. Therefore, we elected not to investigate further in this section. We, however, will revisit admission control policies in the performance evaluation of our m-level architecture (Section 9.2).

9.2   m-Level Architecture

In this section, we will generalize our two-level to a multi-level architecture. With the m-level architecture, requests are submitted to a distributed server (DS) consisting of a number of centralized SM servers connected to each other via dedicated communication links [155], [16], [123]. Upon the arrival of a new request, the system has to locate a node (server) that has the requested object and the required resources to deliver the object. To deliver the SM object, with a consumption rate R_c, from the source N_s to the user N_0, the system needs to establish a *path* (Figure 9.8).

Definition 9.2.1: A *path* is a sequence of nodes $\langle N_s, N_{s-1}..., N_0 \rangle$ where N_s is the source node (i.e., the node delivering the object O_k) and N_0 is the user node (i.e., the node displaying O_k). ∎

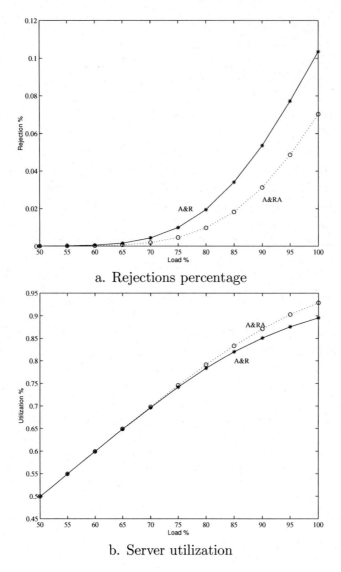

a. Rejections percentage

b. Server utilization

Figure 9.7. Resource Management Policies (50% buffer size).

The streaming policy will reserve R_c on all the links (and possibly the nodes) participating in a path. In this section, however, we describe an m-level super-streaming policy that utilizes a higher bandwidth on the links in order to expedite the object delivery. Since the display is restricted by R_c, the policy once again needs some buffer space (disk and/or memory) at the user side and/or other intermediate nodes to store the portion which is

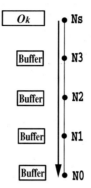

Figure 9.8. Delivery path.

delivered but not yet displayed.

Resource Management

With two-level architecture, assigning resources to streams is not a very complex task. With the m-level architecture, however, investigating the amount of resources a delivery can utilize is not trivial. Therefore, we propose a more sophisticated resource management policy. The policy is designed with the following rules of thumb in mind:

1. Over-committing resources can have a negative impact on the system performance and thus should be avoided.

2. Upward renegotiation of resources (i.e., a request demanding more resources after initiating the delivery) might fail and thus should be avoided. The request, however, can release some of the committed resources during the delivery.

3. The policy should strive to buffer as much data as possible as close as possible to the user. By buffering data at nodes close to the user, the policy can reduce the number of links needed to deliver the buffered data to the user hence reducing the overall communication cost of the delivery.

Given a path, the resource management policy goes through six steps (see Figure 9.9) before starting the delivery.

Identifying Bottleneck Nodes   Given a delivery path $\langle N_s, N_{s-1}..., N_0 \rangle$, each node N_i functions both as a receiver and transmitter of a continuous media stream (except N_s and N_0). If the receiving rate of a node is higher than

---

1. Determine if the policy is applicable.

2. Identify all the bottleneck nodes (i.e., nodes along the path, including N_0, that can be utilized to buffer data for later display).

3. Compute the size of the buffer required at each bottleneck node.

4. Compute the utilized bandwidth at each link and each bottleneck node. As a result, a subset of the bottleneck nodes will be selected as buffering nodes.

5. Generate a retrieval plan that specifies which portion of the object should be buffered at which buffering node.

6. Commit the required resources and initiate the delivery.

Figure 9.9. The Resource Management Policy.

its transmitting rate, the node is identified as a *bottleneck* node. To match the receiving rate with the transmitting rate, the bottleneck node must store the portion of the object that is received but not yet transmitted. In this section, we describe an algorithm that identifies the bottleneck nodes (step 2 of Figure 9.9). A by-product of this algorithm is to examine the applicability of the policy (step 1 of Figure 9.9). We start by defining some terms.

Definition 9.2.2: *Flow-out* of node N_i, $f_{out}(N_i)$, is the amount of available bandwidth on the link connecting N_i to its child (N_{i-1}) in a selected path. ∎

Definition 9.2.3: *Usable flow* of node N_i is the maximum available bandwidth from N_s to N_i or,

$$f_{usable}(N_i) = Min_{j=i+1}^{s} f_{out}(N_j) \tag{9.7}$$

∎

Definition 9.2.4: N_i is a *bottleneck* node if $f_{usable}(N_i) > f_{out}(N_i)$. ∎

The policy needs to identify all the nodes N_i with $f_{usable}(N_i)$ exceeding $f_{out}(N_i)$. N_s can transfer to N_i more data than what N_i can send out to nodes bellow it in the path (N_j where $j < i$). These nodes are identified as bottleneck nodes that can be utilized by the policy to buffer part of the object.

A naive algorithm may traverse the path investigating whether each node is a bottleneck node using Equation 9.7. The complexity of this approach is

current_flow $= f_{out}(N_s)$
for $i = s$ to 0 do
    $f_{usable}(N_i) =$ current_flow
    $BN(N_i) =$ false
    if current_flow $> f_{out}(N_i)$
        /* flow-in exceeds flow-out */
        $BN(N_i) =$ true
        /* node is flagged as a bottleneck */
        current_flow $= f_{out}(N_i)$
        /* flow-in for the next bottleneck */

Figure 9.10. Identifying bottleneck nodes.

$O(s^2)$. Figure 9.10, however, shows a top-down algorithm that performs this task in $O(s)$. If f_{usable} of the first bottleneck node in the path (i.e., the closest bottleneck node to N_s) equals the display rate of the object, the path does not have any extra bandwidth hence making super-streaming inapplicable.

Buffer Size Computation   In this section, we present two data flow paradigms for super-streams: *Sequential Data Flow* (SDF) and *Parallel Data Flow* (PDF). The computation of steps 3 and 4 of Figure 9.9 varies depending on the employed paradigm. We only describe the computation for PDF.

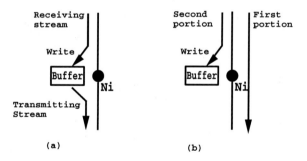

Figure 9.11. SDF vs. PDF.

The *SDF* paradigm is the intuitive way of supporting super-streams (see Figure 9.11a). The buffer is concurrently being read and written by transmitting and receiving streams, respectively. Since N_i is a bottleneck node, the transmitting rate is less than the receiving rate. The buffer size is hence a function of the difference between the two rates. The disadvantage of SDF is that N_i (the server) needs to support an aggregation of transmitting and receiving rates.

An alternative approach is to employ the PDF paradigm. To illustrate, let us first assume that there is only a single bottleneck node N_i in the path (see Figure 9.11b). In this case, PDF conceptually breaks the object into two portions. The first portion is delivered to the user directly, bypassing the bottleneck node. The second portion of the object is written to the buffer simultaneously with the delivery of the first portion. By careful computation of the size of each portion of the object, PDF can guarantee that the second portion becomes entirely resident in the buffer prior to the completion of the delivery of the first portion. Consequently, the flow can continue with no interruptions by delivering the second portion of the object to the user from the buffer. Trivially, the buffer size is equal to the size of the second portion and is computed by modifying Equation 9.1 as:

$$buffer\_size = \frac{(f_{achievable}(N_i) - f_{out}(N_i)) * O_k.size}{f_{achievable}(N_i)} \tag{9.8}$$

where $f_{achievable}$ is the maximum rate that N_i can handle and is dominated by either the rate of the incoming link or the server bandwidth $(SR_{bw}(N_i))$. That is,

Definition 9.2.5: *Achievable flow* of a bottleneck node N_i is the maximum amount of bandwidth that N_i can receive without data loss or,

$$f_{achievable}(N_i) = Min(f_{usable}(N_i), SR_{bw}(N_i) + f_{out}(N_i)) \tag{9.9}$$

■

The buffer size computed by Equation 9.8 is the optimal buffer size. However, this optimal buffer size might exceed the server available storage capacity $(SR_{cap}(N_i))$. In this case, PDF fixes the size of the second portion as $SR_{cap}(N_i)$ and computes the size of the first portion accordingly. Since $SR_{cap}(N_i)$ is less than the buffer size computed by Equation 9.8, the delivery of first portion completes only after the buffer is filled with the second portion. The only problem is that the second portion will now reside in the buffer for a longer duration of time. This is because the size of the first portion is now larger and hence it requires a longer time to deliver. We can remedy this situation by reducing the transmission rate of the second portion correspondingly. Due to lack of space, we will not investigate this optimization any further.

If there is more than one bottleneck node in a path, then the above procedure will be applied recursively to the first portion of the object. To illustrate, consider Figure 9.12 with path $\langle N_s, N_3, N_2, N_1, N_0 \rangle$ where N_3 and N_1 are bottleneck nodes. Assume the three consecutive portions of the object

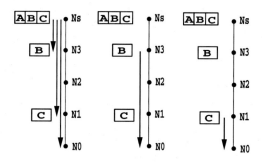

Figure 9.12. Recursive object partitioning.

Subobject_size = $O_k.size$
current_achievable_flow = $f_{usable}(N_0)$
for node = N_0 to N_s do
    $f_{achievable}(N_i)$ = current_achievable_flow
    if $BN(N_i)$ == true /* if a bottleneck node*/
        if $(f_{achievable}(N_i) - f_{out}(N_i)) > SR_{bw}(N_i)$
            $f_{achievable}(N_i) = SR_{bw}(N_i) + f_{out}(N_i)$ /* equation 3 */
        buffer_size = $[(f_{achievable}(N_i) - f_{out}(N_i))$ * Subobject_size]
            / $f_{achievable}(N_i)$ /* equation 2 */
        current_achievable_flow = $f_{achievable}(N_i)$
        if buffer_size > $SR_{cap}(N_i)$
            $SR_{achievable}N_i = SR_{cap}(N_i)$ /*buffer size to utilize */
        else $SR_{achievable}N_i$ = buffer_size
        Subobject_size = Subobject_size - node.achievable_buffer

Figure 9.13. Buffer size calculation with PDF.

are A, B and C. PDF is a bottom-up algorithm. Therefore, it starts with bottleneck N_1 and breaks the object into two portions: AB and C. AB will be scheduled to be delivered directly to the user while C will be delivered to the buffer at N_1. Subsequently, the recursion applies to AB and bottleneck N_3. That is, A is delivered directly to the user and B becomes a resident of N_3's buffer. This recursive partitioning of the object (Figure 9.12) will result in early releasing of the resources as opposed to any alternative partitioning method.

Figure 9.13 depicts a complete algorithm to compute both the required buffer size and $f_{achievable}$ per each bottleneck node in the path (step 3 and 4 in Figure 9.9) in $O(s)$. This algorithm will also generate a delivery plan (step 5 in Figure 9.9) consistent with the above mentioned recursive partitioning. After deciding on the subobjects assignment, the resources are committed and the super-stream is initiated (step 6 in Figure 9.9).

Admission Control

Unlike our two-level architecture, clients in the m-level architecture compete for a number of resources, making admission control more challenging. Once again a newly arriving request may need resources that are utilized by super-streams to buffer part(s) of objects. The system may employ a *non-preemptive* policy and reject the request or a *preemptive* policy and attempt to accept the request by downgrading other super-streams. However, investigating if downgrading one or more super-streams will free enough resources to admit the new request and deciding which super-stream to downgrade if more than one exists is not trivial.

In our performance evaluation, we have measured the performance of the system with both the *preemptive* and *non-preemptive* policies. In our implementation, when a new request arrives and the system does not have enough resources to accept it, the *preemptive* admission control policy attempts to allocate resources by downgrading the most recently admitted super-streaming delivery. Currently, we are implementing the *preemptive* admission control policy with different heuristics to select which super-stream to downgrade if more than one exists.

Multiple Delivery Rates

Applying our multiple delivery rates in the m-level architecture is slightly different. With the m-level architecture, the buffer size calculation algorithm (Section 9.2) partitions the object into n portions (not necessarily equal). When the start of portion q falls into an occupied slot, the system compensates by shifting the start of this portion to the first available slot. However, to avoid overflowing the buffer at any of the bottleneck nodes, this shift must be propagated to all the portions $q - i > 1$. Hence, the size of only the first portion is increased.

Performance Evaluation

We conducted a number of simulation experiments to obtain some insights about the performance of DS with both super-streaming and streaming. We elected to build a simple model that consist of four nodes simulating a path in DS (similar to Figure 9.8). By doing so, we were able to neutralize other issues related to DS (for more on these issues see [155]) and focus on the object delivery problem. To better evaluate super-streaming in the m-level architecture, we forced the system to deliver the object to the user at the

Node	Server Capacity(GB)	Server Bandwidth(Mb/s)	Link Bandwidth(Mb/s)
1	8	100	50
2	12	200	100
3	20	200	100
4	50	200	$\emptyset$

Table 9.5. System Parameters.

consumption rate R_c. By doing so, we can ensure that any improvement in performance is not due to client side buffering but rather due to applying super-streaming at intermediate nodes. The parameters of the model are detailed in Table 9.5.

A static object population was assumed: all objects were stored at the highest node in the path, and no new objects are inserted during the execution of the simulation. Popular objects are replicated at other nodes according to the LFU replacement policy presented in [155]. In addition, objects had random size and bandwidth requirement. A request only arrives at the head-end node (N_1 in Figure 9.8), and subsequently, it is either served immediately, or rejected if its required resources are not available.

Poisson distribution was used to simulate the requests' inter-arrival period. To simulate object selections, Zipf's law [123] was employed. Each experiment was repeated multiple times with different seeds for the random number generators. To simulate multiple requests sharing intermediate nodes and links, we devised an artificial load generator.

We will present two performance comparisons to give a flavor of the improvement achieved by applying our super-streaming paradigm. In Figure 9.14, the x-axis represents the system load (100% indicates that the system is operating at maximum capability). The y-axis represents the number of requests rejected by the system employing streaming and super-streaming (with 20% of the server's capacity used for buffering and the rest for storing regular objects). As expected, super-streaming employing a non-preemptive admission control policy rejected more request than streaming. On the other hand, super-streaming employing a preemptive admission control policy observed approximately 30% to 10% less rejections than streaming under medium system load ($30\% \leq load \leq 70\%$).

At higher system load ($85\% \leq load$), however, super-streaming starts to reject more requests. By designating some of the head-end server capacity as a buffer space, the system reduces the number of objects stored in this head-end. As a result, more requests are forced to retrieve objects from

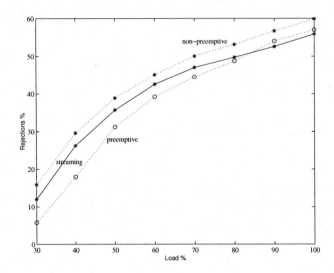

Figure 9.14. Throughput (super-streaming vs. streaming).

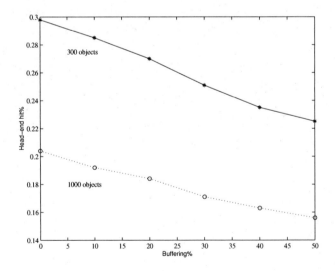

Figure 9.15. Head-end hits.

higher nodes in the path, thus increasing the overall network requirement of the system. At medium load, the improvement in performance gained by applying super-streaming compensates for this increase. At higher load, however, the system resources are approaching saturation which leaves no extra resources for super-streaming to utilize. Therefore, the number of requests rejected by super-streaming becomes higher than that of streaming.

The effect of buffer size on the number of requests served by the head-end is demonstrated in Figure 9.15. The figure shows the head-end hit percentage (the percentage of requests served by the head-end) as a function of buffering percentage (the percentage of server capacity designated for buffering) for our system storing 300 and 1000 objects. As the buffering percentage increases, the number of objects stored in the head-end decreases, thus reducing the number of requests served by it. However, the rate of reduction is lower for the system with 1000 objects since the popularity of the objects stored in the head-end is inversely proportional to the total number of objects stored in the system (Zipf's law).

9.3 Conclusion and Future Work

We proposed a super-streaming paradigm which allows SM servers to utilize resources, that are otherwise unused, to support a display. We demonstrated how this paradigm can be employed in both the two-level and the m-level architectures. Furthermore, we presented a multiple delivery rate technique that is not only essential for our paradigm but can be utilized by other applications. We developed an analytical and a simulation model that demonstrated the superiority of super-streaming over streaming. With the analytical model, we were able to show the effect of delivery ratio r on the performance of super-streaming. Moreover, with the simulation study, we demonstrated how the increase in client buffer size impacted the system performance.

With super-streaming, two alternative resource management policies were proposed. Since $A\&RA$ utilized the resource previously occupied by departing requests, it outperformed $A\&R$ (rejecting 35% less request at $Load=85\%$). In addition, we presented preemptive and non-preemptive admission control policies to provide for fair treatment of request. Our simulation result demonstrated that the preemptive admission control policy significantly outperformed the non-preemptive policy.

Study of the super-streaming paradigm can be extended in several ways. First, to optimize the performance of super-streaming, admission control policies should be equipped with new effective heuristics. Next, the multiple delivery rate technique should be improved to support non-integer multiple rates. Finally, investigating other applications with super-streaming assumed as the delivery technique opens the door to study of numerous interesting research challenges; for example, studying the effect of temporal relationships introduced by digital editing applications.

Chapter 10

YIMA CASE STUDY

10.1 Introduction

In this chapter, we report on the design, implementation and evaluation of a scalable real-time streaming architecture that would enable applications such as news-on-demand, distance learning, e-commerce, corporate training, and scientific visualization on a large scale. A growing number of applications store, maintain, and retrieve large volumes of real-time data, where the data are required to be available online. We denote these data types collectively as *"streaming media,"* or SM for short. Streaming media is distinguished from traditional textual and record-based media in two ways. First, the retrieval and display of streaming media are subject to real-time constraints. If the real-time constraints are not satisfied, the display may suffer from disruptions and delays termed *hiccups*. Second, streaming media objects are large in size. A two-hour MPEG-2 video with a 4 Megabit per second (Mb/s) bandwidth requirement is 3.6 Gigabytes (GB) in size. Popular examples of SM are video and audio objects, while less familiar examples are haptic, avatar and application coordination data [156].

The first research papers on the design of streaming media servers appeared about a decade ago (e.g., [9], [132]), followed by many papers on this topic during the past decade (here is an example for every year [58], [133], [50], [54], [24], [119], [153], [56], [103], [101], [184]). Our research during this past decade started in the early 90s [73] and continued through today [80]. Some of the projects described in these papers resulted in prototype servers, such as Streaming-RAID [175], Oracle Media Server [100], UMN system [90], Tiger [19], Fellini [110], Mitra [76] and RIO [116]. These first generation streaming media servers were primarily focused on the design of different data placement paradigms, buffer management mechanisms, and retrieval scheduling techniques to optimize for throughput and/or startup latency time.

There are two major shortcomings with these research prototypes. First, since they were implemented concurrently during the same time frame, each one of them could not take advantage of the findings and proposed techniques of the other projects. For example, UMN, Fellini and Mitra independently proposed a seemingly identical design for their data placement technique based on round-robin assignments of blocks to disk clusters. While this approach was already superior in throughput to RAID striping (used by Streaming-RAID), it resulted in a higher worst case startup latency time. In contrast, RIO's data placement, which is based on random block assignment, is more flexible and results in the same throughput as round-robin with a shorter expected startup latency time. On the other hand, UMN's simple scheduling policy resulted in a good performance without complicating the code as opposed to the constrained scheduling policies of Mitra (round-robin), Fellini (cycle-based) and RIO (round-based). We extended UMN's scheduling (to deadline-driven) and adapted the disk cluster (in Mitra's and Fellini's terms) or *logical volume striping* (in UMN's vocabulary) storage design. We extended RIO's random data placement (to pseudo-random placement for easier bookkeeping and storage scale-up) and instead of the expensive shared-memory architecture of UMN (based on SGI's Onyx), we employed a shared-nothing approach on commodity personal computer hardware.

The second shortcoming is that almost all of these research prototypes were completed before the industry's standardizations for streaming streaming media over the IP networks. Hence, each prototype has its own proprietary media content format, client (and codec) implementation and communication/network protocol. As a matter of fact, some of these prototypes did not even focus on these aspects and never reported on their assumed network and client configurations. They mainly assumed a very fast network with constant-bit-rate media types in their corresponding research publications. Practically, these environment assumptions and the specific content type are not realistic.

Several commercial implementations of streaming media servers are now available in the marketplace. We group them into two broad categories: 1) single-node, consumer oriented systems (e.g., low-cost systems serving a limited number of users) and 2) multi-node, professional oriented systems (e.g., high-end broadcasting and dedicated video-on-demand systems). Table 10.1 lists some examples for both types. RealNetworks, Apple Computer, and Microsoft product offerings fit into the first category, while SeaChange and nCUBE offer solutions that are oriented towards the second category. Table 10.1 is a non-exhaustive list summarizing some of the more popular and recognizable industry products.

Video Server	Yima	RealNetworks, Microsoft WM, Apple QT,	SeaChange	nCUBE
	Prototype	Consumer Level	Professional Level	
Authoring suite		Yes	Yes	Yes
MPEG-1 & 2	Yes	Yes^a	Yes	Yes
MPEG-4	Yes	Yes^b	Yes	?
HDTV MPEG-2	Yes		?	?
Multi-node clusters	Yes		Yes	Yes
Multi-node fault-tolerance^c	Yes		Yes	Yes
Synchronized streams^d	Yes			
Selective retransmissions	Yes	Yes	?	?
Hardware	Std. PC	Std. PC	Std. PC	Hypercube
Software	Linux	Windows / Mac^e	Windows	Proprietary

^{a}These systems support many other codecs that are commonly used in PC/Mac environments.

^{b}Supported in Windows Media Player.

^{c}Multi-node Fault-Tolerance is defined as the capability of one node taking over from another node without replicating the data.

^{d}By synchronized streams we mean the simultaneous playback of multiple independent audio and video streams for, say, panoramic systems.

^{e}RealSystem Producer products are also available for several Unix variants.

Table 10.1. A comparison of streaming media servers. Note that Yima is a prototype system and does not achieve the refinement of the commercial solutions. However, we use it to demonstrate several advanced concepts. (A ? indicates that we were unable to find conclusive information about the item.)

Commercial systems often use proprietary technology and algorithms, therefore, their design choices and development details are not publicly available and objective comparisons are difficult to achieve[1]. Because these details are not known, it is also unclear as to how much research resulting from academic work have been incorporated into these systems. For example, although a good body of work has been developed on how to configure a video server (e.g., block size, number of disks) given an application's requirements (e.g., tolerable latency, required throughput), as reviewed in [74], there is no indication that any of the commercial products utilize these results. However, those details that we are aware of are referred to throughout this chapter when comparing them with our design choices.

In this chapter we report on our streaming media architecture called Yima[2]. The focus of our implementation has been on providing a high-

---

[1]One notable exception is Apple Computer, Inc., which has published the source code of its Darwin Streaming Server.

[2]Yima, in ancient Iranian religion, is the first man, the progenitor of the human race,

performance, scalable system that incorporates the latest research results and is fully compatible with open industry standards.

Specifically, some of the features of the Yima architecture are as follows. It is designed as a completely distributed system (with no single point of failure or bottleneck) on a multi-node, multi-disk platform. It separates physical disks (used to store the data) from the concept of logical disks (used for retrieval scheduling) to support fault tolerance [189] and heterogeneous disk subsystems[3] [188]. Data blocks are pseudo-randomly placed on all nodes and the non-deterministic scheduling is performed locally on each node. Yima includes a method to reorganize data blocks in real-time for online adding/removing disk drives [80] (Section 10.4). Also included are a flexible rate-control mechanism between clients and the server to support both variable and constant bit rate media types (see Section 10.3) as well as optimization techniques such as Super-streaming [154]. There are also techniques to resolve stream contentions at the server for ensuring inter-stream synchronization as proposed in [157].

Yima follows open industry standards as proposed by the Internet Streaming Media Alliance (ISMA; www.ism-alliance.org). Content-wise, Yima supports the MPEG-4 file format, MP4, and can stream MPEG-1, MPEG-2, MPEG-4 and Quicktime video formats. It supports the RTP and RTSP communication standards for IP based networks. The clients can be off-the-shelf QuickTime players as well as our own proprietary clients [160] that handle advanced multi-channel audio and video playback and HDTV displays. Although our proprietary clients can support specialized playback of multiple media types, they still follow communication and format standards.

A general description of a Yima server and its various clients are provided in [160]. In that paper, we also compared our initial architectural design with the new fully distributed design. In [187] we reported on our experiments in streaming MPEG-4 content from a Yima server located at the USC's campus to a residential location connected via ADSL [187]. In [80], we proposed alternative schemes to redistribute media blocks when disks are added to or removed from the Yima server. These schemes were then compared via a simulation study.

In this chapter, however, we describe the details of our fully distributed architecture and report on several experiments we conducted in real-world settings. We show when different components of a node (i.e., CPU, network and disk) become a bottleneck. We have also implemented the superior

---

and son of the sun.

[3]Note that our concept of *logical* disks is very different from the concept of logical volume striping used in the UMN system.

redistribution scheme [80], Randomized SCADDAR, in Yima and evaluated it in these real-world settings. Finally, we describe our novel client-controlled transmission rate smoothing protocol and explain its implementation in Yima. Several experiments have also been conducted to evaluate our smoothing protocol.

The remainder of this chapter is organized as follows. After describing the overall system architecture in Section 10.2 we will elaborate in greater detail on three innovative aspects of Yima: 1) a client-controlled transmission rate smoothing protocol to support variable bit rate media (Section 10.3), 2) efficient online scalability of storage where disks can be added or removed without stream interruption (Section 10.4), and 3) the scalable multi-node architecture with independent, local scheduling and data transmission (Section 10.5). Finally, in Section 10.6, we summarize this chapter and lay out our future plans in this area.

10.2 System Architecture

There are two basic techniques to assign the data blocks of a media object, in a load balanced manner, to the magnetic disk drives that form the storage system: in a *round-robin* sequence [12], or in a *random* manner [144]. Traditionally, the round-robin placement utilizes a cycle-based approach for scheduling of resources to guarantee a continuous display, while the random placement utilizes a deadline-driven approach. In general, the round-robin approach provides high throughput with little wasted bandwidth for video objects that are retrieved sequentially. This approach can employ optimized disk scheduling algorithms (such as *elevator* [152]) and object replication and request migration [66] techniques to reduce the inherently high startup latency. The random approach allows for fewer optimizations to be applied, potentially resulting in less throughput. However, there are several benefits that outweigh this drawback, as described in [145], such as 1) support for multiple delivery rates with a single server block size, 2) support for interactive applications, and 3) support for data reorganization during disk scaling.

One potential disadvantage of random data placement is the need for a large amount of metadata: the location of each block must be stored and managed in a centralized repository (e.g., tuples of the form $\langle node_x, disk_y \rangle$). Yima avoids this overhead by utilizing a *pseudo-random* block placement. With pseudo-random number generators, a seed value initiates a sequence of random numbers which can be reproduced by using the same seed. File objects are split into fixed-size blocks and each block is assigned to a random

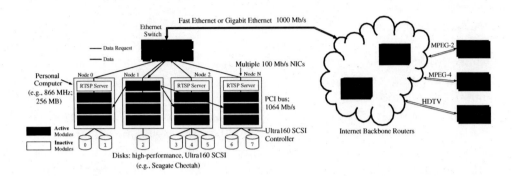

Figure 10.1. The Yima multi-node hardware architecture. Each node is based on a standard PC and connects to one or more disk drives and the network.

disk. Block retrieval is similar. Hence, Yima needs to store only the seed for each file object, instead of locations for every block, to compute the random number sequence.

The design of Yima is based on a *bipartite* model. From a client's viewpoint, the scheduler, the RTSP and the RTP server modules are all centralized on a single master node. Yima expands on decentralization by keeping only the RTSP module centralized (again from the client's viewpoint) and parallelizing the scheduling and RTP functions as shown in Figure 10.1. Hence, every node retrieves, schedules, and sends data blocks that are stored locally directly to the requesting client, thereby eliminating a potential bottleneck caused by routing all data through a single node. The elimination of this bottleneck and the distribution of the scheduler reduces the inter-node traffic to only control related messages, which is orders of magnitude less than the streaming data traffic. The term "bipartite" relates to the two groups, a server group and a client group (in the general case of multiple clients), such that data flows only between the groups and not between members of a group. Although the advantages of the bipartite design are clear, its realization introduces several new challenges. First, since clients are receiving data from multiple servers, a global order of all packets per session needs to be imposed and communication between the client and servers needs to be carefully designed. Second, an RTSP server node needs to be maintained for client requests along with a distributed scheduler and RTP server for each node. Lastly, a flow control mechanism is needed to prevent client buffer overflow or starvation.

Each client maintains contact with one RTSP module for the duration of a session to relay control related information (such as PAUSE and RESUME commands). A session is defined as a complete RTSP transaction for a streaming media stream, starting with the DESCRIBE and PLAY commands and ending with a TEARDOWN command. When a client requests a data stream using RTSP, it is directed to a server node running an RTSP module. For load-balancing purposes each server node may run an RTSP module. For each client, the decision of which RTSP server to contact can be based on either a round-robin DNS or a load-balancing switch. Moreover, if an RTSP server fails, sessions are not lost — instead they are reassigned to another RTSP server and the delivery of data is not interrupted.

In order to avoid bursty traffic and to accommodate variable bitrate media, the client sends slowdown or speedup signals to adjust the data transmission rate from Yima. By periodically sending these signals to the Yima server, the client can receive a smooth flow of data by monitoring the amount of data in its buffer. If the amount of buffer data decreases (or increases), the client will issue speedup (or slowdown) requests. Thus, the amount of buffer data can remain close to constant to support consumption of variable bitrate media. This mechanism will complicate the server scheduler logic, but bursty traffic is greatly reduced as shown in Section 10.3.

10.3   Variable Bitrate Smoothing

Many popular compression algorithms use variable bitrate (VBR) media stream encoding. VBR algorithms allocate more bits per time to complex parts of a stream and less bits to simple parts to keep the visual and aural quality at near constant levels. For example, an action sequence in a movie may require more bits per second than the credits that are displayed at the end. As a result, different transmission rates may be required over the length of a media stream to avoid starvation or overflow of the client buffer. As a contradictory requirement we would like to minimize the variability of the data transmitted through a network. High variability produces uneven resource utilization and may lead to congestion and exacerbate display disruptions.

To achieve scalability and high resource utilization, both at the server and client sides, it was our goal to reduce the variability of the transmitted data. We designed and implemented a novel technique that adjusts the multimedia traffic based on an end-to-end rate control mechanism in conjunction with an intelligent buffer management scheme. Unlike previous studies [114], [92], we consider multiple signaling thresholds and adaptively predict the future bandwidth requirements. With this *Multi-Threshold Flow Control* (MTFC) scheme, VBR streams are accommodated without a priori knowledge of the stream bitrate. Furthermore, because the MTFC algorithm encompasses server, network and clients, it adapts itself to changing network conditions. Display disruptions are minimized even with few client resources (e.g., a small buffer size).

10.3.1 Approach

MTFC is an adaptive technique that works in a real-world, dynamic environment with minimal prior knowledge of the multimedia streams needing to be served. It is designed to satisfy the following desirable characteristics:

- **Online operation**: This is required for live streaming and is also desirable for stored streams.

- **Content independence**: An algorithm that is not tied to any particular encoding technique will continue to work when new compression algorithms are introduced.

- **Minimizing feedback control signaling**: The overhead of online signaling should be negligible to compete with offline methods that do not need any signaling.

- **Rate smoothing**: The peak data rate as well as the number of rate changes should be lowered compared with the original, unsmoothed stream. This will improve network transmissions and resource utilization.

The MTFC algorithm incorporates the following: 1) a multi-threshold client buffer model, 2) a consumption prediction model, and 3) a rate change Δr computation algorithm. The three components work together as follows. The client playout buffer incorporates multiple thresholds. When the data level sufficiently deviates from the buffer midpoint, a correction message is sent to the server to adjust the sending rate. The sending rate change Δr is computed based on the value of the threshold that was crossed and the estimated data consumption in the near future. MTFC is parameterized in many different ways. For example, thresholds can either be equi-spaced or not (e.g., exponential) and we have tested several consumption prediction algorithms. The full details of the MTFC can be found in [186].

10.3.2 Performance Evaluation

MTFC is fully implemented in our architecture and we have tested its effectiveness in both LAN and WAN environments. The following is a subset of the test results. Figure 10.2a visually illustrates the effectiveness of our

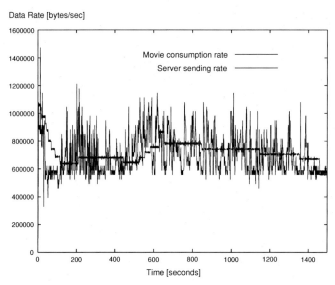

Figure 10.2a: Real consumption rate versus smoothed server sending rate for a 25 minute segment of a typical movie. The smoothing parameters used are as follows: 32 MB playout buffer size, 17 thresholds and 180 seconds prediction window size.

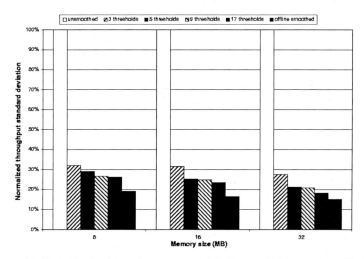

Figure 10.2b: Reduction in rate variability of the movie *Twister* with different client buffer sizes and number of thresholds. The transmission schedule becomes smoother as the number of thresholds increases and also with increased buffer size.

Figure 10.2. Smoothing effectiveness of the MTFC algorithm with the movie *Twister*.

MTFC algorithm. Shown are the unsmoothed and the smoothed transmission rate of the movie *Twister* with a playout buffer size $B = 32$ MB, $m = 17$ thresholds and a consumption prediction window size $w_{pred} = 180$ seconds. The variability is clearly reduced as well as the peak rate.

To objectively quantify the effectiveness of our technique with many different parameter sets, we measured the standard deviation of the transmission schedule. Figure 10.2b presents the reduction in standard deviation achieved by MTFC, across client buffer sizes ranging from 8 MB to 32 MB and with the number of thresholds ranging from 3 to 17. In all cases, the standard deviation is reduced substantially, by 28-42% for a 8 MB client buffer, 30-47% for a 16 MB client buffer, and 38-59% for a 32 MB client buffer. This figure illustrates what is intuitively clear: an increase in the client buffer size yields smoother traffic. More importantly, it also shows that a higher number of thresholds results in smoother traffic.

10.4   Data Reorganization

An important goal for any computer cluster is to achieve a balanced load distribution across all the nodes. Both round-robin and random data placement techniques, described in Section 10.2, distribute data retrievals evenly across all disks drives over time. A new challenge arises when additional nodes are added to a server in order to a) increase the overall storage capacity of the system, or b) increase the aggregate disk bandwidth to support more concurrent streams. If the existing data are not redistributed then the system evolves into a *partitioned* server where each retrieval request will impose a load on only part of the cluster. One might try to even out the load by storing new media files on only the added nodes (or at least skewed towards these nodes), but because future client retrieval patterns are usually not precisely known this solution becomes ineffective.

Redistributing the data across all the disks (new and old) will once again yield a balanced load on all nodes. Reorganizing blocks placed in a random manner requires much less overhead when compared to redistributing blocks placed using a constrained placement technique. For example, with round-robin striping, when adding or removing a disk, almost all the data blocks need to be relocated. More precisely, z_j blocks will move to the old and new disks and $B - z_j$ blocks will stay put as defined in Eq. 10.1:

$$z_j = B - \frac{B-1}{LCM} \times N_{j-1} - \begin{cases} (B-1 \bmod N_{j-1}) + 1 & \text{if } (B-1 \bmod LCM) < N_{j-1} \\ N_{j-1} & \text{otherwise} \end{cases}$$

$$(10.1)$$

where B is the total number of object blocks, D_j is the number of disks upon the jth scaling operation, and LCM is the least common multiple of D_{j-1} and D_j. Instead, with a randomized placement, only a fraction of all data blocks need to be moved. To illustrate, increasing a storage system from four to five disks requires that only 20% of the data from each disk be moved to the new disk. Assuming a well-performing pseudo-random number generator, z_j blocks move to the newly added disk(s) and $B - z_j$ blocks stay on their disks where z_j is approximated as:

$$z_j \approx \frac{|N_j - N_{j-1}|}{\max(N_j, N_{j-1})} \times B \qquad (10.2)$$

z_j is approximated since it is the theoretical percentage of block moves. z_j should be observed when using a 32-bit pseudo-random number generator and the number of disks and blocks increases. The number of block moves is simulated in Figure 10.3a as disks are scaled one-by-one from 1 to 200. Similarly for disk removal, only the data blocks on the disk that are scheduled for removal must be relocated. Redistribution with random placement must ensure that data blocks are still randomly placed after disk scaling (adding or removing disks) to preserve the load balance.

As described in Section 10.2, we use a pseudo-random number generator to place data blocks across all the disks of Yima. Each data block is assigned a random number, X, where $X \bmod D$ is the disk location and D is the number of disks. If the number of disks does not change, then we are able to reproduce the random location of each block. However, upon disk scaling, the number of disks changes and some blocks need to be moved to different disks in order to maintain a balanced load. Blocks that are now on different disks cannot be located using the previous random number sequence; instead, a new random sequence must be derived from the previous one. In Yima we use a composition of random functions to determine this new sequence. Our approach, termed SCAling Disks for Data Arranged Randomly (SCADDAR), preserves the pseudo-random properties of the sequence and results in minimal block movements after disk scaling while it computes the new locations for data blocks with little overhead [80]. This algorithm can support scaling of disks while Yima is online through either an eager or lazy method; the eager method uses a separate process to move blocks, while the lazy method moves blocks as they are accessed.

We call the new random numbers, which accurately reflect the new block locations after disk scaling, X_j, where j is the scaling operation (j is 0 for the initial random numbers generated by the pseudo-random number generator). We wish to map X_0 to X_j for every block at every scaling operation, j, such

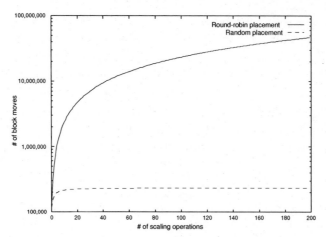

Figure 10.3a: Block movement (72 GB disks, 256 KB blocks).

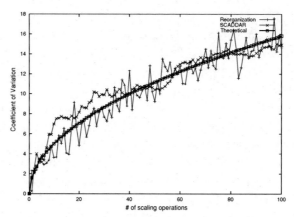

Figure 10.3b: Coefficient of Variation.

Figure 10.3. Block movement and coefficient of variation.

that $X_j \bmod N_j$ results in the new block locations. We have formulated Eqs. 10.3 and 10.4 to compute the new X_j's for an addition of disks and a removal of disks, respectively:

$$X_j = \begin{cases} p\_r(X_{j-1}) \times N_j + r_{j-1} & \text{if } (p\_r(X_{j-1}) \bmod N_j) < N_{j-1} \text{ (a)} \\ p\_r(X_{j-1}) \times N_j \\ \quad + (p\_r(X_{j-1}) \bmod N_j) & \text{otherwise (b)} \end{cases}$$

$$(10.3)$$

$$X_j = \begin{cases} p\_r(X_{j-1}) \times N_j + \text{new}(r_{j-1}) & \text{if } r_{j-1} \text{ is not removed (a)} \\ p\_r(X_{j-1}) & \text{otherwise (b)} \end{cases} \quad (10.4)$$

In order to find X_j (and ultimately the block location at $X_j \bmod N_j$) we must first compute $X_0, X_1, X_2, \ldots, X_{j-1}$. Eqs. 10.3 and 10.4 compute X_j by using X_{j-1} as the seed to the pseudo-random number generator so we can ensure that X_j and X_{j-1} are independent. We base this on the assumption that the pseudo-random function performs well.

We can achieve our goal of a load balanced storage system where similar loads are imposed on all the disks on two conditions: *uniform* distribution of blocks and *random* placement of these blocks. A uniform distribution of blocks on disks means that all disks contain the same (or similar) number of blocks. So, we need a metric to show that SCADDAR achieves a balanced load. We would like to use the uniformity of the distribution as a metric, hence the standard deviation of the number of blocks across disks is a suitable choice. However, because the averages of blocks per disk will differ when scaling disks, we normalize the standard deviation and use the coefficient of variation (standard deviation divided by average) instead. One may argue that a uniform distribution may not necessarily result in a *random* distribution since, given 100 blocks, the first 10 blocks may reside on the first disk, the second 10 blocks may reside on the second disk, and so on. However, given a *perfect* pseudo-random number generator (one that outputs independent X_j's that are all uniformly distributed between 0 and R), SCADDAR is statistically indistinguishable from complete reorganization[4]. This fact can be easily proven by demonstrating a coupling between the two schemes; due to lack of space, we omit the proofs from this chapter version. However, since real life pseudo-random number generators are unlikely to be perfect, we cannot formally analyze the properties of SCADDAR so we use simulation to compare it with complete reorganization instead. Thus, SCADDAR does result in a satisfactory random distribution. For the rest of this section, we use the uniformity of block distribution as an indication of load balancing.

We performed 100 disk scaling operations with 4000 data blocks. Figure 10.3b shows the coefficient of variations of three methods when scaling from 1 to 100 disks (one-by-one). The first curve shows complete block reorganization after adding each disk. Although complete reorganization requires a great amount of block movement, the uniform distribution of blocks

---

[4]This does not mean that SCADDAR and complete reorganization will always give identical results; the two processes are identical in *distribution*.

is ideal. The second curve shows SCADDAR. This follows the trend of the first curve suggesting that the uniform distribution is maintained at a near-ideal level while minimizing block movement. The third curve shows the theoretical coefficient of variation as defined in Def. 10.4.1. The theoretical coefficient of variation is derived from the theoretical standard deviation of Bernoulli trials. The SCADDAR curve also follows a similar trend as the theoretical curve.

Definition 10.4.1: Theoretical coefficient of variation is a percentage and defined as $\sqrt{\frac{D-1}{B}} \times 100$. ∎

When adding disks to a disk array, the average disk bandwidth usage across the array will decrease after data are redistributed using SCADDAR. The resulting load balanced placement leads to more available bandwidth which can be used to support additional streams. We measure the bandwidth usage of a disk within a disk array as the array is scaled. In Figure 10.4a, we show the disk bandwidth of one disk among the array when the array size is scaled from 2 to 7 disks (one disk added every 120 seconds beginning at 130 seconds). Ten clients are each receiving 5.33 Mbps streams across the duration of the scaling operations. The bandwidth continues to decrease and follows the expected bandwidth (shown as the solid horizontal lines) as the disks are scaled. While the bandwidth usage of each disk decreases, we observe in Figure 10.4b that the total bandwidth of all disks remains fairly level and follows the expected total bandwidth (53.3 Mbps); fluctuations are due to the variable bitrate of the media.

10.5   Scalability Experiments

The goal of every cluster architecture is to achieve close to a linear performance scale-up when system resources are increased. However, achieving this goal in a real-world implementation is very challenging. In this section we present the results of two sets of experiments. First, we compared a single node server with two different network interface transmission speeds: 100 Mbps versus 1 Gbps. In the second set of experiments we scaled a server cluster from 1 to 2 to 4 nodes. We start by describing our measurement methodology. Table 10.2 lists all the terms used in this section.

10.5.1   Methodology of Measurement

One of the challenges when stress-testing a high-performance streaming media server is the potential support of a large number of clients. For a realistic test environment, these clients should not only be simulated, but rather

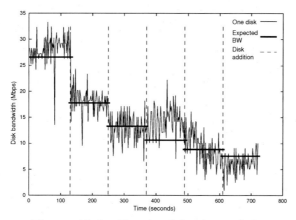

Figure 10.4a: Bandwidth of one disk.

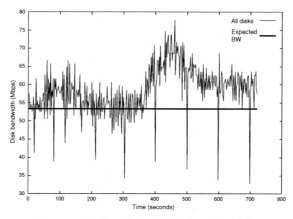

Figure 10.4b: Total disk bandwidth.

Figure 10.4. Disk bandwidth when scaling from 2 to 7 disks. One disk is added every 120 seconds (ten clients each receiving 5.33 Mbps).

should be real viewer programs that run on various machines across a network. To keep the number of client machines managable we actually ran several client programs on a single machine. Since decompressing multiple MPEG-2 compressed DVD-quality streams require a high CPU load, we changed our client software to not actually decompress the media streams. Such a client is identical to a real client in every respect, except that it does not render any video or audio. Instead, this *dummy* client consumes data according to a movie trace data file, which was the pre-recorded consumption behavior of a real client with respect to a particular movie. Thus, by chang-

Term	Definition
N	The number of concurrent clients supported by Yima server
N_{max}	The maximum number of sustainable, concurrent clients
μ_{idle}	Idle CPU in percentage
μ_{system}	System (or kernel) CPU load in percentage
μ_{user}	User CPU load in percentage
B_{avgNet}	Average network bandwidth per client (Mb/s)
B_{net}	Network bandwidth (Mb/s)
B_{disk}	The amount of movie data accessed from disk per second (termed disk bandwidth) (MB/s)
B_{cache}	The amount of movie data accessed from server cache per second (termed cache bandwidth) (MB/s)
$B_{avgNet}[i]$	The B_{avgNet} measured for i-th server node in a multinode experiment
$B_{net}[i]$	The B_{net} measured for i-th server node in a multinode experiment
$B_{disk}[i]$	The B_{disk} measured for i-th server node in a multinode experiment
$B_{cache}[i]$	The B_{cache} measured for i-th server node in a multinode experiment
$R_{\Delta r}$	The number of rate changes per second

Table 10.2. List of terms used repeatedly in this section and their respective definitions.

ing the movie trace file, each dummy client can emulate a DVD stream (5 Mbps, VBR), a HDTV stream (20 Mbps, CBR), or an MPEG-4 stream (800 Kbps, VBR). For all the experiments in this section, we chose trace data from the DVD movie *Twister* as the consumption load. The dummy client used the following smoothing parameters: 16 MB playout buffer, 9 thresholds and 90-second prediction window. For each experiment, we started clients in a staggered manner (one after the other every 2 to 3 minutes).

On the server side, we recorded the following statistics every 2 seconds: CPU load (μ_{idle}, μ_{system} and μ_{user}), disk bandwidth (B_{disk}), cache bandwidth (B_{cache}), $R_{\Delta r}$, the total network bandwidth (B_{net}) for all clients, the number of clients served, and B_{avgNet}. On the client side, the following statistics were collected every second: the stream consumption rate (this could be used as the input movie trace data for dummy clients), the stream receiving rate, the amount data in the buffer, the time and value of each rate change during that period, and the sequence number for each received data packet.

Our servers were run without any admission control policies because we wanted to find the maximum sustainable throughput using multiple client sessions. Therefore, client starvation would occur when the number of sessions increased beyond a certain point. We defined that threshold as the maximum number of sustainable, concurrent sessions. Specifically, this threshold marks the point where certain server system resources reach full utilization and become a bottleneck, such as the available network bandwidth, the disk bandwidth and the CPU load. Beyond this threshold, the server is unable to provide enough data for all clients.

First, we first describe the Yima server performance with two different network connections, and we then evaluate Yima in a cluster scale-up experiment.

10.5.2 Network Scale-up Experiments

Experimental setup: We tested a single node server with two different network connections: 100 Mb/s and 1 Gb/s Ethernet. Figure 10.1 illustrates our experimental setup. In both cases, the server consisted of a single Pentium III 933 MHz PC with 256 MB of memory. The PC is connected to an Ethernet switch (model Cabletron 6000) via a 100 Mb/s network interface for the first experiment and a 1 Gb/s network interface for the second experiment. Movies are stored on a 73 GB Seagate Cheetah disk drive (model ST373405LC). The disk is attached through an Ultra2 low-voltage differential (LVD) SCSI connection that can provide 80 MB/s throughput. RedHat Linux 7.2 is used as the operating system. The clients are based on several Pentium III 933 MHz PCs, which are connected to the same Ethernet switch via 100 Mb/s network interfaces. Each PC could support 10 concurrent MPEG-2 DVD dummy clients (5.3 Mbps for each client). The total number of client PCs involved was determined by the number of clients needed. For both experiments, dummy clients were started every 2 to 3 minutes and use the trace data file pre-recorded from a real DVD client playing the movie *Twister* (see Figure 10.2a).

Experimental results: Figure 10.5 shows the server measurement results for both sets of experiments (100 Mbps, 1 Gbps) in two columns. Figures 10.5c and 10.5d present B_{avgNet} with respect to the number of clients, N. Figure 10.5c shows that, for a 100 Mb/s network connection, B_{avgNet} remains steady (between 5.3 and 6 Mbps) when N is less than 13; after 13 clients, B_{avgNet} decreases steadily and falls below 5.3 Mbps (depicted as a dashed horizontal line), which is the average network bandwidth of our test movie. Note that the horizontal dashed line intersects with the B_{avgNet} curve at approximately 12.8 clients. Thus, we consider 12 as the maximum number of clients, N_{max}, supportable by a 100 Mb/s networking interface. An analoguous result can be observed in Figure 10.5d, indicating that $N_{max} = 35$ with a 1 Gb/s network connection.

Figures 10.5a and 10.5b show the CPU utilization as a function of N

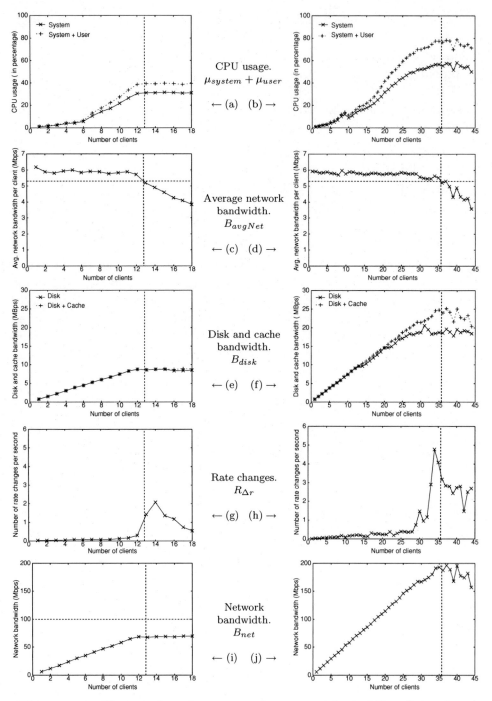

Figure 10.5. Yima single node server performance with 100 Mbps (left column) versus 1 Gbps (right column) network connection.

for 100 Mb/s and 1 Gb/s network connections. Both figures contain two
curves: μ_{system} and $\mu_{system}+\mu_{user}$. As expected, the CPU load (both μ_{system}
and μ_{user}) increases steadily as N increases. With the 100 Mb/s network
connection, the CPU load reaches its maximum at 40% with 12 clients, which
is exactly N_{max} suggested by Figure 10.5c (vertical dashed line). Similarly,
for 1Gb/s, the CPU load levels off at 80% where $N_{max} = 35$ clients. Note that
in both experiments, μ_{system} accounts for more than 2/3 of the maximum
CPU load.

Yima implements a simple yet flexible caching mechanism in the file I/O
module (Figure 10.1). Movie data are loaded from disks as blocks (e.g.,
1 MB). These blocks are organized into a shared block list maintained by
the file I/O module in memory. For each client session, there are at least
two corresponding blocks on this list. One is the block currently used for
streaming, and the other is the prefetched, next block. Some blocks may be
the shared because the same block is used by more than one client session
simultaneously. Therefore, a session counter is implemented for each block.
When a client session requests a block, the file I/O module checks the shared
block list first. If the block is found, then the corresponding block counter
will be incremented and the block made available; otherwise, the requested
block will be fetched from disk and added to the shared block list (with its
counter set to one). We define the cache bandwidth, B_{cache}, as the amount
of data accessed from the shared block list (server cache) per second.

Figures 10.5e and 10.5f show B_{disk} and B_{cache} as a function of N for 100
Mb/s and 1 Gb/s network connections. In both experiments, the $B_{disk} +$
B_{cache} curves increase linearly until N reaches its respective N_{max} (12 for
100 Mb/s and 35 for 1 Gb/s), and they level off beyond those points. For the
100 Mb/s network connection, $B_{disk} + B_{cache}$ level off at around 8.5 MBps,
which equals the 68 Mb/s peak rate, B_{net}, in Figure 10.5i with $N = N_{max}$.
Similarly, for the 1 Gb/s network connection, $B_{disk} + B_{cache}$ level off at 25
MBps, which corresponds to the 200 Mb/s maximum, B_{net}, in Figure 10.5j
with $N = N_{max} = 35$. In both cases, B_{cache} contributes little to $B_{disk}+B_{cache}$
when N is less than 15. For $N > 15$, caching becomes increasingly effective.
For example, with 1 Gb/s network connection, B_{cache} accounts for 20% of
30% to $B_{disk} + B_{cache}$ with N between 35 and 40. This is because for higher
N, the probability that the same cached block is accessed by more than one
client increases. Intuitively, caching is more effective with large N.

Figures 10.5i and 10.5j show the relationship of B_{net} and N for both
network connections. Both figures nicely complement Figures 10.5e and
10.5f. With the 100 Mbps connection, B_{net} increases steadily with respect
to N until it levels off at 68 Mbps with N_{max} (12 clients). For the 1 Gb/s

connection, the result is similar except that B_{net} levels off at 200 Mbps with $N = 35$ (N_{max} for 1 Gb/s setup). Notice that the horizontal dashed line in Figure 10.5i represents the theoretical bandwidth limit for the 100 Mb/s setup.

Figures 10.5g and 10.5h show $R_{\Delta r}$ with respect to N for 100 Mb/s and 1 Gb/s network connections. Both figures suggest a similar trend: there exists a threshold T where, if N is less than T, $R_{\Delta r}$ is quite small (approximately 1 per second); otherwise, $R_{\Delta r}$ increases significantly to 2 for 100 Mb/s connection and 5 for the 1 Gb/s connection. With the 100 Mb/s setup, T is reached at approximately 12 clients. For the 1 Gb/s case, the limit is 33 clients. In general, T roughly matches N_{max} for both experiments. Note that in both cases, for $N > T$, at some point $R_{\Delta r}$ begins to decrease. This is due to client starvation. Under these circumstances such clients send a maximum stream transmission rate change request. Because this maximum cannot be increased, no further rate changes are sent.

Note that in both the 100 Mb/s and 1 Gb/s experiments, N_{max} is reached when some system resources become a bottleneck. For the 100 Mb/s setup, Figure 10.5a and Figure 10.5e suggest that the CPU and disk bandwidth are not the bottleneck, because neither of them reaches more than 50% utilization. On the other hand, Figure 10.5i indicates that the network bandwidth, B_{net}, reaches approximately 70% utilization for $N = 12$ (N_{max} for 100 Mb/s setup), and hence is most likely the bottleneck of the system. For the 1 Gb/s experiment, Figure 10.5f and Figure 10.5j show that the disk and network bandwidth are not the bottleneck. Conversely, Figure 10.5b shows that the CPU is the bottleneck of the system because it is heavily utilized ($\mu_{system} + \mu_{user}$ is around 80%) for $N = 35$ (N_{max} for the 1 Gb/s setup).

10.5.3 Server Scale Up Experiments

Experimental setup: To evaluate the cluster scalability of Yima server, we conducted 3 sets of experiments. Figure 10.1 illustrates our experimental setup. The server cluster consists of the same type of rack-mountable Pentium III 866 MHz PCs with 256 MB of memory. We increased the number of server PCs from 1 to 2 to 4, respectively. The server PCs are connected to an Ethernet switch (model Cabletron 6000) via 100 Mb/s network interfaces. The movies are striped over several 18 GB Seagate Cheetah disk drives (model ST118202LC, one per server node). The disks are attached through an Ultra2 low-voltage differential (LVD) SCSI connection that can provide 80 MB/s throughput. RedHat Linux 7.0 is used as the operating system and the client setup is the same as in Section 10.5.2.

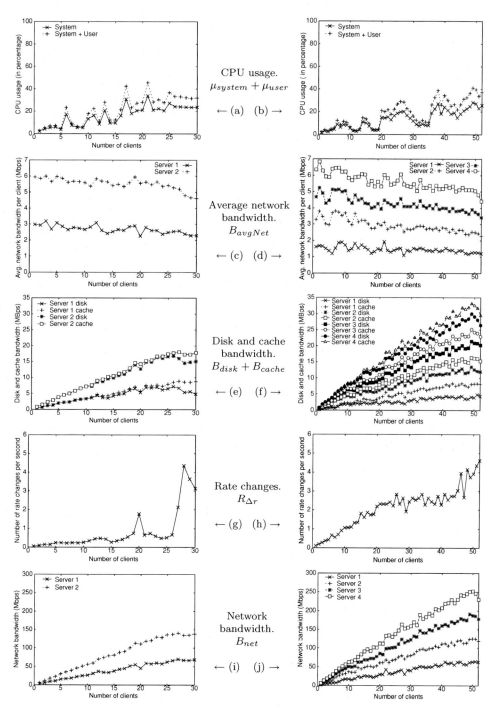

CPU usage.
$\mu_{system} + \mu_{user}$

$\leftarrow$ (a) (b) $\rightarrow$

Average network
bandwidth.
B_{avgNet}

$\leftarrow$ (c) (d) $\rightarrow$

Disk and cache
bandwidth.
$B_{disk} + B_{cache}$

$\leftarrow$ (e) (f) $\rightarrow$

Rate changes.
$R_{\Delta r}$

$\leftarrow$ (g) (h) $\rightarrow$

Network
bandwidth.
B_{net}

$\leftarrow$ (i) (j) $\rightarrow$

Figure 10.6. Yima two node (left column) versus four node (right column) server performance.

Experimental results: The results for a single node server have already been reported in Section 10.5.2. So, we will not repeat them here, but refer to them where appropriate. Figure 10.6 shows the server measurement results for the 2 node and 4 node experiments in two columns.

Figures 10.6c and 10.6d present the measured bandwidth B_{avgNet} as a function of N. Figure 10.6c shows two curves representing the two nodes: $B_{avgNet}[1]$ and $B_{avgNet}[1] + B_{avgNet}[2]$. Similarly, Figure 10.6d shows four curves: $B_{avgNet}[1]$, $B_{avgNet}[1]+B_{avgNet}[2]$, $B_{avgNet}[1]+B_{avgNet}[2]+B_{avgNet}[3]$ and $B_{avgNet}[1] + B_{avgNet}[2] + B_{avgNet}[3] + B_{avgNet}[4]$. Note that the each server node contributes roughly the same portion to the total bandwidth B_{avgNet}, i.e., 50% in case of the 2 node system and 25% for the 4 node cluster. This illustrates how well the nodes are load balanced within our architecture. Recall that the same software modules are running on every server node, and the movie data blocks are evenly distributed by the random data placement technique. Similarly to Figure 10.5c and 10.5d, the maximum number of supported clients can be derived as $N_{max} = 25$ for 2 nodes and $N_{max} = 48$ for 4 nodes. Including the previous results from 1 node (see 100 Mb/s experimental results in Figure 10.5) with 2 and 4 nodes, the maximum number of client streams N_{max} are 12, 25, and 48 respectively, which represents an almost ideal linear scale-up.

Figures 10.6a and 10.6b show the average CPU utilization on 2 and 4 server nodes as a function of N. In both figures, μ_{system} and $\mu_{system} + \mu_{user}$ are depicted as two curves with similar trends. For 2 nodes the CPU load $(\mu_{system} + \mu_{user})$ increases gradually from 3% with $N = 1$ to approximately 38% with $N = 25$ (N_{max} for this setup), and then levels off. With 4 nodes, the CPU load increases from 2% with $N = 1$ to 40% with $N = 48$ (N_{max} for this setup).

Figures 10.6e and 10.6f show $B_{disk} + B_{cache}$. The 4 curves presented in Figure 10.6e cumulatively show the disk and cache bandwidth for 2 nodes: $B_{disk}[1]$, $B_{disk}[1] + B_{cache}[1]$, $B_{disk}[2] + B_{disk}[1] + B_{cache}[1]$, and $B_{disk}[2] + B_{cache}[2] + B_{disk}[1] + B_{cache}[1]$. The curves exhibit the same trend as shown in Figure 10.5e and 10.5f for a single node. $B_{disk} + B_{cache}$ reach a peak value of 17 MB/s with $N = N_{max}$ for the 2 node setup and 32 MB/s for the 4 node experiment. Note that $B_{disk} + B_{cache}$ for 4 nodes is nearly doubled compared with 2 nodes, which is double that of the 1 node setup. In both cases, each server contributes approximately the same portion to the total of $B_{disk} + B_{cache}$, illustrating the balanced load in the Yima cluster. Furthermore, similar to Figure 10.5e and 10.5f, caching effects are more pronounced with large N in both the 2 and 4 node experiments.

Figures 10.6i and 10.6j show the achieved network throughput B_{net}.

Again, Figure 10.6i and 10.6j nicely complement Figures 10.6e and 10.6f. For example, the peak rate, B_{net}, of 136 Mb/s for the 2 node setup is equal to the 17 MB/s peak rate of $B_{disk} + B_{cache}$. Each node contributes equally to the total served network throughput.

Finally, Figures 10.6g and 10.6h show the number of rate changes, $R_{\Delta r}$, that are sent to the server cluster by all clients. Similarly to the 1 node experiment, for the 2 node server $R_{\Delta r}$ is very small (approximately 1 per second) when N is less than 26, and increases significantly above this threshold. For the 4 node experiment, a steady increase is recorded when N is less than 26; after that it remains constant at 2.5 for N between 27 and 45, and finally $R_{\Delta r}$ increases for N beyond 45. Note that for all experiments, with $N < N_{max}$, the rate change messages $R_{\Delta r}$ generate negligible network traffic and server processing load. Therefore, our MTFC smoothing technique (see Section 10.3) is well suited for a scalable cluster architecture.

Overall, the experimental results presented here demonstrate that our current architecture scales linearly to four nodes while at the same time achieves an impressive performance on each individual node. Furthermore, the load is nicely balanced and remains such, even if additional nodes or disks are added to the system (with SCADDAR). We expect that high-performance Yima systems can be built with 8 and more nodes. When higher performing CPUs are used (beyond our dated 933 MHz Pentium IIIs) each node should be able to eventually reach 300 to 700 Mb/s. With such a configuration almost any currently available network could be saturated (e.g., 8×700 Mb/s = 5.6 Gb/s effective bandwidth).

10.6   Conclusions

Yima is a second generation scalable real-time streaming architecture, which incorporates results from first generation research prototypes, and is compatible with industry standards. By utilizing a pseudo-random block assignment technique, we are able to reorganize data blocks efficiently when new disks are added or removed while in a normal mode of operation. Yima has a rate-control *mechanism* that can be utilized by a control *policy* such as Super-streaming [154], to speed-up and slowdown streams. We illustrated the usefulness of this mechanism in supporting variable-bitrate streams. Finally, the fully distributed design of Yima yields a linear scale up in throughput. We conducted several experiments to verify and evaluate the above design choices in realistic setups.

Our experiments demonstrate the graceful scale-up of Yima with SCAD-DAR as disks are added to the storage subsystem. We showed that the average load across all disks uniformly decreases as disks are added. The experiments also illustrated the effectiveness of the MTFC smoothing technique in providing a hiccup-free display of variable-bitrate streams. We showed that the technique is flexible and lightweight enough so that by increasing the number of thresholds, smoother traffic is achieved. Finally, we demonstrated that with a Yima single node setup, up to 12 streams can be supported, each with a 5.3 Mbps bandwidth requirement, before the 100 Mbps network interface becomes the bottleneck. We installed Apple Computer's Darwin Streaming Server on the same hardware and modified the software to support the same media format; we could only support 9 streams with Quicktime. The superiority of Yima is attributed to our optimized lightweight scheduler and RTP/RTSP servers.

To demonstrate that the bottleneck was indeed the network interface card (NIC) with the single node Yima server, we upgraded the NIC to a 1 Gbps model and achieved up to 35 streams on a single node server before the CPU became the bottleneck. We observed a linear scale-up in throughput with 2 node and 4 node configurations of Yima. A total of 48 streams are supported on the 4 node configuration when each server carries a 100 Mbps network card.

Work on Yima continues and we plan to extend the architecture in three ways. During 2001 we co-located a 4-node Yima server at one of our industry partner's data centers and performed various tests. Currently we have a server located across the continental United States that is connected to our campus via an Internet2 link (http://www.internet2.org/). We are using this setup to stream multi-channel audio and high-definition video content to our campus as part of a demonstration event. The experiment is called *Remote Media Immersion*. It provides us with a valuable environment to perform intensive real-world tests on Yima and also collect detailed data on hiccup and packet-loss rates as well as explore synchronization issues. Second, we plan to extend our experiments to support distributed clients from more than one Yima server. We already have four different hardware setups hosting Yima. By co-locating two Yima server clusters at off-campus locations and the other two servers in different buildings within our campus, we have an initial setup to start our distributed experiments. We have some preliminary approaches to manage distributed streaming media servers [158] that we would like to implement, experiment with and extend. Finally, we performed some studies on supporting other media types, in particular the haptic [159] data type. Our next step would be to store and stream haptic data.

Appendix A

PERSONAL EDITION SINGLE DISK SM SERVER SOFTWARE

Powered by

A.1 Introduction

Yima is a streaming media architecture composed of *server, client,* and *utility* software components. Both the server and the client are currently Linux based. The full version of Yima and its internal operation are described in Chapter 10. As part of this book a simplified version of Yima is distributed: the *Yima Personal Edition* or *Yima PE* for short.

Yima PE is missing some of the features of its big sibling. Notably absent are multi-node support and retransmission capabilities in the server component. The provided client only recognizes and plays WAVE[1] audio streams. The idea behind the Personal Edition is to have a fully functional streaming media system that can be used for experiments and teaching while at the same time keeping the source code at a manageable size. This chapter describes the installation and use of Yima PE. It is a guide on how to compile and run the various components of the system as well as how to use the different configuration parameters.

---

[1]Uncompressed, 16-bit linear PCM, 44,100 Hz stereo files are supported in the current software. This is the same format that one would typically find on an audio CD.

The Yima PE software is included on a CD-ROM with this book. For updates and other online information about Yima PE and this book please refer to the following web site:

`http://streamingmedia.usc.edu`

This guide makes several assumptions about the user's level of expertise:

- The user is familiar with the Linux operating system.

- The user has access to two Linux computers that are connected via an IP-based network (e.g., a LAN or the Internet). The software has been tested mostly on Red Hat Linux versions 7.x, however, any recent Linux distribution should work fine.

- The user has read Chapter 10; "Yima Case Study" and is familiar with some of the architectural features of Yima.

- The user has a CD-ROM or some other file archive of the software called *Yima Personal Edition*.

The rest of this chapter is organized as follows. We will start with an overview of the software components that are included with Yima PE and describe the contents and organization of the CD-ROM (or the file archive, if the user acquired the software as a `tar` or `zip` file). Subsequently, we will describe the installation and operation of both the server and the client components. Finally, we will elaborate on how to prepare new content for the system and make it available for streaming.

Please also refer to the `README` file included with the software for any changes and updates that were made after this book was printed.

A.2   Yima PE Components

Yima PE consists of the following software components:

1. A server executable called `yimaserver` to manage the storage, retrieval, and scheduling of the streaming media files. The server is also responsible for the communication with the clients. The server implementation is multi-threaded and some of its tasks include:

    - Media block storage and retrieval scheduling.
    - Media block caching.
    - RTP packetizing (see RFC 1889 [148]). The data is pre-packetized with the `yimasplit` utility (see item 3).

- RTP packet scheduling and transmission.

- RTSP communication with the clients (see RFC 2326 [149]).

2. A client executable called `yimaclient` capable of receiving and playing uncompressed (linear PCM) WAVE audio files. This software requires a computer with a soundcard that is supported under Linux. If you are having difficulty with sound support under Linux you may want to check the following web site `http://www.alsa-project.org` for more information.

3. A `yimasplit` utility to prepare media files for storage and streaming. A regular WAVE file needs to be pre-processed by this utility before the `yimaserver` process can serve the file to clients. The `yimasplit` program pre-packetizes the original file data into RTP packets and then combines the packets into fixed sized media blocks.

A.2.1 CD-ROM Contents

The Yima PE CD-ROM is organized with the following directory structure:

```
/YimaPE_v1.0
          /Server
          /Client
          /Splitter
          /Streams
```

The `Server` directory contains a tar archive with the source code of the complete Yima PE server software source tree. Section A.3 describes how to build the `yimaserver` executable.

The `Client` directory contains the software client capable of playing streamed WAVE files. The software is distributed as a tar archive that includes the complete software source tree. Section A.4 describes how to build the `yimaclient` executable.

The `Splitter` directory contains the `yimasplit` tool to make original WAVE files streamable. Again this is a tar archive with source code. See Section A.5 for installation instructions.

The `Streams` directory contains a short sample PCM audio WAVE file ("yimaintro.wav") with the following format: stereo, 16-bits/sample, 44,100 samples/second (1.4 Mb/s bandwidth). The `yimaintro.wav` file can be processed with the `yimasplit` program to test the correct operation of the installation (see Section A.5).

A.3   Server Installation and Operation

To install the Yima PE server please complete the following steps. Step 1 may be skipped if you already have a functioning Linux installation.

1. Install a recent Linux distribution on your computer and create an appropriate user account (for example "yima"). The software has been tested on Red Hat Linux 7.x. Boot the machine and login.

2. Open a terminal (shell) window.

3. Mount the CD-ROM and copy the `YimaPE_Server_v1.0.tar` source file from the CD-ROM (from the `YimaPE_v1.0/Server` directory) to the user home directory.

4. Untar the `YimaPE_Server_v1.0.tar` file. A source directory structure will be created:

   ```
   $tar xvf YimaPE_Server_v1.0.tar
   ```

5. Change to the subdirectory `YimaPE_v1.0/Server/qpthread-1.3.1`.

6. Install the `qpthread` library version 1.3.1 by following the directions contained in the `README` file. QpThread is a third-party software and is not covered by the Yima PE license.
   (See http://lin.fsid.cvut.cz/~kra/index.html for more information.)

7. Set the shell environment variable `LD_LIBRARY_PATH` to include the qpthread library directory path (refer to the man pages of the shell that you are using, e.g., bash, csh, etc., on how to do this). You will need to do this with superuser privileges so that it applies to all users.

8. Create a directory that is used to store the movie data. This directory can be located on a different path or disk, for example. We will refer to this directory as the *media directory*.

9. Change to the `YimaPE_v1.0/Server` directory in the yima user home directory.

   Note: The tar file contains a precompiled executable file of the server called `yimaserver`. If you want to rebuild the software, then type `make` in the `Server` directory.

10. Copy the `config_Sample` file to a file called `config` (while still being in the `YimaPE_v1.0/Server` directory):

    ```
    $cp config_Sample config
    ```

11. Edit the server `config` file: on line 2 change `/home/yima/Movies` to the path of the media directory that you designated on this machine (see item 8).

At this point the server is ready to run and you can test it by becoming superuser and invoking the server executable.

1. Change to become superuser:

   ```
   $su
    password:
   ```

2. Start the Yima PE server:

   ```
   $./yimaserver
   ```

The server needs superuser privileges because it listens on TCP port 554 for incoming RTSP connections from clients[2]. If everything went well then you should see:

```
<YimaPE 1.0> begin scheduler
<YimaPE 1.0> begin rtsps
```

If you see an error message like the following:

```
(RTSPS/frontend/interface): Permission denied
Abort.
(RTSPS/backend): frontend suddenly closed the link!
(RTSPS/backend/link): Success
Abort.
(RTSPS/frontend): socket suddenly closed by RTSP server!
: Success
```

then the server executable is not running with superuser privileges and the RTSP port 554 could not be opened.

If you see an error message like this:

```
./yimaserver: error while loading shared libraries:
./yimaserver: undefined symbol: __tiQ25qpthr10QpCondBase
```

then the dynamic qpthread library could not be found and loaded. Make sure that the `LD_LIBRARY_PATH` environment variable points to the location where `libqpthr.so` can be found.

---

[2]If you change this port number in both the client and server code to a number > 1024 then no superuser privileges are required. However, port 554 is the standard port for RTSP connections.

Even though the server can run at this point, we first need to prepare some media content for streaming to make it useful. See Section A.5 on how to prepare a WAVE file for streaming.

A.4   Client Installation and Operation

To install the Yima client please complete the following steps. Step 1 may be skipped if you already have a functioning Linux installation.

1. Install a recent Linux distribution on your computer and create an appropriate user account (for example "yima"). Boot the machine and log in.

2. Open a terminal (shell) window.

3. Mount the CD-ROM and copy the `YimaPE_Client_v1.0.tar` source file from the CD-ROM (from the `YimaPE_v1.0/Client` directory) to the user home directory.

4. Untar the `YimaPE_Client_v1.0.tar` file. A source directory structure will be created:

        $tar xvf YimaPE_Client_v1.0.tar

   Note: The tar file contains a pre-built executable file of the client, `yimaclient`. If you want to rebuild the software, then type `make` in the `YimaPE_v1.0/Client` directory.

   The client code makes use of a graphical user interface library called **XForms** which is a third-party software and is not covered by the Yima PE license.
   (See http://world.std.com/~xforms/ for more information.)

The client can be run at this point by executing `yimaclient`. However, we first need to prepare some media content for streaming on the server and then we will need to tell the client where to find this content. So let us continue with Section A.5 on how to prepare a WAVE file for streaming.

A.5   Media Preparation

To prepare a new media file for Yima PE do the following. First you will need to install the splitter utility from the CD-ROM. This utility called `yimasplit` is used to pre-packetize media content and to break it into fixed sized data blocks.

1. Mount the CD-ROM and copy the `YimaPE_Splitter_v1.0.tar` source file from the CD-ROM (from the `YimaPE_v1.0/Splitter` directory) to the user home directory.

2. Untar the `YimaPE_Splitter_v1.0.tar` file. A source directory structure will be created:

   ```
   $tar xvf YimaPE_Splitter_v1.0.tar
   ```

   Note: The tar file contains a pre-built executable file of the splitter, `yimasplit`. If you want to rebuild the software, then type `make` in the `YimaPE_v1.0/Splitter` directory.

3. You may want to add the `YimaPE_v1.0/Splitter` directory to your shell search path expression so that the `yimasplit` can be executed from any directory location. Refer to the man pages of your shell for instructions on how to do this.

Once the `yimasplit` utility is installed and ready we can proceed with preparing our media content.

1. Obtain a WAVE file. A short audio file is contained on the CD-ROM ("yimaintro.wav") in the `YimaPE_v1.0/Streams` directory.

2. Copy the file (e.g., `yimaintro.wav`) from the CD-ROM (or your own file) to some temporary directory. Let us assume that directory is called `/tmp`.

3. Execute the following command in the temporary directory:

   ```
   $/<path_to_yimasplit>/yimasplit yimaintro.wav 25
   ```

   You may need to specify the full path to the `yimasplit` utility as shown in the example. Alternatively, you can add the location of `yimasplit` to your shell search path and then you will be able to execute it without a path.

   The parameters of `yimasplit` are as follows: (1) the movie name ("yimaintro.wav" in the example) and (2) the movie length in seconds ("25" in the example). A subdirectory `BLOCKS` with many files will be created; these files are the media blocks. Also, a `config` file is created that contains information about the movie for the server.

4. Copy all the files in the `BLOCKS` directory into the media directory: the one that you configured in the `config` file for the Yima PE server.

5. Change to the `YimaPE_v1.0/Server` directory and append the movie information from the temporary directory to the server `config` file as follows (assuming that the temporary directory is called `tmp`):

   `$cat /tmp/config >> config`

   Now the server can access this new content called `yimaintro.wav` and can serve it to any clients. Note however that the server program `yimaserver` only reads the `config` file at startup. If any changes are made to that file then the server needs to be stopped (e.g., with control C) and restarted.

6. Finally, we need to add the new content to the clients' list of known media files. Ideally, the media name could be entered at runtime, however, currently the client reads this information from a configuration file. To update the client configuration, edit the `Yima.cfg` file that is contained in the `YimaPE_v1.0/Client` directory. Append the new content information in the following format and update the `movies: X` counter:

   `rtspX://<name_or_IP_address_of_server>/<medianame>`

   For example (X is replaced with an ascending number):

   `rtsp1://192.168.1.100/yimaintro.wav`
   `rtsp2://yima.usc.edu/yimaintro.wav`

Congratulations! The Yima PE system is now ready for streaming. Make sure that the server is running. Execute the client, click on the `[List]` button, select your content by double-clicking and then press the `[Play]` button. If everything went well, you should hear the audio of your media content after a few seconds of buffering delay.

Note that uncompressed, linear PCM WAVE files require about 1.4 Mb/s of bandwidth for their smooth transmission. Therefore, you will not be able to run this demonstration over a regular phone modem, DSL, or cable modem connection. However, if the client is extended with, for example, an MP3 or other audio decompression module so that it can accept compressed streams, then much less transmission bandwidth is required. There will be no changes necessary to the server side. Refer to Appendix C under Chapter 10 for this and other exercises.

Once again, for updates, help and other online information about Yima PE and this book please refer to the following web site:

`http://streamingmedia.usc.edu`

Appendix B

GLOSSARY

$Disk$ Price of a single magnetic disk drive, in dollars ($).

MB Price of main memory, in dollars ($) per megabyte (MByte).

MB_T Price of tape storage system, in dollars ($) per megabyte (MByte). Note: this depends on the total tape storage subsystem cost ($Cost_{Tape}$) and the database size (DB), $\$MB_T = \frac{Cost_{Tape}}{DB}$.

W Cost per megabyte of unused storage space, in dollars ($).

$\#cyl$ Number of cylinders of a disk drive.

$\#disk$ Number of physical disks per type.

$\#seek$ Number of seeks incurred per time period.

λ Requests arrival rate to the system. The number of requests per second.

μ Service rate of a server. (Note: in Chapter 8 and Chapter 9, μ refers to expected value of a random variable.

ρ System utilization, $\rho = \lambda \times s$.

ω Block retrieval time. Time to read a block from a disk including queueing time and servicing time.

σ Standard Deviation of a random variable.

θ Skew factor of the Zipf distribution.

A

$A(j, O_i)$ Location of jth block of object O_i in a multidisk platform (disk id).

$A\&R$ Assign and Release resource management policy.

$A\&RA$ Assign and Reassign resource management policy.

ADSL See xDSL.

B

b Number of blocks per region (REBECA).

B Block size, in MByte. (Note: in Chapter 8 and Chapter 9, B refers to blocking ratio of the requests.)

$\mathcal{B}_i$ Size of a fragment assigned to physical disk i, in MByte.

$\mathcal{B}^l$ Size of a logical block assigned to a logical disk, in MByte.

B_H Bandwidth of a head-end.

B_M See R_C.

$Band_{Effective}$ The effective bandwidth of the tape storage subsystem. It depends on the effectiveness of the data placement technique in reducing T_{Access}, $Band_{Effective} = \frac{O}{T_{Access} + \frac{O}{D_T}}$.

C

CBR Constant Bit Rate.

Cluster A set of disks that forms a logical unit of data storage.

CM Continuous Media (which is synonymous with Streaming Media).

CORBA Common Object Request Broker Architecture (an open, vendor-independent architecture and infrastructure that computer applications use to work together over networks).

$Cost_{Bandwidth}$ Tape bandwidth cost, in dollars ($). It depends on: 1) the bandwidth requirement of the application, $Band_{Requirement}$, 2) the effective bandwidth of a tape drive, $Band_{Effective}$, and 3) the tape drive cost, $Cost_{Drive}$, $Cost_{Bandwidth} = \left\lceil \frac{Band_{Requirement}}{Band_{Effective}} \right\rceil \times Cost_{Drive}$.

$Cost_{Cartridge}$ Tape cartridge cost, in dollars ($).

$Cost_{Storage}$ The total cost of the cartridges (media) necessary to hold the database in its entirety, in dollars ($). It depends on the database size DB, the effectiveness of the data placement technique to pack the objects onto a tape cartridge, and the tape cartridge cost $Cost_{Cartridge}$.

$Cost_{Tape}$ Tape storage cost, in dollars ($),
$Cost_{Tape} = Cost_{Storage} + Cost_{Bandwidth}$.

CP w/ track-sharing Contiguous Placement with track sharing. Places one object after the other on a tape, and hence, multiple objects may share a track.

CP w/o track-sharing Contiguous Placement without track sharing. It is a modification of CP w/ track-sharing. It does not allow two objects to share the same track.

CPS Cost Per Stream, in dollars ($).

CPS' Adjusted cost per stream (including cost for unusable space), in dollars ($).

CPU Central Processing Unit.

CSMS Centralized Streaming Media Server.

Cylinder Stack of tracks across all the surfaces at a common distance from the spindle.

D

d Distance travelled by the disk read/write head, in cylinders (tracks).

d_i^l Logical disk drive i.

d_i^p Physical disk drive i.

deadline A time limit set force to retrieve a specific block.

D Number of physical disk drives.

D^l Number of logical disk drives.

D-VHS Digital Video Home System (digital VHS).

DHT Distributed Hash Table.

$Disk_i$ Number of blocks that are disk resident on the disk clusters at time period i, $Disk_i = Min(Disk_{i-1} + Min(k, PCR), n)$.

DS Distributed Server.

DSMS Decentralized Streaming Media Server.

DVD Digital Versatile Disk is a standard for optical discs that features the same form factor as CD-ROMs but holds 4.7 GB of data and more.

E

E[L] Expected startup latency.

EDF Earliest Deadline First scheduling policy.

ERO Early Release Optimistic (a reservation protocol that avoids over-reservation).

ERP Early Release Pessimistic (a reservation protocol that avoids under-reservation).

F

$f_{achievable}$ Maximum rate that node N_i can handle.

$f_{out}(N_i)$ Flow-out of node N_i.

$f_{usable}(N_i)$ Usable flow of node N_i.

G

$\mathcal{G}_i$ Parity group i; a data stripe allocated across a set of G logical disks and protected by a parity code.

G Parity group size.

Gb/s Gigabits per second.

GB/s Gigabytes per second.

GByte Gigabyte.

GSS Group Sweeping Scheme (a disk scheduling algorithm).

H

hiccups Deviation from the real-time display requirements of a SM object, such as artifacts, disruptions, and jitters.

HDTV High Definition Television.

I

I_B Improvement of the blocking ratio in RedHi as compared to the pure hierarchy.

I-Frame MPEG intra-code picture. These frames are encoded using intra-frame compression that is similar to JPEG. Thus, they can be decoded without other frames.

$IdleTime$ The idle time of the tape storage system when multiplexed among k requests, $IdleTime = \frac{z \times size(B)}{R_C} - \sum_{i=1}^{k} T_{Reposition}(o_i) - \frac{z \times k \times size(B)}{R_T}$.

IDF Incomplete Data Flow.

IETF Internet Engineering Task Force.

ISP Internet Service Provider.

J

JPEG Joint Photographic Expert Group.

K

K Number of physical disks that constitute a logical disk.

L

l See $T_{Latency}^{Max}$.

L Disk utilization (also referred to as *load*) of an individual logical disk. (Note: in Chapter 8 and Chapter 9, L refers to set of the network links.)

$L_{avg-min}$ See $T_{Latency}^{avg-min}$.

L_{max} Maximum possible number of requests during a simulation.

LAN Local Area Network.

LBN Logical Block Number.

Local-Streaming Paradigm Streaming of SM objects from a local machine, e.g., PC or terminal.

Lossless Compression A type of compression with which the reconstructed information is mathematically equivalent to the original information, thus no loss or alteration of information occurs during the compression and the decompression processes.

Lossy Compression A type of compression that achieves very high degree of compression (compression ratios up to 200) by removing less important information.

LRU Least Recently Used (a caching algorithm that favors the objects which are accessed more recently for caching).

LT Logical-track.

M

m Number of different physical disk types.

M Total amount of memory needed to support N streams, in MByte.

MAMP Media Asset Mapping Problem.

MaxMem Maximum memory required to retrieve an object from tape using the non-multiplexing technique, $MaxMem = n - \frac{n \times Min(k, PCR)}{PCR}$.

Mb/s Megabits per second.

MB/s Megabytes per second.

MByte Megabyte.

Mem_i Number of blocks that are memory resident at time period i, $Mem_i = n - (Ter_i + Disk_i)$.

MM Multimedia.

MM-DBMS Multimedia - Database Management System.

MPEG Motion Picture Encoding Group.

MPEG-2 Moving Pictures Excerpt Group, Version 2 (a streaming media compression format).

MTBF Mean Time Between Failure (not to be confused with MTTF, Mean Time To Failure).

MTTDL Mean time to data loss.

MTTF Mean time to failure; mean lifetime of an individual, physical disk (or link) with failure rate λ.

MTTR Mean time to repair of a physical disk (or link) with repair rate μ.

MTTSL Mean time to service loss.

N

n Number of blocks that comprise an object.

n_i Number of seeks incurred because of sub-fragmentation.

$\mathcal{N}$ Number of simultaneous displays supported by a single logical disk.

N Number of simultaneous displays supported by the system (throughput). (Note: in Chapter 8 and Chapter 9, N refers to set of the network nodes.)

$N_{r/w}$ Number of read/write heads in the recording assembly.

N_s Number of blocks (segments) per track, $N_s = \frac{S_t}{S_s}$.

N_t Number of tracks per tape cartridge.

N_D Number of simultaneous displays supported by a single disk drive.

N_H Number of head-ends.

NS Network Simulator (a testbed to simulate and test networked systems and network protocols; see http://www.isi.edu/nsnam/ns/ for more information).

O

O Object size, in MBytes.

$Object(size)$ See O.

OC-3 One of the Optical Carriers with bandwidth equal to 155.52 Mbps.

OC-x Any of the Optical Carriers; higher "x" indicates higher bandwidth (for example, OC-1 = 51.85 Mbps, and OC-3 = 155.52 Mbps).

P

path_freeBW Measures availability of the bandwidth with a network path.

p_i Number of logical disks that map to physical disk type i.

p_0 Number of logical disks that map to the slowest physical disk type.

P_{rej} Probability of request rejection.

PBN Physical Block Number.

PCR Production Consumption Ratio $PCR = \frac{R_T}{R_C}$, where C is the display rate of the object.

PDF Parallel Data Flow.

Pos Tape load/unload position, i.e., beginning of the tape, Pos=beginning, or middle of the tape, Pos=midpoint.

POP Point Of Presence.

PRAM Parallel Random Access Machines/Memories.

Progressive-Download Display of the SM object starts as soon as enough of the data is buffered locally, while the remainder of the file is being retrieved and stored on the local storage (at the end of the display the entire SM object is stored locally).

Q

q The time required to render an object that consists of n blocks (from tape) disk resident, $q = \lceil \frac{n}{Min(k,PCR)} \rceil$.

QOS Quality Of Service.

R

r Delivery ratio.

round Time period during which a block per requested object is retrieved.

R Number of regions in REBECA.

R_{Active} Active region of the disk (REBECA).

R_C SM object display rate, in Mb/s.

R_D Magnetic disk drive transfer rate, in Mb/s.

R_{D_i} Transfer rate of physical disk i, in MB/s.

R_T Tape drive transfer rate, in MB/sec, which is the aggregate transfer rate of the $N_{r/w}$ heads.

$R(t)$ Reliability function.

$R(Z_i)$ Zone i transfer rate.

REBECA REgion BasEd bloCk Allocation.

RedHi Redundant Hierarchy.

Remote-Streaming Paradigm SM object's data blocks are retrieved from a remote server over a network and processed by the client as soon as they are received; data blocks are not written to the local storage (although they might be buffered). This paradigm is referred to as streaming paradigm throughout the textbook.

Rotational Latency The time required to wait for the desired data (sector) to rotate under the disk read/write head, in sec.

RR Round-robin.

RSVP Resource reSerVation Protocol.

RTCP RTP Control Protocol.

RTP Real Time Transport Protocol.

RTSP Real Time Streaming Protocol.

RTT Round Trip Time (time required for a packet to travel the network distance from a source to a destination and back).

S

s Average service time for requests, in sec.

S Storage space usable for SM, in MByte.

S_c Tape cartridge capacity, in MByte.

S_s Tape data block (segment) size, in MByte.

S_t Tape track size, in MByte, $S_t = \frac{S_c}{N_t}$.

S_M See O.

S_P Duration of a sub-period, in sec.

SCSI Small Computer System Interface standard.

SDF Sequential Data Flow.

SDSL See xDSL.

Sector The smallest storage unit on a magnetic disk drive (which commonly holds 512 bytes of user data).

Seek Time The time required to reposition the disk read/write head from the current track to the target track, in sec.

SM Streaming Media.

SR_{bw} Server bandwidth, in Mb/s.

SR_{cap} Server storage capacity, in MByte.

SSTF Shortest Seek Time First.

Startup Latency The time elapsed from when a display request arrives at the system until the onset of the display of the requested object, in sec.

Store-and-Display Paradigm SM objects are first downloaded in their entirety from a remote server, before initiating the display process.

STREAM 1 Data flow from tape storage to memory.

STREAM 2 Data flow from memory to disk storage.

STREAM 3 Data flow from disk storage to memory.

STREAM 4 Data flow from memory to the display station referencing the SM object.

Streaming Paradigm SM object's data blocks are retrieved from a remote server over a network and processed by the client as soon as they are received; data blocks are not written to the local storage (although they might be buffered).

Striping Process of dividing a body of data into blocks and spreading the data blocks across multiple disks.

T

t_{no} Number of tracks partially occupied by an object using *CP w/ sharing*, $t_{no} = \lceil \frac{O}{S_t} \rceil$.

t'_{no} Number of tracks required per object by *CP w/o sharing* and *WARP* ($t'_{no} = t_{no}$ when t_{no} is an even number, otherwise $t'_{no} = t_{no} + 1$).

T_M Duration (i.e., display time) of an object.

T_P Time period, in sec, i.e., time required to display a block (B) ($T_P = \frac{B}{R_C}$).

T_S Duration of each simulation execution.

T_{Access} Time required to access an object on a tape device, in sec, $T_{Access} = T_{Switch} + T_{Reposition}$.

$T_{Latency}$ Time elapsed from when the request arrives until the onset of its display, in sec.

$T_{Latency}^{avg-min}$ Average-minimum latency ($\frac{T_p}{2}$), in sec.

$T_{Latency}^{Max}$ Maximum startup latency (without queuing), in sec.

$T_{Reposition}$ Time spent in tape and recording assembly repositioning, in sec.

T_{Rewind} Time required to rewind from one end of the tape to the other end, in sec.

$T_{Search}(i, j)$ Time required to search a tape cartridge from block location i to location j, in sec.

T_{Seek} The seek time of magnetic disk drive, in sec.

$T_{Seek_y}(x)$ Seek time to traverse x cylinders $(0 < x \leq \#cyl_y)$ with seek model of disk type y, in sec, where $\#cyl_y$ denotes the total number of cylinders of disk type y.

$T_{Service}$ Disk service time, in sec, and it is composed of the data transfer time (desirable) and the mechanical positioning delays (undesirable).

$T_{Service_y}$ Total time to retrieve a block of data from a disk of type y, in sec ($T_{Service_y} = T_{Transfer_y} + T_{Seek_y}$).

T_{Switch} Time required to unload the tape in the idle drive and load the tape with the referenced data, in sec.

T_{Track} Time required to align the tape recording assembly from one track (or set of tracks) to another track (or set of tracks), with possibly some minor block adjustments, in sec.

$T_{Transfer}$ Time to transfer an object from disk, in sec.

$T_{Transfer_T}$ Time required to complete the transfer of an object, $T_{Transfer_T} = \frac{O}{R_T}$.

$T_{Transfer_y}$ Time to transfer a block of data from a disk of type y, in sec.

T_{W_Seek} The worst seek time estimate, in sec. This estimate is different for different techniques. For example, T_{W_Seek} for Simple Disk Scheduling Technique is $= Seek(\#cyl)$, while for SCAN scheduling technique it is $= Seek(\frac{\#cyl}{N})$.

$TC_{STEP1}(X)$ Time required for STEP 1 of the pipelining technique, in sec, $TC_{STEP1}(X) = \lceil \frac{n}{PCR} \rceil - n + 1$.

Ter_i Number of blocks that are tape resident at time period i, $Ter_i = Max(n - (i \times PCR), 0)$.

TG Track group are a set of tape tracks that are read or written in parallel, using the $N_{r/w}$ tape read/write heads.

TP Track-Pairing.

Throughput Number of simultaneous displays.

Track Multiple sectors that form a concentric circle.

TWG Track wrap group is two TG's in opposite directions, used for the placement of an object on a tape.

U

UDP User Datagram Protocol (a connectionless transport protocol on the Internet).

V

VBR Variable Bit Rate.

VDSL See xDSL.

VOD Video-On-Demand.

W

wasteful Percentage of time that a disk performs wasteful work.

W Percentage of storage space unusable for SM.

WARP Warp ARound data Placement. It exploits the special characteristics of the serpentine tape technology to optimize the data placement. The objective of *WARP* is to eliminate T_{Search} entirely by: 1) placing the object on an even number of tracks, and 2) adjusting the amount of data on each track group (TG), such that every two tracks holds *almost* the same amount of data, i.e., same amount of data is allocated to each track wrap group (TWG).

X

X Streaming media (SM) object.

X_i Logical block i of object X. (Note: in Chapter 8 and Chapter 9, X_i refers to Object i.)

$X_{i.j}$ Fragment j of logical block i (of object X).

xDSL Any descendant of the Digital Subscriber Line protocol including SDSL (Symmetric DSL), ADSL (Asymmetric DSL), and VDSL (Very high bit rate DSL).

Y

y Number of sub-fragments that constitute a fragment.

Z

Zone A contiguous collection of disk cylinders whose tracks have the same storage capacity.

Zone-Bit-Recording (ZBR) A recording technique that allows for a tighter packaging of data, by recording more data on the outer tracks than on the inner ones (tracks are physically longer towards the outside of a platter).

Appendix C

EXERCISES

Chapter 1

1. What is the purpose of buffering at the display station, as shown in Figure 1.3?

2. The NTSC standard requires 30 frames per second (FPS) to provide a smooth continuous display. What happens if the frame rate drops below 30 fps? How low can the rate drop? Does your answer depend on the type of video objects being displayed (e.g., action movie)?

3. Is the MPEG-1 standard a constant bit rate (CBR) standard or is it a variable bit rate (VBR) standard? How about the MPEG-2 standard?

4. Dropped frames (blocks) can cause a retransmission of the blocks with an extra load on the network and the server. Explain.

5. Dropped frames (blocks) can cause hiccups in the display. Explain.

6. Why do display stations buffer SM objects before starting their display?

7. What are the main advantages and disadvantages of compression methods for SM systems? (Note: Consider all aspects of the system.)

8. Why are MPEG standards more appropriate than JPEG standards for SM display?

9. What are the important issues to consider when streaming multimedia data on the Internet?

10. Explain why Real-time Transport Protocol (RTP) is based on User Datagram Protocol (UDP).

Chapter 2

1. Find the parameters shown in Table 2.1 for a modern disk drive of your choice, and compare this disk with the ones in the table. Keep your references.

2. Increasing spindle speed from 7,200 RPM to 10,000 RPM (15,000 RPM) with all other parameters being equal, will cause an increase in data transfer rate of 39% (108%). Explain.

3. Identify two technological improvements that can improve disk transfer rate.

4. Relatively speaking, the wasteful overhead of modern disks is on the rise. Explain.

5. A limitation of deadline-driven scheduling is that the seek time cannot be optimized. Illustrate this scheduling technique and explain this limitation.

6. A possible approach to minimizing the seek time is to have the SM objects contiguously placed on the magnetic disks. This way, no seek times are incurred when retrieving different blocks of the same object. Explain the limitations of this approach.

7. Why does Equation 2.4 assume the maximum seek time?

8. Why is $l_{SCAN} = 2 \times T_p$, while $l_{Simple} = T_p$? Does this mean that $L_{SCAN} = 2 \times l_{Simple}$?

9. In Equation 2.33, for $R > 2$, $l = (2 \times R + 1) \times T_p$. Explain this equation.

10. REBECA is a tradeoff between throughput and latency. Explain.

11. With REBECA, under what circumstances is round-robin placement of SM objects appropriate?

12. Using the Track Pairing (TP) technique on a Cheetah 4LP disk causes a 4.2% waste of disk space. How much disk space is wasted on a modern disk drive?

13. Quantify the percentage of disk space wasted by FIXB and VARB for the commercial disks of Table 2.1.

14. Explain why it is difficult to implement VARB.

Chapter 3

1. You are constructing a multi-disk SM server using ten disk drives with 200 Mb/s of data transfer rate. Video clips are encoded with 8 Mb/s of constant bit rate. The worst seek time of the disk is 15 milliseconds. The target time period is one second.

   (a) When we construct ten clusters with each disk being a cluster, what percent of total disk bandwidth can be wasted at maximum?

   (b) When we construct a cluster with all disks, what percent of total disk bandwidth can be wasted at maximum?

   (c) Which approach can support a greater number of simultaneous displays when disk utilization approaches 1?

   (d) Argue for a small number of clusters in the system. Argue for a large number of clusters in the system.

2. In general, using a big block size, a system can reduce the wasteful portion of disk operations caused by seeks. Consequently, we can achieve a higher throughput. However, a big block size can significantly increase the average startup latency in a multi-disk system. Explain why.

3. Some approaches, such as round-robin data placement and cycle-based scheduling, may incur a long startup latency as the number of disks in the system increases. What would you do in order to reduce the startup latency?

4. Replication can also improve the reliability of the system as discussed in Chapter 6. Explain how full, selective, partial replication schemes enhance the system reliability in the case of a single disk failure in a multi-disk system.

5. Based on the discussion in Section 3.5.1, how can you control the frequency of online reconfiguration processes?

Chapter 4

1. Discuss what factors will affect the block retrieval time in the deadline-driven scheduling with random data placement.

2. Argue for the *Wait* approach in handling a hiccup situation. Argue for the *Skip* approach.

3. N buffering with Prefetching might waste system resources when a display is terminated before the normal end of display. What resources will be wasted? How?

4. Prefetching can increase the system load more than necessary for a short moment by issuing $N - 1$ block requests together. Can this be a problem in Movie-On-Demand application? Can this cause problems in some cases?

Chapter 5

1. In general, can we use all the disk space and all the disk bandwidth of a streaming media server with a heterogeneous disk subsystem?

2. With the Disk Grouping and Staggered Grouping techniques, each block fragment size $\mathcal{B}_i$ is approximately proportional to physical disk i's bandwidth. However, the fragment sizes are not exactly proportional. Why not? Explain.

3. With Disk Grouping the physical disks can be ordered (and activated) in any order within a logical disk. With Staggered Grouping they must be ordered from fastest to slowest. Explain why this is the case.

4. Which of the two techniques, Disk Grouping or Staggered Grouping, generally requires less memory? Explain why.

5. The three techniques, Disk Grouping, Staggered Grouping and Disk Merging all form logical disks on top of a collection of physical disk drives. Which technique will generally produce the most logical disks (assuming the same physical storage system)? Explain.

6. With Disk Merging the number of streams supported per logical disk can arbitrarily be chosen within a certain range. For example, consider Table 5.2. The total throughput for both shown configurations is $N = 96$ streams. In one case, each logical disk only supports one stream (i.e., $N_D = 1$) while in the second case, each logical disk supports four streams ($N_D = 4$). Why are the block sizes and wasted disk space different for both configurations? Explain.

7. Most of the discussion about Disk Merging is assuming that one or more logical disks are mapped to each physical disk, i.e., $p_0 \geq 1$. However, configurations with $p_0 < 1$ are certainly possible. In this case several physical disks may form a logical disk. How is this different from Disk Grouping? Explain.

8. In the comparison of Deterministic Disk Merging (DDM) and Random Disk Merging (RDM) of Table 5.8, DDM performs worse because the number of seeks is increased. Why is that the case? Explain.

Chapter 6

1. Why can we assume that the mean-time-between-failures (MTBF) is similar to the mean-time-to-failure (MTTF)? Explain.

2. Current commercial disk drives have a MTTF of approximately 1,000,000 hours. This is equal to more than 100 years. However, a significant number of disk drives will fail after 5 to 10 years of operation. Are the claimed MTTF's incorrect? Hint: the answer to this question is not directly contained in the material of this book. Research such fault tolerance concepts as *bathtub curve*, *infant mortality* and *wear out* (end of life).

3. Why are we considering the mean-time-to-service-loss (MTTSL) and not the mean-time-to-data-loss (MTTDL)? Explain.

4. What are the general tradeoffs between mirroring and parity-based fault tolerance? Explain.

5. Design an algorithm that assigns logical disks to parity groups for Disk Merging.

6. Write a program that can simulate the disk failures in a disk storage system that uses parity groups to protect against service loss. The program input will be the storage system configuration and the MTTF of the different disk models (see Table 6.3 for some ideas). The output will be the MTTSL of the complete storage system. Reproduce the results shown in Figure 6.12.

Chapter 7

1. What is the current data growth, disk storage capacity growth, and tape storage capacity growth? Identify the sources for your data.

2. What is your outlook on the future of tape storage as a:

   - Backup storage device
   - Near-online storage device
   - Online-storage device

in the next five years? Use the data from the previous question.

3. Do you expect tape storage to remain the tertiary storage of choice, or do you expect CD-ROMs, DVDs, holographic storage, magnetic disks, or other types of storage to replace it?

4. Find the parameters shown in Table 7.1 for a helical-scan and a serpentine scan tape drives of your choice.

5. Mean-time-to-failure (MTTF) of a tape cartridge is a very important factor in a SM server design that includes tape devices either as a backup storage or as a near-online storage. First, explain the importance of MTTF of tape cartridges. Second, compare the MTTF of helical-scan and serpentine-scan tape cartridges.

6. The multiplexing technique is impractical due to the following reasons: 1) it wastes tape bandwidth (a valuable resource), and 2) it negatively impacts the storage reliability. Explain.

7. Is it possible to apply the pipelining technique to alternative types of tertiary storage (such as CD-ROMs and DVDs)?

8. Why is neither PDF nor IDF appropriate when $PCR < 1$?

9. Why is neither PDF nor IDF appropriate when using the non-multiplexing approach while $PCR > 1$?

10. Prediction-based data placement techniques suffer a number of disadvantages. Identify and explain these disadvantages.

11. Is it possible to apply *WARP* to *helical-scan* tapes? Explain your answer.

12. Explain how $\$MB_T$ of *WARP* could be lower than $\$MB_T$ of *CP w/ sharing* (note: *WARP* always wastes more tape storage than *CP w/ sharing*).

13. Calculate the cost effectiveness (i.e., $\$MB_T$) of *WARP* and *CP w/ sharing* for a tape storage system based on IBM 3950, for objects ranging in size from 500 MByte up to 2 GByte, for the following systems:

    System 1: $DB = 1$ TByte, and $Band_{Effective} = 5$ MB/s.

    System 2: $DB = 1$ TByte, and $Band_{Effective} = 10$ MB/s.

    System 3: $DB = 10$ TByte, and $Band_{Effective} = 5$ MB/s.

System 4: $DB = 10$ TByte, and $Band_{Effective} = 10$ MB/s.

Chapter 8

1. What are the reasons for full decentralization of the meta-data with stream servers? Explain.

2. Explain the potential benefits of redundancy with the RedHi topology. Also, describe the potential difficulties with implementing the RedHi topology in the real world as compared to a pure hierarchy.

3. How do you compare the RedHi topology with a Fat-tree topology? Are these two topologies equivalent? Explain in details.

   (Refer to the following for more information about Fat-trees: C.E. Leiserson. Fat-trees: Universal networks for hardware efficient supercomputing. *IEEE Transactions on Computers*, C-34(10): 892-901, Oct. 1985.)

4. Assume the RedHi configuration defined in Table 8.5. Assume the single copy of the object "A" is located at the left-side root node and a request for object A is initiated at the rightmost head-end. Explain the life cycle of the request. Indicate your assumptions where required.

5. Assume the RedHi configuration defined in Table 8.5. Suppose all nodes, irrespective of their object inventories, always forward a request packet based on the selected propagation policy.

   a. For the three propagation policies, namely Parent-only, Inclusive, and Sibling, derive rigorous results for the percentage of the network covered by a request initiated at the rightmost head-end.

   b. Generalize your results from part (a) to a general RedHi topology.

6. Implement the RedHi topology defined in Table 8.5 using Yima Personal Edition (provided as a supplement to this book) as server nodes. Extend the Yima nodes to implement the three propagation policies described in Section 8.2.2. Perform experiments to verify your results from previous exercises.

7. A single point of failure

   a. can protect against a single failure.

   b. can bring down an entire network with one fault.

    c. can protect against multiple failures.

    d. does not exist in a well-designed system.

8. What is fault tolerance?

    a. A system component that can bring down the system.

    b. The ability to protect against faults.

    c. A faulty system.

    d. A system which has many areas with potential failures.

9. RedHi achieves load balancing and fault tolerance by

    a. adding redundant links.

    b. increasing storage space.

    c. adding more nodes.

    d. using Zipf's law for movie selection.

10. Poisson distribution is used in our case to:

    a. determine the placement of movie objects.

    b. determine the placement of server nodes.

    c. simulate the frequency of movie requests.

    d. simulate packet drops.

11. Why does RedHi perform almost identically with and without a link failure?

    a. Data will be rerouted through the redundant links.

    b. Poisson distribution forces data packets to arrive at their destinations faster.

    c. Because link failures only affect a small number of head-end nodes leaving overall performance identical.

    d. Failed links are diagnosed quickly with RedHi and, therefore, automatically repaired assuming the problem is not a physical problem.

12. Which is the correct order of the request phases?

    a. locate, select, reserve, initiate.

    b. reserve, locate, select, initiate.

c. initiate, locate, reserve, select.

d. select, locate, initiate, reserve.

13. True or false: Multiple independent CM servers form a more fault tolerant system as compared to a networked (distributed) server with multiple nodes because it removes the single point of failure.

Chapter 9

1. How do you compare the multiple-delivery-rate techniques proposed for 2-level and m-level architectures? Explain.

2. Assume the round-robin placement policy is used to assign fragments of an object X to n disks of a streaming media server.

   a. Consider a specific placement scenario (e.g., number of fragments for object X, the location of its first fragment) and then explain how reading m object fragments at a time may complicate the disk scheduling.

   b. Assume the multiple-delivery-rate approach proposed in Section 9.1.1 and explain how it simplifies the disk scheduling.

3. In Section 9.1.4, we assume objects have identical size. Instead, assume a Zipf distribution for the object size and repeat the analysis and compare your results with the results derived in Section 9.1.4.

4. The Internet is

   a. a network with highly predictable traffic.

   b. a stable network.

   c. best used to deliver streaming media.

   d. a network with unpredictable traffic.

5. At the client side, one difference between super-streaming and streaming is

   a. super-streaming allows clients to receive higher quality video.

   b. super-streaming guarantees the elimination of any video glitches.

   c. super-streaming requires the client to have a larger buffer size.

   d. super-streaming uses a higher compression standard.

6. Which of the following is not a reason against full downloads?

   a. It hogs resources.

   b. It requires a large client buffer.

   c. Fewer clients can be served.

   d. It might reduce the chance of glitches.

7. Admission control for super-streaming

   a. is a strict set of guidelines that cannot be changed.

   b. always blocks new requests if there are unavailable resources.

   c. can be flexible in deciding which resources to release depending on the application.

   d. always involves only one node.

8. In the simulation, as the client buffer size _____, the rejection rate _____.

   a. increases, decreases.

   b. increases, increases.

   c. decreases, decreases.

   d. stays the same, increases.

9. A bottleneck node is:

   a. a node which has less storage capacity than any other node at a certain path.

   b. a node which has fewer redundant links than any other node at a certain path.

   c. a node which has fewer redundant links than any other node between it and the source node.

   d. a node whose available bandwidth is less than the maximum available bandwidth of all nodes between it and the source node.

10. True or false:

   a. A node can store stream data in its buffer parallel to streaming the data to the next node.

b. Super-streaming is shown analytically to be superior to streaming because the probability of reaching a state where the server rejects requests is higher with streaming as compared to super-streaming.

Chapter 10

Appendix A describes the Yima Personal Edition (Yima PE) streaming media software that is included with this book. The following exercises are based on that software.

1. The Yima PE software can stream uncompressed, linear PCM WAVE files that require about 1.4 Mb/s of bandwidth for their smooth transmission. That is more bandwidth than a regular phone modem, DSL, or cable modem connection can support. Implement one of the following:

   a. Extend the client so that it can understand other uncompressed audio formats that require less bandwidth. For example, a mono audio stream with 8-bits per sample and 11 kHz sampling rate only requires 88 Kb/s for its transmission.

   b. Extend the client with, for example, an MP3, Ogg or other audio decompression module so that it can accept and play compressed streams. As a result, much less transmission bandwidth will be required. You may want to find a decompressor library for which the source code is available.

   Hint: There will be no changes necessary to the server side.

2. The RTP media transport protocol is based on the unreliable UDP protocol. Therefore, any lost packet will not be recovered and result in "hiccups" at the client side. Implement one or more of the following:

   a. A *retransmission* scheme that will request (client) and retransmit (server) lost packets.

   b. A *forward error correction* (FEC) scheme that lets the server include additional, redundant information in the transmission such that some lost packets can be reconstructed by the client.

   c. An *error concealment* method that interpolates missing data from data in packets before and after the lost packet.

3. Implement a statistics function at the client side that outputs how much data was lost during a session in two different ways:

   a. the total number of RTP packets lost, and

   b. the percentage of data (bytes) lost with respect to the total amount of data transmitted.

4. Implement a video decompressor in the client software such that video files can now be streamed.

BIBLIOGRAPHY

[1] NS, the Network Simulator. Information about NS is availabale at http://www.isi.edu/nsnam/ns/.

[2] *Electronic Engineering Times*, March 1993.

[3] C.C. Aggarwal, J.L. Wolf, and P.S. Yu. On Optimal Batching Policies for Video-on-Demand Storage Servers. *Proceedings of the Third IEEE International Conference on Multimedia Computing and Systems*, pages 253 –258, 1996.

[4] D. Aksoy, S. Zdonik, and M. J. Franklin. Data Staging for On-Demand Broadcast. *VLDB 2001, International Conference on Very Large Databases*, September 2001.

[5] J. Al-Marri and S. Ghandeharizadeh. An Evaluation of Alternative Disk Scheduling Techniques in Support of Variable Bit Rate Continuous Media. In *EDBT98*, Valencia, Spain, March 1998.

[6] A.O. Allen. *Probability, Statistics and Queueing Theory, 2nd Ed.* Academic Press, 1990.

[7] K.C. Almeroth and M.H. Ammar. The Use of Multicast Delivery to Provide a Scalable and Interactive Video-on-Demand Service. *IEEE Journal on Selected Areas in Communications*, 14(6):1110 –1122, August 1996.

[8] American National Standard of Accredited Standards Commitee X3, 11 West 42nd Street, New York, NY 10036. *Small Computer System Interface (SCSI-2), ANSI X3.131 - 199x*, March 1994.

[9] D. Anderson and G. Homsy. A Continuous Media I/O Server and Its Synchronization. *IEEE Computer*, October 1991.

[10] B. Awerbuch and D. Peleg. Concurrent Online Tracking of Mobile Users. *Proceedings of the Conference on Communications, Architectures, and Protocols; ACM SIGCOMM'91*, pages 221–233, September 1991.

[11] S.A. Barnett and G.J. Anido. A Cost Comparison of Distributed and Centralized Approaches to Video-on-Demand. *IEEE Journal on Selected Areas in Communications*, 14(6):1173–1183, August 1996.

[12] S. Berson, S. Ghandeharizadeh, R. Muntz, and X. Ju. Staggered Striping in Multimedia Information Systems. In *Proceedings of the ACM SIGMOD International Conference on Management of Data*, pages 79–89, 1994.

[13] S. Berson, L. Golubchik, and R. R. Muntz. A Fault Tolerant Design of a Multimedia Server. In *Proceedings of the ACM SIGMOD International Conference on Management of Data*, pages 364–375, 1995.

[14] E. W. Biersack and C. Bernhardt. A Fault Tolerant Video Server Using Combined Raid 5 and Mirroring. In *Proceedings of Multimedia Computing and Networking 1997 Conference (MMCN'97)*, pages 106–117, San Jose, California, February 1997.

[15] Y. Birk. Track-Pairing: A Novel Data Layout for VOD Servers with Multi-Zone-Recording Disks. In *Proceedings of the International Conference on Multimedia Computing and Systems*, pages 248–255, 1995.

[16] C. Bisdikian and B. Patel. Cost-Based Program Allocation for Distributed Multimedia-on-Demand Systems. *IEEE MultiMedia*, pages 62–72, Fall 1996.

[17] D. Bitton and J. Gray. Disk shadowing. In *Proceedings of the International Conference on Very Large Databases*, September 1988.

[18] S. Blake, D. Black, M. Carlson, E. Davies, Z. Wang, and W. Weiss. An Architecture for Differentiated Services. RFC 2475, December 1998.

[19] W. J. Bolosky, J. S. Barrera, R. P. Draves, R. P. Fitzgerald, G. A. Gibson, M. B. Jones, S. P. Levi, N. P. Myhrvold, and R. F. Rashid. The Tiger Video Fileserver. In *6th Workshop on Network and Operating System Support for Digital Audio and Video*, Zushi, Japan, April 1996.

[20] A. Broido and K. Claffy. Internet Topology: Connectivity of IP Graphs. *Cooperative Association for Internet Data Analysis (CAIDA)*, July 2001. Available online at http://www.cadia.org.

[21] H.J. Burch and F. Ercal. Mapping the Internet. *IEEE Computer*, 32(4):97 – 102, April 1999.

[22] M. Carey, L. Haas, and M. Livny. Tapes Hold Data, Too: Challenges of Tuples on Tertiary Storage. In *Proceedings of the ACM SIGMOD International Conference on Management of Data*, pages 413–417, 1993.

[23] S.G. Chan and F. Tobagi. Providing On-Demand Video Services Using Request Batching. *Proceedings Of IEEE ICC'98*, 3:1716 –1722, 1998.

[24] E. Chang and H. Garcia-Molina. Reducing Initial Latency in a Multimedia Storage System. In *Proceedings of IEEE International Workshop on Multimedia Database Management Systems*, 1996.

[25] E. Chang and H. Garcia-Molina. MEDIC: A Memory and Disk Cache for Multimedia Clients. Technical Report SIDL-WP-1997-0076, Stanford University, 1997.

[26] H.J. Chen and T. Little. Physical Storage Organizations for Time-Dependent Multimedia Data. In *Proceedings of the Foundations of Data Organization and Algorithms (FODO) Conference*, October 1993.

[27] L. T. Chen, D. Rotem, and S. Seshadri. Declustering Databases on Heterogeneous Disk Systems. In *Proceedings of the 21st International Conference on Very Large Databases*, pages 110–121, Zürich, Switzerland, September 1995.

[28] L.T. Chen, R. Drach, M. Keating, S. Louis, D. Rotem, and A. Shoshani. Efficient Organization and Access of Multi-Dimensional Datasets on Tertiary Storage Systems. *Information Systems Journal*, April 1995.

[29] M. Chen and D. D. Kandlur. Stream Conversion to Support Interactive Video Playout. *IEEE Multimedia*, Summer 1996.

[30] P. M. Chen, E. K. Lee, G. A. Gibson, R. H. Katz, and D. A. Patterson. RAID: High-Performance, Reliable Secondary Storage. *ACM Computing Surveys*, 26(2):145–185, June 1994.

[31] S. Christodoulakis, P. Triantafillou, and F. Zioga. Principles of Optimally Placing Data in Tertiary Storage Libraries. In *Proceedings of the International Conference on Very Large Databases*, pages 236–245, 1997.

[32] D. D. Clark, S. Shenker, and L. Zhang. Supporting Real-time Applications in an Integrated Services Packet Network: Architecture and Mechanism. *Proceedings of Conference on Communications Architectures and Protocols*, pages 14 – 26, 1992.

[33] National Storage Industry Consortium. Tape Roadmap. June 1998.

[34] A. Dan, D.M. Dias, R. Mukherjee, D. Sitaram, and R. Tewari. Buffering and Caching in Large Video Servers. In *Proceedings of COMPCON*, 1995.

[35] A. Dan and D. Sitaram. An Online Video Placement Policy based on Bandwidth to Space Ratio (BSR). In *Proceedings of the ACM SIGMOD International Conference on Management of Data*, pages 376–385, San Jose, May 1995.

[36] A. Dan, D. Sitaram, and P. Shahabuddin. Scheduling Policies for an On-Demand Video Server with Batching. *Proceedings of the Second ACM International Conference on Multimedia, MM'94*, pages 15–23, October 1994.

[37] A. E. Dashti. *Data Placement Techniques For Multimedia Hierarchical Storage Systems*. Ph.D. Dissertation, University of Southern California, Los Angeles, CA, August 1999.

[38] A. E. Dashti. WARP: Wrap Around Data Placement for Serpentine Tapes. In *Proceedings of the Joint 18th IEEE Symposium on Mass Storage Systems and the 9th NASA Goddard Conference on Mass Storage Systems and Technologies*, April 2001.

[39] A. E. Dashti and S. Ghandeharizadeh. On Configuring Hierarchical Storage Structures. In *Proceedings of the Joint NASA/IEEE Mass Storage Conference*, March 1998.

[40] A. E. Dashti and C. Shahabi. Data Placement Techniques for Serpentine Tapes. In *Proceedings of the Software Technology Track of the Thirty-Third Hawaii International Conference on System Sciences (HICSS-33)*, January 2000.

[41] A. E. Dashti, C. Shahabi, and R. Zimmermann. Cost Analysis of Serpentine Tape Data Placement Techniques in Support of Continuous Media Display. In *Proceedings of the 10th International Conference on Computing and Information (ICCI 2000)*, November 2000.

[42] The Internet Movie Database. In *http://www.imdb.com*, 2002.

[43] J. Dedek. *Basics of SCSI, Second Edition*. Ancot Corporation, 1994.

[44] P. J. Denning. Effects of scheduling on file memory operations. *AFIPS Spring Joint Computer Conference*, pages 9–21, April 1967.

[45] P. J. Denning. The Working Set Model for Program Behavior. *Communications of the ACM*, 11(5), May 1968.

[46] P. J. Denning and S. C. Schwartz. Properties of the Working Set Model. *Communications of the ACM*, 15(3), March 1972.

[47] DLT.com. *http://www.dlttape.com*.

[48] A. L. Drapeau and R. H. Katz. Striped Tape Arrays. *Twelfth IEEE Symposium on Mass Storage Systems*, April 1993.

[49] Exabyte. *http://www.exabyte.com*.

[50] C. Fedrighi and L. A. Rowe. A Distributed Hierarchical Storage Manager for a Video-on-Demand System. In *Storage and Retrieval for Image and Video Databases II, IS&T/SPIE Symp. on Elec. Imaging Science and Tech.*, pages 185–197, 1994.

[51] K. D. Fisher, W. L. Abbott, J. L. Sonntag, and R. Nesin. PRML Detection Boosts Hard-Disk Drive Capacity. *IEEE Spectrum*, pages 70–76, November 1996.

[52] R. Flynn and W. Tetzlaff. Disk Striping and Block Replication Algorithms for Video File Servers. In *Proceedings of the International Conference on Multimedia Computing and Systems*, pages 590–597, Hiroshima, Japan, June, 17-23 1996.

[53] S. Fortune and J. Willie. Parallelism in Random Access Machines. *Proceedings of the 10th ACM Symposium on Theory of Computing*, pages 114–118, 1978.

[54] C. S. Freedman and D. J. DeWitt. The SPIFFI Scalable Video-on-Demand System. In *Proceedings of the ACM SIGMOD International Conference on Management of Data*, pages 352–363, 1995.

[55] G. R. Ganger. *System-Oriented Evaluation of I/O Subsystem Performance*. Ph.D. Dissertation, University of Michigan, 1995.

[56] L. Gao and D. F. Towsley. Supplying Instantaneous Video-on-Demand Services Using Controlled Multicast. In *ICMCS*, volume 2, pages 117–121, 1999.

[57] A.D. Gelman, H. Kobrinski, L.S. Smoot, S.B. Weinstein, M. Fortier, and D. Lemay. A Store-and-Forward Architecture for Video-on-Demand Service. *ICC'91, IEEE International Conference on Communications*, 2:842 –846, 1991.

[58] D. J. Gemmell and S. Christodoulakis. Principles of Delay Sensitive Multimedia Data Storage and Rtrieval. *ACM Trans. Information Systems*, 10(1):51–90, January 1992.

[59] D. J. Gemmell, H. M. Vin, D. D. Kandlur, P. V. Rangan, and L. A. Rowe. Multimedia Storage Servers: A Tutorial. *IEEE Computer*, May 1995.

[60] D. James Gemmell. Disk Scheduling for Continuous Media. In Soon M. Chung, editor, *Multimedia Information Storage and Management*, chapter 1. Kluwer Academic Publishers, Boston, MA, August 1996.

[61] S. Ghandeharizadeh, A. Dashti, and C. Shahabi. A Pipelining Mechanism to Minimize the Latency Time in Hierarchical Multimedia Storage Managers. *Computer Communications*, March 1995.

[62] S. Ghandeharizadeh, D. Ierardi, D. H. Kim, and R. Zimmermann. Placement of Data in Multi-Zone Disk Drives. In *Second International Baltic Workshop on DB and IS*, June 1996.

[63] S. Ghandeharizadeh, D. Ierardi, and R. Zimmermann. An on-line algorithm to optimize file layout in a dynamic environment. *Information Processing Letters*, (57):75–81, 1996.

[64] S. Ghandeharizadeh, S. H. Kim, and C. Shahabi. On Configuring a Single Disk Continuous Media Server. In *Proceedings of the ACM SIGMETRICS*, 1995.

[65] S. Ghandeharizadeh, S. H. Kim, C. Shahabi, and R. Zimmermann. Placement of Continuous Media in Multi-Zone Disks. In Soon M. Chung, editor, *Multimedia Information Storage and Management*, chapter 2. Kluwer Academic Publishers, Boston, MA, August 1996.

[66] S. Ghandeharizadeh, S. H. Kim, W. Shi, and R. Zimmermann. On Minimizing Startup Latency in Scalable Continuous Media Servers. In

Proceedings of Multimedia Computing and Networking 1997, pages 144–155, February 1997.

[67] S. Ghandeharizadeh and S.H. Kim. Striping in Multi-disk Video Servers. In *High-Density Data Recording and Retrieval Technologies*, pages 88–102. Proc. SPIE 2604, October 1995.

[68] S. Ghandeharizadeh and S.H. Kim. Design of Multi-user Editing Servers for Continuous Media. *Journal of Multimedia Tools and Applications Journal*, 11(1), May 2000.

[69] S. Ghandeharizadeh, S.H. Kim, and C. Shahabi. On Disk Scheduling and Data Placement for Video Servers. Technical report, University of Southern California, Los Angeles, CA, 1996.

[70] S. Ghandeharizadeh, S.H. Kim, W. Shi, and R. Zimmermann. On Minimizing Startup Latency in Scalable Continuous Media Servers. In *Proceedings of Multimedia Computing and Networking*, pages 144–155. Proc. SPIE 3020, February 1997.

[71] S. Ghandeharizadeh and L. Ramos. Continuous retrieval of multimedia data using parallelism. *IEEE Transactions on Knowledge and Data Engineering*, 5(4), August 1993.

[72] S. Ghandeharizadeh, L. Ramos, Z. Asad, and W. Qureshi. Object Placement in Parallel Hypermedia Systems. In *Proceedings of the International Conference on Very Large Databases*, pages 243–254, September 1991.

[73] S. Ghandeharizadeh and C. Shahabi. Management of Physical Replicas in Parallel Multimedia Information Systems. In *Proceedings of the Foundations of Data Organization and Algorithms (FODO) Conference*, October 1993.

[74] S. Ghandeharizadeh and C. Shahabi. Distributed Multimedia Systems. In John G. Webster, editor, *Wiley Encyclopedia of Electrical and Electronics Engineering*. John Wiley and Sons Ltd., 1999.

[75] S. Ghandeharizadeh, J. Stone, and R. Zimmermann. Techniques to Quantify SCSI-2 Disk Subsystem Specifications for Multimedia. Technical Report USC-CS-TR95-610, University of Southern California, Los Angeles, CA, 1995.

[76] S. Ghandeharizadeh, R. Zimmermann, W. Shi, R. Rejaie, D. Ierardi, and T.W. Li. Mitra: A Scalable Continuous Media Server. *Kluwer Multimedia Tools and Applications*, 5(1):79–108, July 1997.

[77] G. Gibson, J. S. Vitter, and J. Wilkes. Report on the Working Group on Storage I/O for Large-Scale Computing. *ACM Computing Surveys*, 28(4), December 1996.

[78] G. A. Gibson. *Redundant Disk Arrays: Reliable, Parallel Secondary Storage*. Ph.D. Dissertation, University of California at Berkeley, Berkeley, CA, December 1991. Also available from MIT Press, 1992.

[79] L. De Giovanni, A.M. Langellotti, L.M. Patitucci, and L. Petrini. Dimensioning of Hierarchical Storage for Video-on-Demand Services. *ICC'94, IEEE International Conference on Communications*, 3:1739 –1743, 1994.

[80] A. Goel, C. Shahabi, S.-Y. D. Yao, and R. Zimmermann. SCADDAR: An Efficient Randomized Technique to Reorganize Continuous Media Blocks. In *Proceedings of the 18th International Conference on Data Engineering*, February 2002.

[81] F. Goodenough. DSP Technique Nearly Doubles Disk Capacity. *Electronic Design*, pages 53–58, February 1993.

[82] E. Grochowski. IBM leadership in disk storage technology. *IBM Almaden Research Center. http://www.storage.ibm.com/technolo/grochows*, 1999.

[83] E. Grochowski. Internal Data Rate Trend & Storage Price Projections, 2001. IBM Almaden Research Center, San Jose, CA. URL: http://www.almaden.ibm.com/sst/.

[84] M. Hall. *Combinatorial Theory (2nd Edition)*. Wiley-Interscience, 1986.

[85] S. R. Heltzer, J. M. Menon, and M. F. Mitoma. Logical Data Tracks Extending Among a Plurality of Zones of Physical Tracks of one or More Disk Devices., April 1993. U.S. Patent No. 5,202,799.

[86] B. Hillyer and A. Silberschatz. On the Modeling and Performance Characteristics of a Serpentine Tape Drive. In *Proceedings of the ACM SIGMETRICS International Conference on Measurement and Modeling of Computer Systems*, pages 170–179, May 1996.

[87] B. Hillyer and A. Silberschatz. Random I/O Scheduling in Online Tertiary Storage Systems. In *Proceedings of the ACM SIGMOD International Conference on Management of Data*, pages 195–204, June 1996.

[88] B. Hillyer and A. Silberschatz. *Storage Technology: Status, Issues, and Opportunities.* http://www.bell-labs.com/user/hillyer/, June 1996.

[89] B. Hillyer and A. Silberschatz. Scheduling Non-Contiguous Tape Retrievals. In *Joint NASA/IEEE Mass Storage Conference*, pages 113–124, March 1998.

[90] J. Hsieh, J. Liu, D. Du, T. Ruwart, and M. Lin. Experimental Performance of a Mass Storage System for Video-On-Demand. *Special Issue of Multimedia Systems and Technology of Journal of Parallel and Distributed Computing (JPDC)*, 30(2):147–167, November 1995.

[91] K.A. Hua, Y. Cai, and Simon Sheu. Patching: A Multicast Technique for True Video-on-Demand Services. *Proceedings of the Sixth ACM International Conference on Multimedia, MM'98*, pages 191–200, September 1998.

[92] J.Y. Hui, E. Karasan, J. Li, and J. Zhang. Client-Server Synchronization and Buffering for Variable Rate Multimedia Retrievals. *IEEE Journal on Selected Areas in Communications, 14(1)*, pages 226–237, January 1996.

[93] T. Ibaraki and N. Katoh. *Resource Allocation Problems - Algorithmic Approaches.* The MIT Press, 1988.

[94] O. H. Ibarra and C. E. Kim. Fast Approximation Algorithms for the Knapsack and Subset Sum Problems. *Journal of the ACM*, 22(4):463–468, 1975.

[95] IBM. *http://www.storage.ibm.com.*

[96] K. Kalpakis, K. Dasgupta, and O. Wolfson. Optimal Placement of Replicas in Trees with Read, Write, and Storage Costs. *IEEE Transactions on Parallel and Distributed Systems*, 12(6):628 –637, June 2001.

[97] S.H. Kim and S. Ghandeharizadeh. Design of Multi-user Editing Servers for Continuous Media. In *International Workshop on Research Issues in Data Engineering*, February 1998.

[98] T.Y. Kim, B.H. Roh, and J.K. Kim. Bandwidth Renegotiation with Traffic Smoothing and Joint Rate Control for VBR MPEG Video over ATM. *IEEE Transactions on Circuits and Systems for Video Technology*, 10(5):693 –703, August 2000.

[99] L. Kleinrock. *Queueing Systems Volume I: Theory*, page 105. Wiley-Interscience, 1975.

[100] A. Laursen, J. Olkin, and M. Porter. Oracle Media Server: Providing Consumer Based Interactive Access to Multimedia Data. In *Proceedings of the ACM SIGMOD International Conference on Management of Data*, pages 470–477, 1994.

[101] S.H. Lee, K.Y. Whang, Y.S. Moon, and I.Y. Song. Dynamic Buffer Allocation in Video-on-Demand Systems. In *Proceedings of the ACM SIGMOD International Conference on Management of Data*, 2001.

[102] C. Leopold. *Parallel and Distributed Computing: A Survey of Models, Paradigms, and Approaches*. Wiley Series on Parallel and Distributed Computing, 1st edition. John Wiley, New York, 2001.

[103] P. W.K. Lie, J. C.S. Lui, and L. Golubchik. Threshold-Based Dynamic Replication in Large-Scale Video-on-Demand Systems. *Multimedia Tools and Applications*, 11(1):35–62, 2000.

[104] C.L. Liu and J.W. Layland. Scheduling Algorithms for Multiprogramming in a Hard Real-Time Environment. *Journal of ACM*, 20(1):46–61, Jan 1973.

[105] P. Lougher and D. Shepherd. The Design of a Storage Server for Continuous Media. *The Computer Journal (special issue on multimedia)*, 36(1), February 1993.

[106] LTO. *http://www.lto-technology.com*.

[107] R. Lüling. Static and Dynamic Mapping of Media Assets on a Network of Distributed Multimedia Information Servers. *Proceedings of 19th IEEE International Conference on Distributed Computing Systems*, pages 253–260, 1999.

[108] D. Malon and E. May. Critical Database Technologies for High Energy Physics. In *Proceedings of the International Conference on Very Large Databases*, pages 580–584, 1997.

[109] Martin, P.S. Narayan, B. Ozden, R. Rastogi, and A. Silberschatz. The Fellini multimedia storage. *Journal of Digital Libraries*, 1997.

[110] C. Martin, P. S. Narayan, B. Özden, R. Rastogi, and A. Silberschatz. The Fellini Multimedia Storage Server. In Soon M. Chung, editor, *Multimedia Information Storage and Management*, chapter 5. Kluwer Academic Publishers, Boston, MA, August 1996.

[111] P. McGowan and J. Hickey, editors. *Quantum DLT Handbook*. Quantum Corperation, 1998.

[112] J.M. McManus and K.W. Ross. Video-on-Demand over ATM: Constant-Rate Transmission and Transport. *IEEE Journal on Selected Areas in Communications*, 14(6):1087 – 1098, August 1996.

[113] A. G. Merten. *Some quantitative techniques for file organization*. Ph.D. Dissertation, University of Wisconsin, Madison, Wisconsin, 1970. Technical Report No. 15.

[114] M. Mielke and A. Zhang. A Multi-Level Buffering and Feedback Scheme for Distributed Multimedia Presentation Systems. In *Proceedings of Seventh International Conference on Computer Communications and Networks (IC3N'98)*, Lafayette, LA, October 1998.

[115] Antoine Mourad. Reliable Disk Striping in Video-On-Demand Servers. In *Proceedings of the 2nd IASTED/ISMM International Conference on Distributed Multimedia Systems and Applications*, pages 113–118, Stanford, CA, August 1995.

[116] R. Muntz, J. Santos, and S. Berson. RIO: A Real-time Multimedia Object Server. *ACM Sigmetrics Performance Evaluation Review*, 25(2):29–35, September 1997.

[117] R. R. Muntz and J. C.S. Lui. Performance Analysis of Disk Arrays Under Failure. In *Proceedings of the 16th Very Large Databases Conference*, pages 162–173, Brisbane, Australia, 1990.

[118] NCITS. *Magnetic Tape Format for Information Interchange, 128 Track Parallel Serpentine 12.65 mm, 3400 bpmm, Run Length Limited Recording*. National Committee for Information Technolgy Standards, 1997.

[119] G. Nerjes, P. Muth, and G. Weikum. Stochastic Service Guarantees for Continuous Data on Multi-Zone Disks. In *Proceedings of the Principles of Database Systems Conference*, pages 154–160, 1997.

[120] R.T. Ng and J. Yang. Maximizing Buffer and Disk Utilizations for News On-Demand. In *Proceedings of the International Conference on Very Large Databases*, September 1994.

[121] S. W. Ng. Advances in Disk Technology: Performance Issues. *IEEE Computer*, pages 75–81, May 1998.

[122] NSIC. *NSIC Tape Roadmap*. National Storage Industry Consortium, 1998.

[123] J. Nussbaumer, B. Patel, F. Schaffa, and J. Sterbenz. Network Requirements for Interactive Video-on-Demand. *IEEE Journal on Selected Areas in Communications*, 13(5):779–787, June 1995.

[124] National Institute of Standards and Technology. National Storage Industry Consortium. In *Workshop on Digital Data Storage*, March 1994.

[125] B. Özden, R. Rastogi, P. Shenoy, and A. Silberschatz. Fault-tolerant Architectures for Continuous Media Servers. In *Proceedings of the ACM SIGMOD International Conference on Management of Data*, pages 79–90, June 1996.

[126] B. Özden, R. Rastogi, and A. Silberschatz. Disk Striping in Video Server Environments. In *Proceedings of the International Conference on Multimedia Computing and Systems*, pages 580–589, Hiroshima, Japan, June 1995.

[127] C.H. Papadimitriou, S. Ramanathan, and P.V. Rangan. Information Caching for Delivery of Personalized Video Programs on Home Entertainment Channels. *Proceedings of the International Conference on Multimedia Computing and Systems*, pages 214 –223, 1994.

[128] D. Patterson, G. Gibson, and R. Katz. A Case for Redundant Arrays of Inexpensive Disks (RAID). In *Proceedings of the ACM SIGMOD International Conference on Management of Data*, May 1988.

[129] F. Periera and T. Ebrahimi, editors. *The MPEG-4 Book*. Prentice Hall IMSC Press Multimedia Series, 2002.

[130] R. Perlman. Fault-tolerant Broadcast of Routing Information. *Computer Networks*, 7:395–405, December 1983.

[131] J.L. Peterson and A. Silberschatz. *Operating Systems Concepts, 2nd Ed.* Addison-Wesley, 1985.

[132] V.G. Polimenis. The Design of a File System that Supports Multimedia. Technical Report TR-91-020, ICSI, 1991.

[133] S. Ramanathan and P. V. Rangan. Feedback Techniques for Intra-Media Continuity and Inter-Media Synchronization in Distributed Multimedia Systems. *The Computer Journal*, 36(1):19–31, 1993.

[134] R. Ramarao and V. Ramamoorthy. Architectural Design of On-Demand Video Delivery Systems: the Spatio-Temporal Storage Allocation Problem. *Proceedings of IEEE International Conference on Communications, ICC'91*, 1:506 –510, 1991.

[135] P. Rangan and H. Vin. Efficient Storage Techniques for Digital Continuous Media. *IEEE Transactions on Knowledge and Data Engineering*, 5(4), August 1993.

[136] P. Rangan, H. Vin, and S. Ramanathan. Designing an On-Demand Multimeida Service. *IEEE Communications Magazine*, 30(7), July 1992.

[137] P.V. Rangan and H.M. Vin. Designing File Systems for Digital Video and Audio. In *12th ACM Symposium on Operating Systems*, 1991.

[138] A. L. Narasimha Reddy and James C. Wyllie. I/O Issues in a Multimedia System. *IEEE Computer*, 27(3):69–74, March 1994.

[139] C. Ruemmler and J. Wilkes. An Introduction to Disk Drive Modeling. *IEEE Computer*, pages 17–28, March 1994.

[140] J.D. Ryoo and S.S. Panwar. Algorithms for Determining File Distribution in Networks with Multimedia Servers. *Proceedings of IEEE International Conference on Communications, ICC'99*, 2:875 –879, 1999.

[141] J. D. Salehi, Z.L. Zhang, J. F. Kurose, and D. Towsley. Supporting Stored Video: Reducing Rate Variability and End-to-End Resource Requirements Through Optimal Smoothing. *Proceedings of the ACM SIGMETRICS Conference on Measurement and Modeling of Computer Systems*, pages 222–231, May 1996.

[142] K. Salem and H. Garcia-Molina. Disk Striping. In *Proceedings of International Conference on Database Engineering*, February 1986.

[143] O. Sandta and R. Midstraum. Low-cost Access Time Model for Serpentine Tape Drives. In *Proceedings of the Joint NASA/IEEE Mass Storage Conference*, pages 116–127. IEEE, March 1999.

[144] J. R. Santos and R. R. Muntz. Performance Analysis of the RIO Multimedia Storage System with Heterogeneous Disk Configurations. In *ACM Multimedia Conference*, Bristol, UK, 1998.

[145] J. R. Santos, R. R. Muntz, and B. Ribeiro-Neto. Comparing Random Data Allocation and Data Striping in Multimedia Servers. In *SIGMETRICS*, Santa Clara, CA, June 2000.

[146] S. Sarawagi and M. Stonebraker. Efficient Organization of Large Multidemnsional Arrays. In *Proceedings of the IEEE 10th International Conference on Data Engineering*, pages 328–336. IEEE, 1994.

[147] F. Schaffa and J.P. Nussbaumer. On Bandwidth and Storage Tradeoffs in Multimedia Distribution Networks. *INFOCOM'95; Proceedings of Fourteenth Annual Joint Conference of the IEEE Computer and Communications Societies*, 3:1020 –1026, 1995.

[148] H. Schulzrinne, S. Casner, R. Frederick, and V. Jacopson. RTP: A Transport Protocol for Real Time Applications RFC 1889. In *Internet Engineering Task Force*, January 1996.

[149] H. Schulzrinne, A. Rao, and R. Lanphier. Real Time Streaming Protocol (RTSP). RFC 2326, April 1998.

[150] N. D. Schwartz. *The Tech Boom will Keep on Rocking*. Fortune, February 1999.

[151] P. H. Seaman, R. A. Lind, and T. L. Wilson. An analysis of auxiliary-storage activity. *IBM Systems Journal*, 5(3):158–170, 1996.

[152] M. Seltzer, P. Chen, and J. Ousterhout. Disk Scheduling Revisited. In *Proceedings of the 1990 Winter USENIX Conference*, pages 313–324, Washington DC, Usenix Association, 1990.

[153] H. Shachnai and P. S. Yu. Exploring Wait Tolerance in Effective Batching for Video-on-Demand Scheduling. *Multimedia Systems*, 6(6):382–394, 1998.

[154] C. Shahabi and M. Alshayeji. Super-streaming: A New Object Delivery Paradigm for Continuous Media Servers. *Journal of Multimedia Tools and Applications*, 11(1), May 2000.

[155] C. Shahabi, M. Alshayeji, and S. Wang. A Redundant Hierarchical Structure for a Distributed Continuous Media Server. *Fourth European*

Workshop on Interactive Distributed Multimedia Systems and Telecommunication Services IDMS'97, September 1997.

[156] C. Shahabi, G. Barish, B. Ellenberger, N. Jiang, M. Kolahdouzan, A. Nam, and R. Zimmermann. Immersidata Management: Challenges in Management of Data Generated within an Immersive Environment. In *Proceedings of the International Workshop on Multimedia Information Systems*, October 1999.

[157] C. Shahabi, S. Ghandeharizadeh, and S. Chaudhuri. On Scheduling Atomic and Composite Continuous Media Objects. *Transactions on Knowledge and Data Engineering*, 14(2):447–455, 2002.

[158] C. Shahabi and F. B. Kashani. Decentralized Resource Management for a Distributed Continuous Media Server. *IEEE Transactions on Parallel and Distributed Systems (TPDS)*, 13(6), June 2002.

[159] C. Shahabi, M. R. Kolahdouzan, G. Barish, R. Zimmermann, D. Yao, K. Fu, and L. Zhang. Alternative Techniques for the Efficient Acquisition of Haptic Data. In *ACM SIGMETRICS/Performance*, 2001.

[160] C. Shahabi, R. Zimmermann, K. Fu, and S.-Y. D. Yao. Yima: A Second Generation Continuous Media Server. *IEEE Computer*, 35(6):56–64, June 2002.

[161] D. P. Siewiorek and R. S. Swarz. *The Theory and Practice of Reliable Systems Design*. Digital Press, Bedford, MA, 1982.

[162] A. Silberschatz, M. Stonebraker, and J. Ullmen. Database Systems: Achievements and Opportunities. *Communications of the ACM*, October 1991.

[163] A. Silberschatz, M. Stonebraker, and J. Ullmen. Database Systems: Achievements and Opportunities. *NSF Workshop on the Future of Database Systems Research*, May 1995.

[164] A.J. Smith. Bibliography on Paging and Related Topics. *Operating Systems Reviews*, 12:39–56, 1978.

[165] Sony. *http://www.sony.com*.

[166] M.F. Mitoma S.R. Heltzer, J.M. Menon. Logical data tracks extending among a plurality of zones of physical tracks of one or more disk devices. In *U.S. Patent No. 5,202,799*, April 1993.

[167] M. Stonebraker. Managing Persistent Objects in a Multi-level Store. In *Proceedings of the ACM SIGMOD International Conference on Management of Data*, 1991.

[168] T.J. Teorey. A Comparative Analysis of Disk Scheduling Policies. *Communications of ACM*, 15(3), 1972.

[169] T.J. Teory. Properties of Disk Scheduling Policies in Multiprogrammed Computer Systems. In *Proc. AFIPS Fall Joint Computer Conf.*, pages 1–11, 1972.

[170] R. Tewari, D. M. Dias, R. Mukherjee, and H. Vin. High Availability for Clustered Multimedia Servers. In *Proceedings of International Conference on Database Engineering*, pages 645–654, New Orleans, LA, February 1996.

[171] R. Tewari, R. P. King, D. Kandlur, and D. M. Dias. Placement of Multimedia Blocks on Zoned Disks. In *Proceedings of IS&T/SPIE Multimedia Computing and Networking*, San Jose, CA, January 1996.

[172] R. Tewari, R. Mukherjee, D.M. Dias, and H.M. Vin. Design and Performance Tradeoffs in Clustered Video Servers. In *Proceedings of IEEE ICMCS*, June 1995.

[173] T. Thompson. The Elements of Design: Packing a Platter. *Byte*, 21(8):80NA2, 80NA6, August 1996.

[174] B. Tierney, B. Johnston, H. Herzog, G. Hoo, G. Jin, J. Lee, T. Chen, and D. Rotem. Distributed Parallel Data Storage Systems: A Scalable Approach to High Speed Image Servers. In *Second ACM Conference on Multimedia*, pages 399–405, San Francisco, CA, October 1994.

[175] F.A. Tobagi, J. Pang, R. Baird, and M. Gang. Streaming RAID-A Disk Array Management System for Video Files. In *First ACM Conference on Multimedia*, August 1993.

[176] H. M. Vin, S. S. Rao, and P. Goyal. Optimizing the Placement of Multimedia Objects on Disk Arrays. In *Proceedings of the IEEE International Conference on Multimedia Computing and Systems, ICMCS'95*, Washington, D.C., May 1995.

[177] P. Wayner. Digital Hard Drives. *Byte Magazine*, pages 91–96, March 1994.

[178] N. C. Wilhelm. A General Model for the Performance of Disk Systems. *Journal of the ACM*, 24(1):14–31, January 1977.

[179] J.L. Wolf, P.S. Yu, and H. Shachnai. DASD Dancing: A Disk Load Balancing Optimization Scheme for Video-on-Demand Computer Systems. In *Proceedings of the ACM SIGMETRICS*, Ottawa, Canada, May 1995.

[180] B. L. Worthington, G. R. Ganger, and Y. N. Patt. Scheduling Algorithms for Modern Disk Drives. In *Proceedings of the ACM SIGMETRICS*, pages 241–251, May 1994.

[181] B. L. Worthington, G. R. Ganger, Y. N. Patt, and J. Wilkes. On-Line Extraction of SCSI Disk Drive Parameters. In *Proceedings of the ACM SIGMETRICS*, May 1995.

[182] P. S. Yu, M. S. Chen, and D. D. Kandlur. Design and Analysis of a Grouped Sweeping Scheme for Multimedia Storage Management. In *Proceedings of the Third International Workshop on Network and Operating System Support for Digital Audio and Video*, November 1992.

[183] P.S. Yu, M-S. Chen, and D.D. Kandlur. Grouped sweeping scheduling for DASD-based multimedia storage management. *Multimedia Systems*, 1(1):99–109, January 1993.

[184] A. Zhang, Y. Song, and M. Mielke. Netmedia: Streaming Multimedia Presentations in Distributed Environments. *IEEE Multimedia*, 9(1), January - March 2002.

[185] R. Zimmermann. *Continuous Media Placement and Scheduling in Heterogeneous Disk Storage Systems*. Ph.D. Dissertation, University of Southern California, Los Angeles, CA, December 1998.

[186] R. Zimmermann, K. Fu, and C. Shahabi. A Multi-Threshold Online Smoothing Technique for Variable Rate Multimedia Streams. Technical report, University of Southern California, 2002. URL http://www.cs.usc.edu/tech-reports/technical_reports.html.

[187] R. Zimmermann, K. Fu, C. Shahabi, D. Yao, and H. Zhu. Yima: Design and Evaluation of a Streaming Media System for Residential Broadband Services. *Proceedings of VLDB 2001 Workshop on Databases in Telecommunications (DBTel 2001)*, September 2001.

[188] R. Zimmermann and S. Ghandeharizadeh. Continuous Display Using Heterogeneous Disk-Subsystems. In *Proceedings of the Fifth ACM Multimedia Conference*, pages 227–236, Seattle, WA, November 1997.

[189] R. Zimmermann and S. Ghandeharizadeh. HERA: Heterogeneous Extension of RAID. In *Proceedings of the 2000 International Conference on Parallel and Distributed Processing Techniques and Applications (PDPTA 2000)*, Las Vegas, NV, June 26-29 2000.

[190] G. K. Zipf. *Human Behavior and the Principle of Least Effort*. Addison-Wesley, Reading MA, 1949.

INDEX

InformIT

Articles

Online Books

Catalog

LICENSE AGREEMENT AND LIMITED WARRANTY

READ THE FOLLOWING TERMS AND CONDITIONS CAREFULLY BEFORE OPENING THIS SOFTWARE MEDIA PACKAGE. THIS LEGAL DOCUMENT IS AN AGREE-MENT BETWEEN YOU AND PRENTICE-HALL, INC. (THE "COMPANY"). BY OPENING THIS SEALED SOFTWARE MEDIA PACKAGE, YOU ARE AGREEING TO BE BOUND BY THESE TERMS AND CONDITIONS. IF YOU DO NOT AGREE WITH THESE TERMS AND CONDI-TIONS, DO NOT OPEN THE SOFTWARE MEDIA PACKAGE. PROMPTLY RETURN THE UNOPENED SOFTWARE MEDIA PACKAGE AND ALL ACCOMPANYING ITEMS TO THE PLACE YOU OBTAINED THEM FOR A FULL REFUND OF ANY SUMS YOU HAVE PAID.

1.   **GRANT OF LICENSE:** In consideration of your payment of the license fee, which is part of the price you paid for this product, and your agreement to abide by the terms and conditions of this Agreement, the Company grants to you a nonexclusive right to use and display the copy of the enclosed software program (hereinafter the "SOFTWARE") on a single computer (i.e., with a single CPU) at a single location so long as you comply with the terms of this Agreement. The Company reserves all rights not expressly granted to you under this Agreement.

2.   **OWNERSHIP OF SOFTWARE:** You own only the magnetic or physical media (the enclosed SOFTWARE) on which the SOFTWARE is recorded or fixed, but the Company retains all the rights, title, and ownership to the SOFTWARE recorded on the original SOFTWARE copy(ies) and all subsequent copies of the SOFTWARE, regardless of the form or media on which the original or other copies may exist. This license is not a sale of the original SOFTWARE or any copy to you.

3.   **COPY RESTRICTIONS:** This SOFTWARE and the accompanying printed materials and user manual (the "Documentation") are the subject of copyright. You may not copy the Docu-mentation or the SOFTWARE, except that you may make a single copy of the SOFTWARE for backup or archival purposes only. You may be held legally responsible for any copying or copyright infringement which is caused or encouraged by your failure to abide by the terms of this restriction.

4.   **USE RESTRICTIONS:** You may not network the SOFTWARE or otherwise use it on more than one computer or computer terminal at the same time. You may physically transfer the SOFTWARE from one computer to another provided that the SOFTWARE is used on only one com-puter at a time. You may not distribute copies of the SOFTWARE or Documentation to others. You may not reverse engineer, disassemble, decompile, modify, adapt, translate, or create derivative works based on the SOFTWARE or the Documentation without the prior written consent of the Company.

5.   **TRANSFER RESTRICTIONS:** The enclosed SOFTWARE is licensed only to you and may not be transferred to any one else without the prior written consent of the Company. Any unauthorized transfer of the SOFTWARE shall result in the immediate termination of this Agree-ment.

6.   **TERMINATION:** This license is effective until terminated. This license will termi-nate automatically without notice from the Company and become null and void if you fail to comply with any provisions or limitations of this license. Upon termination, you shall destroy the Documen-tation and all copies of the SOFTWARE. All provisions of this Agreement as to warranties, limita-tion of liability, remedies or damages, and our ownership rights shall survive termination.

7.   **MISCELLANEOUS:** This Agreement shall be construed in accordance with the laws of the United States of America and the State of New York and shall benefit the Company, its affili-ates, and assignees.

8.   **LIMITED WARRANTY AND DISCLAIMER OF WARRANTY:** The Company warrants that the SOFTWARE, when properly used in accordance with the Documentation, will operate in substantial conformity with the description of the SOFTWARE set forth in the Documen-tation. The Company does not warrant that the SOFTWARE will meet your requirements or that the operation of the SOFTWARE will be uninterrupted or error-free. The Company warrants that the

media on which the SOFTWARE is delivered shall be free from defects in materials and workmanship under normal use for a period of thirty (30) days from the date of your purchase. Your only remedy and the Company's only obligation under these limited warranties is, at the Company's option, return of the warranted item for a refund of any amounts paid by you or replacement of the item. Any replacement of SOFTWARE or media under the warranties shall not extend the original warranty period. The limited warranty set forth above shall not apply to any SOFTWARE which the Company determines in good faith has been subject to misuse, neglect, improper installation, repair, alteration, or damage by you. EXCEPT FOR THE EXPRESSED WARRANTIES SET FORTH ABOVE, THE COMPANY DISCLAIMS ALL WARRANTIES, EXPRESS OR IMPLIED, INCLUDING WITHOUT LIMITATION, THE IMPLIED WARRANTIES OF MERCHANTABILITY AND FITNESS FOR A PARTICULAR PURPOSE. EXCEPT FOR THE EXPRESS WARRANTY SET FORTH ABOVE, THE COMPANY DOES NOT WARRANT, GUARANTEE, OR MAKE ANY REPRESENTATION REGARDING THE USE OR THE RESULTS OF THE USE OF THE SOFTWARE IN TERMS OF ITS CORRECTNESS, ACCURACY, RELIABILITY, CURRENTNESS, OR OTHERWISE.

IN NO EVENT, SHALL THE COMPANY OR ITS EMPLOYEES, AGENTS, SUPPLIERS, OR CONTRACTORS BE LIABLE FOR ANY INCIDENTAL, INDIRECT, SPECIAL, OR CONSEQUENTIAL DAMAGES ARISING OUT OF OR IN CONNECTION WITH THE LICENSE GRANTED UNDER THIS AGREEMENT, OR FOR LOSS OF USE, LOSS OF DATA, LOSS OF INCOME OR PROFIT, OR OTHER LOSSES, SUSTAINED AS A RESULT OF INJURY TO ANY PERSON, OR LOSS OF OR DAMAGE TO PROPERTY, OR CLAIMS OF THIRD PARTIES, EVEN IF THE COMPANY OR AN AUTHORIZED REPRESENTATIVE OF THE COMPANY HAS BEEN ADVISED OF THE POSSIBILITY OF SUCH DAMAGES. IN NO EVENT SHALL LIABILITY OF THE COMPANY FOR DAMAGES WITH RESPECT TO THE SOFTWARE EXCEED THE AMOUNTS ACTUALLY PAID BY YOU, IF ANY, FOR THE SOFTWARE.

SOME JURISDICTIONS DO NOT ALLOW THE LIMITATION OF IMPLIED WARRANTIES OR LIABILITY FOR INCIDENTAL, INDIRECT, SPECIAL, OR CONSEQUENTIAL DAMAGES, SO THE ABOVE LIMITATIONS MAY NOT ALWAYS APPLY. THE WARRANTIES IN THIS AGREEMENT GIVE YOU SPECIFIC LEGAL RIGHTS AND YOU MAY ALSO HAVE OTHER RIGHTS WHICH VARY IN ACCORDANCE WITH LOCAL LAW.

ACKNOWLEDGMENT

YOU ACKNOWLEDGE THAT YOU HAVE READ THIS AGREEMENT, UNDERSTAND IT, AND AGREE TO BE BOUND BY ITS TERMS AND CONDITIONS. YOU ALSO AGREE THAT THIS AGREEMENT IS THE COMPLETE AND EXCLUSIVE STATEMENT OF THE AGREEMENT BETWEEN YOU AND THE COMPANY AND SUPERSEDES ALL PROPOSALS OR PRIOR AGREEMENTS, ORAL, OR WRITTEN, AND ANY OTHER COMMUNICATIONS BETWEEN YOU AND THE COMPANY OR ANY REPRESENTATIVE OF THE COMPANY RELATING TO THE SUBJECT MATTER OF THIS AGREEMENT.

Should you have any questions concerning this Agreement or if you wish to contact the Company for any reason, please contact in writing at the address below.

Robin Short
Prentice Hall PTR
One Lake Street
Upper Saddle River, New Jersey 07458

YIMA PERSONAL EDITION — PERSONAL USE LICENSE AGREEMENT

This Agreement is made by and between the University of Southern California (hereinafter "USC") a California nonprofit corporation with its principal place of business at University Park, Los Angeles, California 90089, and you, a private individual (hereinafter "Developer").

BY USING THE YIMA PERSONAL EDITION SOFTWARE CONTAINED IN THIS CDROM OR THIS BOOK, YOU AGREE TO THE TERMS OF THIS AGREEMENT. IF YOU REQUIRE BROADER RIGHTS FOR YOUR INTENDED USE, PLEASE CONTACT OTL@USC.EDU TO INQUIRE FOR MORE INFORMATION.

Whereas, USC is the owner of Yima; and Whereas, Developer desires to use Yima for personal educational purposes ("Personal Use"); Therefore, USC and Developer hereby agree as follows:

1. DEFINITIONS

1.1 "Yima" and "Yima Personal Edition Software" mean a computer-based streaming media system consisting of Source Code and Executable Code appearing in any of three forms: (i) in printed form in the pages of this book; (ii) in electronic form in the CDROM attached to this book; or (iii) in online or electronic form, available for download over the Internet.

1.2 "Effective Date" means the date as of which you transfer Yima, or any portion of Yima, onto any computer by any means.

1.3 "Source Code" means code of a computer program that is not executable by a computer system directly but must be converted into machine language by compilers, assemblers, and/or interpreters, as well as documentation, release notes, or other specifications which describe the content, organization, and structure of the Source Code included therein to the extent that such documentation is available.

1.4 "Executable Code" means machine-readable, object, linkable or executable code.

1.5 "Derivative Work" means a work which is based upon one or more pre-existing copyrightable or copyrighted material, such as a revision, modification, translation, abridgement, condensation, expansion, collection, compilation, or any other form in which such pre-existing works may be recast, transformed, or adapted such that, in the absence of this Agreement, the preparation, copying, use, distribution, and/or display thereof would constitute an infringement of the pre-existing material.

1.6 "Authors" shall mean the University of Southern California, Cyrus Shahabi and/or Roger Zimmermann.

2. LICENSE GRANT

2.1 USC hereby grants to Developer a nontransferable, non-exclusive license to: (i) use and copy the Source Code for your personal educational use only; (ii) compile the Source Code into Executable Code for your personal educational use only; and (iii) create Derivative Works of the Source Code and compile such Derivative Works into Executable Codes for your personal educational use only.

2.2 All other rights are expressly reserved by USC.

3. FURNISHED AS-IS, WITHOUT WARRANTY

3.1 YIMA IS FURNISHED TO DEVELOPER AS-IS. USC MAKES NO REPRESENTATIONS OR WARRANTIES, EXPRESS OR IMPLIED. BY WAY OF EXAMPLE, BUT NOT LIMITATION, USC MAKES NO REPRESENTATIONS OR WARRANTIES OF MERCHANTABILITY OR FITNESS FOR ANY PARTICULAR PURPOSE, OR THAT THE USE OF YIMA WILL NOT INFRINGE ANY PATENTS, COPYRIGHTS, TRADEMARKS OR OTHER RIGHTS OF ANY THIRD PARTIES. USC SHALL NOT BE HELD LIABLE FOR ANY LIABILITY OR FOR ANY DIRECT, INDIRECT OR CONSEQUENTIAL DAMAGES WITH RESPECT TO ANY CLAIM BY DEVELOPER OR ANY THIRD PARTY ON ACCOUNT OF OR ARISING FROM THIS AGREEMENT OR ANY USE OF YIMA BY DEVELOPER.

3.2 Nothing in this Agreement shall be construed as (i) a warranty or representation that anything made, used, sold or otherwise disposed of under this Agreement is or will be free from infringement of patents, copyrights and trademarks of third parties; or (ii) conferring rights to Developer to use the name of USC or the Authors.

4. INDEMNIFICATION

4.1   Developer shall defend, indemnify and hold harmless USC and its trustees, officers, professional staff, employees and agents and their respective successors, heirs and assigns (the "Indemnitees"), against all liability, demand, damage, loss, or expense incurred by or imposed upon the Indemnitees or any one of them in connection with any claims, suits, actions, demands or judgments arising out of any theory of product liability (including but not limited to, actions in the form of tort, warrantee, or strict liability) for death, personal injury, illness, or property damage arising from Developer's use or other disposition of Yima.

5. TERM AND TERMINATION

5.1   The term of this Agreement shall commence on the Effective Date and shall continue in accordance with the provisions of this Paragraph.

5.2   Developer may terminate this Agreement at any time without notice to USC.

5.3   USC may terminate this Agreement at any time upon its making a reasonable effort to provide notice to Developer, whether such notice is provided directly to Developer, or provided indirectly through such means as publication of notice made on a web site or in a daily newspaper. Such termination shall become effective immediately.

5.4   Upon termination, Developer shall destroy all copies of Yima, provided that Developer may retain one copy of Yima for archive purposes only.

5.5   Surviving any termination are (i) any cause of action or claim of USC, accrued or to accrue, because of any breach or default by Developer; and (ii) the provisions of Paragraphs 3 and 4.

6. MISCELLANEOUS

6.1   Developer will function solely as an independent contractor and not as agent of USC.

6.2   The headings used herein are intended solely for ease of reference, and are not intended to describe, construe or interpret this Agreement.

6.3   A waiver of any breach of any provision of this Agreement shall not be construed as a continuing waiver of said breach or a waiver of any other breaches of the same or other provisions of this Agreement.

6.4   Developer shall not engage in any activity in connection with this Agreement that is in violation of any applicable U.S. law.

6.5   Developer may deliver notices to USC by postal mail to University of Southern California, Office of Technology Licensing, 716 S. Hope Street, Ste. 313, Los Angeles, CA 90007.

6.6   If any provision of this Agreement is determined to be invalid or unenforceable under any controlling body of law, such invalidity or unenforceability shall not in any way affect the validity of enforceability of the remaining provisions hereof.

6.7   Neither party shall use the name, trade name, trademark or other designation of the other party in connection with any products, promotion or advertising without the prior written permission of the other party.

6.8   This Agreement shall be deemed to be executed and to be performed in the State of California, and shall be construed in accordance with the laws of the State of California as to all matters, including but not limited to matters of validity, construction, effect and performance.

6.9   This Agreement constitutes the entire agreement between the parties concerning the subject matter hereof. No amendment of this Agreement shall be binding on the parties unless mutually agreed to and executed in writing by each of the parties.

BY USING THE YIMA PERSONAL EDITION SOFTWARE CONTAINED IN THIS CDROM OR THIS BOOK, YOU AGREE TO THE TERMS OF THIS AGREEMENT. IF YOU REQUIRE BROADER RIGHTS FOR YOUR INTENDED USE, PLEASE CONTACT OTL@USC.EDU TO INQUIRE FOR MORE INFORMATION.